Quality Control - An Anthology of Cases

Edited by Leo D. Kounis

Published in London, United Kingdom

Quality Control - An Anthology of Cases
http://dx.doi.org/10.5772/intechopen.97961
Edited by Leo D. Kounis

Contributors
Shashikiran Venkatesha, Ranjani Parthasarathi, Mohsen Modarres, Shadha ALAmri, Fatima AlAbri, Tripti Sharma, James Stribling, Susan Lanberg, Blaine Snyder, Jerry Diamond, Benjamin Jessup, Hugh Sullivan, Abduletif Hebo, Nefise Gönül Şengöz, Marcel Randermann, Roland Jochem, Timo Hinrichs, Ming-Fung Francis Siu, Siwei Chang, Lindiwe Julia Ncube, Marieta Georgieva Stefanova, Valeryi Semenov, Samuel I. Egwunatum, Ovie I. Akpokodje, Andrew I. Awo-Osagie, Ahmed Mateen Buttar, Muhammad Majid

First published in London, United Kingdom, 2023 by IntechOpen
IntechOpen is the global imprint of INTECHOPEN LIMITED, registered in England and Wales, registration number: 11086078, 5 Princes Gate Court, London, SW7 2QJ, United Kingdom

British Library Cataloguing-in-Publication Data
A catalogue record for this book is available from the British Library

Additional hard and PDF copies can be obtained from orders@intechopen.com

Quality Control - An Anthology of Cases
Edited by Leo D. Kounis
p. cm.
Print ISBN 978-1-80355-990-2
Online ISBN 978-1-80355-991-9
eBook (PDF) ISBN 978-1-80355-992-6

For EU product safety concerns:
IN TECH d.o.o., Prolaz Marije Krucifikse Kozulić 3, 51000 Rijeka, Croatia,
info@intechopen.com or visit our website at intechopen.com.

Meet the editor

Leo D. Kounis is head of the Department of the Communication and Informatics Battalion of the Hellenic Ministry of Defense, Hellenic National Defense General Staff. He obtained his BEng (Hons) in Manufacturing Systems Engineering, his MSc in Quality Engineering, and his Ph.D. in Systems Reliability from the University of Hertfordshire, UK. Dr. Kounis has worked as a senior quality engineer in several private companies in Greece and as a part-time lecturer and scientific advisor in academia. His research interests include quality, transportation, and sustainable energy. He has edited and published numerous scientific papers.

Contents

Preface

The Fourth Industrial Revolution that started less than a decade ago in Germany is taking the world by storm. In a report, the Organisation for Economic Co-operation and Development (OECD) defines Industry 4.0 as: "... the use in industrial production of recent, and often interconnected, digital technologies that enable new and more efficient processes, and which in some cases yield new goods and services. The associated technologies are many, from developments in machine learning and data science, which permit increasingly autonomous and intelligent systems, to low-cost sensors which underpin the IoT, to new control devices that make second-generation industrial robotics possible" (OECD, 2017, p. 27). OECD thus addresses twelve so-called enabling technologies that are driving the core parts of the latest industrial revolution. These technologies range from big data and analytics to autonomous robots and synthetic biology to augmented reality and cloud computing, among others, assisted by the Internet of Things (IoT) and covering anything and everything in between.

For all technologies to "shake hands" with one another, digitalization is regarded as the main driver, with emphasis on the word "smart" in terms of producing sustainable products. In a 2016 study, the European Parliament 2016, [1] defined the primary features of this revolution:

- Interoperability
- Virtualization
- Decentralization
- Real-time capability
- Service orientation
- Modularity

It may be argued that the COVID-19 pandemic may inadvertently have contributed to an accelerated transition into digitalization in several sectors. The latter holds true for both developed and developing countries. In the former category, Greece experienced an increase of 18%, with Ireland, Hungary, and Romania following suit with 15%, whereas in the latter group, the United Arab Emirates led the way with an increase of 36%, followed by Thailand, Bahrain, and Uzbekistan. The 2021 United Nations Conference on Trade and Development (UNCTAD) report claims that "...online purchases increased from 53% ...to 60%." These percentages, however, are the result of statistics obtained from sixty-six countries [https://unctad.org/news/covid-19-boost-e-commerce-sustained-2021-new-unctad-figures-show].

To further facilitate and safeguard quality levels, the term "quality 4.0" has been coined. Prof. J. Oakland stated that "...if properly defined and understood", quality 4.0 has the potential to link present-day advances with technological breakthroughs to foster customer expectations, requirements, and experiences.

This book provides a comprehensive review of quality control, illustrating its application in several different cases and scenarios. Chapter 1 introduces techniques relating to computing performance measures in aquatic environments to assist users in the monitoring process for evaluating environmental outcomes.

Nigeria possesses the largest road network in South Africa covering 195.000 km, of which approximately 30% (60.000Km) is paved [2]. Non-conformance to established guidelines and requirements results in failures that may either be gradual or sudden. Chapter 2 introduces the underlying physical processes involved in the gradual degradation of road pavements and recommends a set of actions.

The global textile industry, predominantly driven by China, was appraised at approximately US $920 billion in 2018 and is estimated to exceed US $1,230 billion by 2024. Turkey ranks fifth among the top ten textile exporters globally, achieving an export value of US $11.9 billion in 2018. Chapter 3 presents a novel statistical approach pertaining to multiple-stream processes based on a spinning mill exercise.

According to a report by the United Nations Conference on Trade and Development (UNCTAD), 67.54% of Ethiopia's leather products were exported to the United States in 2018. In Europe, Germany, and Italy are regarded as the main leather import countries, representing 11.35% and 3.63% of exports, respectively. The export of raw hides and skins to Italy and China generates more than US $446 million as foreign currency [United Nations, UNCTAD, National Green Export Review of Ethiopia: Leather and Sesame Seeds, 2018]. It is thus regarded as one of Ethiopia's main economic drivers. Despite considerable bottlenecks, further growth opportunities do exist. Chapter 4 discusses quality control aspects within the garment manufacturing sub-sector.

Additive manufacturing (AM), also referred to as additive layer manufacturing (ALM) is yet another term for three-dimensional printing. The use of fused deposition modelling (FDM) has found its way into a variety of industrial applications. However, problems still occur in FDM-specific processes and focus on the failure rate of printed items. To overcome this, Chapter 5 introduces a quality gate reference process including defined requirement criteria capable of being applied both in quality control and in situ process monitoring.

Major disruptions in logistics may lead to detrimental consequences in the supply chain. Although international trade recovered in 2021, the OECD argues in a 2022 report that "...the heterogeneity of trade impacts and changes in trade flows..." calls for the implication of risk mitigation measures and strategies. Early recognition and identification of potential problem areas minimize overall costs for returning the system to its normal operation. Chapter 6 analyzes the contribution of prevention costs to total costs.

Quality control equally applies to the food industry. Of particular interest is Africa because although food safety management systems are in place, the risk of microbiological and chemical contamination levels in most food products in Africa is still high. Chapter 7 describes the ISO22000 series of quality standards and recommends its implementation.

Network security is of paramount importance. The intrusion detection system (IDS) is a widely applied network tool for fostering network security management. Chapter 8 postulates that its use may be perplexing to users and elaborates on software interfaces and assessment procedures to appraise novel heuristics. In doing so, it presents new paths for assessing programmed usability.

Complementary metal oxide semiconductor (CMOS) technology is widely applied for manufacturing a variety of products that incorporate integrated circuits. Currently applied methodologies utilize redundancy-based approaches to identify fault tolerance in multi-core scenarios. Based on several implications, Chapter 9 introduces a novel multi-core system.

Concrete build structures are subjected to periodic quality control assessment, as structural problems may have detrimental effects on safety, stability, and integrity. The use of advanced technological means assists in effectively identifying defects while minimizing future repair-specific costs. Chapter 10 introduces a computer vision-based automated crack inspection technique that is applied to a structure in Hong Kong.

The COVID-19 outbreak led to increased use of digital communication means with companies and academia, among others, intensifying work-from-home distance-learning schemes. Chapter 11 discusses the development and usage of electronic teaching technologies in a technical university and addresses the stages and structures thereof.

Numerous scientific papers have addressed the terms quality, quality control, and (total) quality management in businesses worldwide, resulting in the optimization of processes and techniques alike while facilitating the development of new tools and fostering product innovations. Chapter 12 explores the effects of learning capability and innovation on quality management-specific organizational performance.

Finally, Chapter 13 investigates artificial intelligence (AI) deployment for the purposes of safeguarding the Industrial Internet of Things (IIoT). It presents an analysis of the various security challenges confronted by IoT systems as well as a comparative study of recommended AI categories.

Dr. Leo D. Kounis
Hellenic Ministry of Defense,
Telecommunications and Informatics Battalion,
Athens, Greece

References

[1] European Union, 2016. POLICY DEPARTMENT A: ECONOMIC AND SCIENTIFIC POLICY Industry 4.0, Brussels, Policy Department A 2016,94 p. [https://www.europarl.europa.eu/RegData/etudes/STUD/2016/570007/IPOL_STU%282016%29570007_EN.pdf]

[2] Logistics Capacity Assessments (LCAs). [https://dlca.logcluster.org/display/public/DLCA/2.3+Nigeria+Road+Network]

Chapter 1

Performance Measures for Evaluating and Communicating Data Quality in Aquatic Environmental Monitoring in the U.S.

James Stribling, Susan Lanberg, Blaine Snyder, Jerry Diamond, Benjamin Jessup and Hugh Sullivan

Abstract

Quantitative data quality descriptors are important for evaluating and communicating acceptability of information used in environmental monitoring and assessment. In this chapter, we present (1) the rationale for establishing and using performance measures and MQOs in routine quality control planning and analysis, (2) field and laboratory methods for capturing input data required for performance calculations, and (3) approaches for setting data acceptability thresholds and determining the need for corrective actions. Relevant examples are available from local, regional, and national programs in the U.S. charged with monitoring and assessing aquatic biological condition, physical habitat, contaminants, and toxicity testing. We will describe techniques for calculating and determining acceptability of performance measures, such as, among other data quality indicators, precision, accuracy, sensitivity, representativeness, and completeness of field sampling, laboratory processing, and data management and analysis. Data types on which these examples will be based include benthic macroinvertebrates, fish assemblage, tissue body burden, laboratory analytical and toxicity testing, physical habitat, selected geomorphic characteristics, and algal toxins.

Keywords: precision, bias, indicators, error, corrective actions, acceptability

1. Introduction

Science is recognized as treating uncertainty and variability as information that serves as a key component of decision-making, helping formulate new questions, experimental designs, and testing and measurement procedures [1–3]. This is the essence of the scientific method; knowledge itself increases through the process of trial and error. Perhaps the most well-known effort to begin quantifying data variability as part of the decision-making process in technical endeavors led to development of the concept of statistical process control [4]. With a focus on manufacturing, Shewhart's ideas largely originated with the desire to better understand causes of anomalous or unwanted output, and to provide clarity on what might be necessary to improve outcomes. He stated [4] "*Through the use of the scientific method...it has been found possible to set up limits within which the results of routine efforts must lie if they are to be economical. Deviations in the results of a routine process outside such limits indicate that the routine has broken down and will no longer be economical until the cause of the trouble has been removed*". Shewhart [5] further developed statistical techniques and demonstrated their application, helping to broaden the appeal of using control charts to document and track the quality of various data characteristics. The quality of data, and especially environmental data, is tracked through various quality control (QC) processes as discussed in this chapter. As Woodall [6] and others have noted, QC analyses and their interpretation are best handled by those who are knowledgeable about the field of practice.

The 1993 passing of the Government Performance and Results Act (GPRA) in the United States (US) elevated attention to documenting effectiveness, efficiency, and accountability of programs, and resulted in Federal agencies setting goals for program performance [7]. The GPRA focused the need for and use of performance characteristics and quantitative measurement quality objectives (MQO) to strengthen programs at any scale, not just Federal. The Information Quality Act (IQA) of 2000, sometimes referred to as the Data Quality Act, required Federal programs to ensure the "quality, objectivity, utility, and integrity" of publicly available information they produced [8]. It also required agencies to develop techniques for acquiring, reporting, and acting on results, where necessary.

The concept of process or QC is adaptable to any measurement system, requiring only that key points of the process are identified as providing opportunity for taking measurements, and that there is some standard or criterion for comparison. When anomalous or extreme results are detected via comparison to MQO, they would be investigated to determine what might be causing performance deviations.

Routine environmental monitoring requires consistent collection of data and information such that they are of known and acceptable quality. The purpose of this chapter is to describe data requirements, numeric structure, and interpretation thresholds for MQO related to several diverse and important indicators used in aquatic environmental monitoring throughout the US. These include biological, physical, chemical, and toxicological indicators.

2. Quality assurance and control for environmental monitoring

In the US, environmental data are collected by many federal, state, tribal, and local agencies, including the US Environmental Protection Agency (EPA), as well as non-EPA organizations supporting environmental programs on behalf of EPA in accordance with the EPA agency-wide quality system. Other federal agencies such as the US Geological Survey (USGS), the National Oceanographic and Atmospheric Administration (NOAA), and the US Fish and Wildlife Service (USFWS) collect data under different data quality frameworks but have similar requirements for known and acceptable quality. The quality assurance (QA) planning processes established by the EPA are recognized as a high standard that should be *attempted* even in non-EPA projects, such as state-, industry-, or non-profit-funded special projects. The EPA quality system is based on ANSI/ASQC E4-1994, *Specifications and Guidelines for Quality Systems for Environmental Data Collection and Environmental Technology Programs*, a national standard for quality management practices for environmental programs involving the collection and evaluation of environmental data and the design, construction, and operation of environmental technologies [9, 10]. Quality planning documentation prepared for collection of environmental data (by or for EPA) includes descriptions of project-specific data quality objectives (DQO), quality assurance project plans (QAPP), and standard operating procedures (SOP). DQO are integral to the QA planning process. The DQO process includes identifying the decisions to be made based on the information collected, as well as the data quality and quantity acceptance criteria required to make those project decisions [11]. QAPPs are developed and implemented to ensure that data collected for a project are complete and of a quality sufficient for their intended purpose [12]. A QAPP includes a section on DQO, and SOP for relevant field collection and laboratory analysis procedures are often included as QAPP attachments. SOP are developed and followed to ensure that procedures for data collection and analysis are performed consistently within boundaries defined by MQO, thus meeting acceptance criteria.

There are different sources of error, some of which may yield uncertainty and all of which can affect variability observed in data and outcomes. This chapter discusses several commonly used indicators of aquatic environmental condition and the types of performance measures and QC processes used to ensure that data are acceptable to use in a particular environmental program.

3. Indicators of environmental condition

3.1 Biological

Field sampling, laboratory processing, and data analysis procedures for biological indicators are relatively well-established for many monitoring programs. For programs focused on community level indicators of biological integrity, such as the Index of Biological Integrity (IBI) or River Invertebrate Prediction and Classification System (RIVPACS) of observed to expected (O/E) conditions, based on consistent sampling and interpreting of taxa and individual counts. Field sampling for these indicators typically gathers composite samples from multiple habitats distributed throughout some defined area of the stream, river, lake, or estuarine/near-coastal waters. Depending on the program, the sampling area for rivers and streams can be a defined channel length, such as 100 m, or some multiple of the wetted width. Organism groups targeted by this kind of sampling includes, for example, benthic macroinvertebrates (BMI), fish, and algae/diatoms. Laboratory processing for BMI and diatoms includes sorting, subsampling, and taxonomic identifications. Estuarine and near-coastal programs sample benthic invertebrates from a surface area defined by gear type. Example methods documents are [13, 14], and several field and laboratory operations manuals from EPA National Aquatic Resource Surveys (NARS) [15–24].

Efforts to customize performance measures to biological monitoring programs have sought to use the process to isolate potential sources of variability, or error, and determine the need for and nature of corrective actions [14, 25]. A biological assessment protocol is a series of methods encompassing field sampling, laboratory processing (if necessary, and including sorting/subsampling and taxonomic identification), enumeration, data analysis, and assessment endpoints such as a regionally calibrated multimetric IBI. Community-level fish indicators typically do not involve laboratory work as identification and counting is done in the field while on site. [14, 25] propose performance measures and MQO to cover the sequential phases of biological assessments. Key components considered are field sampling precision, and for BMI, sorting/subsampling and taxonomic identification. In the framework they propose, the ability to detect or highlight errors in these phases requires specific activities that provide data to calculate performance measures, the results of which are then compared to the MQO (**Table 1**). Descriptions below are examples of performance measures and how data are acquired for their calculation.

Field sampling precision (requires duplicate samples). Biological samples are taken from duplicate 100 m channel reaches that are immediately adjacent to each other. Laboratory processing and indicator calculation proceeds for each as separate samples. Comparison of results using specific performance measures (relative percent difference [RPD], coefficient of variability [CV], and confidence intervals [CIs]) (**Table 1**) reveals the precision and repeatability of the sampling method and its application.

Sorting/subsampling bias (requires sort residue rechecks). The objective of primary sorting of BMI samples is to remove all organisms from nontarget sample material, such as leaf litter, twigs, sand/silt, and other organic and inorganic detritus. The remaining sample material (sort residue) is checked for specimens missed by the primary sorter, and the performance measure, percent sorting efficiency (PSE) (**Table 1**) calculated as indicative of bias in the process.

Taxonomic precision (requires sample re-identification). Biological samples undergo identification by a primary taxonomist, then reidentification by a separate, independent taxonomist. Identification and count results are directly compared, and differences or error rates are quantified as a measure of taxonomic performance, specifically, precision. Terms calculated are percent taxonomic disagreement (PTD), percent difference in enumeration (PDE), and percent taxonomic completeness (PTC) (**Table 1**). All three terms quantify distinctly different aspects of the taxonomic identification process and relate directly to overall sample characteristics. Further, PTC indicates the proportion of the sample identified to the target hierarchical level (species, genus, tribe, family, or higher), where the absolute value of the difference between primary and QC taxonomist (|PTC|) indicates precision and consistency. Results from QC analyses can be presented in reports or associated with datasets in a straightforward manner (**Table 2**) that allows the data user to understand and move ahead with subsequent analyses.

The sites and samples for which these analyses are done use a *randomly selected* subset of sites, sort residue samples, and samples, respectively. As a rule of thumb, approximately 10% would be selected from the sample lot. The outcomes of these calculations and comparison to MQO can and should be used to (1) help detect potential problems in how the specific activity was implemented, (2) help inform the nature and need for corrective actions; and (3) summarize the overall quality of the full dataset. Subsequent values exceeding the MQO are not automatically taken to be unacceptable data points; rather, such values should receive closer scrutiny to determine reasons for the exceedance and might indicate a need for corrective actions.

The rationale for determining numeric values to be used as MQO should be based on observable data which are relevant to the monitoring program and the indicators that are being tracked as a part of it [25]. As an example, the MQO for PTD is 15 [26] and was arrived at through recognizing that taxonomic comparison (TAXCOMP) results for many samples were <20 and that there were very few <10. The 15% simply splits the difference.

Indicator category	Indicator/ group	Data	Data origin	Performance term	Source
Biological	BMI	Assemblage-level, taxonomy, count	Laboratory processing	Sample sorting: PSE Taxonomic identification: PTD, PDE, PTC, |PTC|	[14, 18, 19, 25–27]
Biological	BMI	Individual metrics, MMI	Field sampling; MMI	Precision (among sites): CV, CI90 Precision (within sites): RPD	[14, 17, 21, 22, 25–27]
Biological	Fish	Individual metrics, MMI	Field sampling; field processing; MMI	Precision (among sites): CV Precision (within sites): RPD Taxonomic identification: PTD Percent completeness: % comp.	[28, 29]
Physical	Physical habitat		Field observations	Precision (among sites): CV, CI90 Precision (within sites): RPD	[13, 14, 25]
Physical	Sediment	Sediment grain size and total organic carbon	Laboratory processing	Precision and accuracy: recovery of spikes in blanks and matrices; MDLs (calculated for lab)	[23]
Physical	Water clarity	Photosynthetically active radiation transmittance at 1 m	field measurements, calculation	Slope of least squares regression [−ln(light UW/light AMB) vs depth]; $R^2 > 0.75$	[24, 30]
Physical	Water clarity	Mean Secchi depth	Field measurements, calculation	Precision: all disappear and reappear values (3 of each) within 0.5 m	[23, 30]
Chemical	Fish	Tissue contaminant load	Laboratory processing	Sample preparation: RPD for duplicate homogenized tissue sample pairs; sample analysis: RSD for initial precision recovery, matrix spike, and matrix spike duplicate samples	[31–34, 51]
Chemical	Residuals and water quality	PFAS—16 analytes	Laboratory analysis of samples collected by facilities	Accuracy of measurements: % recovery for internal standards, LCS % recovery, MS % recovery; precision: RPD for MS/ MSD and FDs	[35]
Harmful algal blooms	Algal toxins	Cylindrospermopsin, microcystins	Laboratory analysis of proficiency test (PT) samples	Accuracy of measurement: % recovery; precision among analytical laboratories: RPD	[18–20, 23, 36–40]

Indicator category	Indicator/ group	Data	Data origin	Performance term	Source
Harmful algal blooms	Algal toxins	Cylindrospermopsin, microcystins	Generally field sampling	False positive rate; false negative rate; sensitivity (detection limit); CV for precision	[41, 42]
Ecotoxicity testing	Sediment toxicity	Acute toxicity of whole sediment sample	Laboratory processing	Minimum mean control corrected % survival	[43, 44]
Ecotoxicity testing	Aquatic toxicity	Counts, weight, % survival, % fertilization	Lab testing; field exposures	Within-test variability; sensitivity to specific contaminants; control precision (CV); PMSD	[45, 46]

BMI, benthic macroinvertebrates; PSE, percent sorting efficiency; PTD, percent taxonomic disagreement; PDE, percent difference in enumeration; PTC, percent taxonomic completeness; |PTC|, absolute value of PTC difference; MMI, multimetric index; CV, coefficient of variability; RPD, relative percent difference; RSD, relative standard deviation; CI90, 90% confidence interval; UW, under water; AMB, ambient; PFAS, per- and polyfluoroalkyl substances; LCS, laboratory control sample; MS, matrix spike; MS/MSD, matrix spike/matrix spike duplicate; FD, field duplicate; MDL, method detection limit; PMSD, percent minimum significant difference.

Table 1.
Selected example performance measures for QC planning and analysis.

Performance characteristic	MQO	Observed
1. Field sampling precision (MMI)	CV < 15%	10.6
	CI90 ≤ 1.0	0.8
2. Sorting/subsampling bias	PSE ≥ 90	96.7
3. Taxonomic precision	Median PTD ≤ 15%	5.4
	Median PDE ≤ 5%	0.5
4. Taxonomic completeness	Median PTC ≥ 90%	91
	Median |PTC| ≤ 5%	1.5

Table 2.
Summary results from QC analyses BMI samples (n = 9) from the Prince George's County (Maryland, USA) biological monitoring program, 2010–2013.

Subsequent TAXCOMP results support using 15%, whether at broad national scales or smaller programs of anywhere from 10 to 50 samples. MQO are also not necessarily intended to be permanently fixed. As a monitoring program or testing procedure matures and more experience is gained, subsequent values often are observed as being consistently lower; a program may determine it would be beneficial to lower the MQO. Among all programs, PTD values are increasingly more commonly observed <10. It is advisable to use improved understanding of variability and its causes to adjust thresholds.

3.2 Physical habitat

3.2.1 Wadeable streams

One approach for characterizing the quality of stream physical habitat is a visual-based procedure [13] that assesses channel conditions in terms of stability, complexity, and availability of habitat for stream biota. There are 10 parameters, seven of which are rated for all streams, and 3 each for low and high gradient streams (Table OS-1[1]). Parameters are graded along a continuum of conditions from the perspective that as a stream becomes physically degraded, it loses physical complexity. Each parameter is rated on a 20-point scale while the observer is on site, then the values are summed for an overall site score. The range for the overall score is 0–200, with low values indicating poor quality habitat incapable of supporting stream biota and high indicating optimal conditions.

Data for input to QC calculations are from assessments done on adjacent 100 m channel reaches, identical to those discussed above for biological sampling. Reaches for which duplicate assessments are performed are randomly selected from the full site load, and pairs of habitat assessment results are used to calculate different performance measures (**Table 1**). As an example of results from such a QC analysis, consider a project that assessed 87 wadeable stream locations in Prince George's County, Maryland USA, and thus had nine (9) pairs of habitat assessment scores (Table OS-2 cdn.intechopen.com/public/259766_osi.zip).

Even though the field technique is qualitative, these numbers demonstrate the consistency of the results, particularly the median relative percent difference (mRPD) and CV. The values of RPD range from 1.4 to 35.3, with the substantial difference at the high end of the range suggesting that either the two reaches are dramatically different in quality, or potentially a data recording error occurred. These numbers characterize quality of the physical habitat data, as well as provide a roadmap for investigating potential anomalous results.

3.2.2 Estuarine/near coastal

Environmental monitoring programs assess abiotic indicators to understand how stressors may impact organisms, as well as how the habitat may be impacted by human disturbance. For example, because light underwater diminishes with depth [47] programs such as the U.S. EPA NARS National Coastal Condition Assessment (NCCA) survey and the Chesapeake Bay Program collect *in situ* water clarity measurements to estimate the impact of cultural eutrophication on light attenuation through the water column [24]. The EPA measures water clarity as Secchi depth at Great Lakes nearshore sites (the average depth of disappearance and reappearance of a 20 cm black and white disk lowered and retrieved through the water column three times), or transmission of photosynthetically active radiation (PAR) by comparing simultaneous ambient and underwater light measurements at incremental depths for estuarine sites. Performance measures for water clarity are intended to ensure accuracy and precision, as well as repeatability and consistency across the wide array of sites encountered in the survey. Secchi depth performance checks are implemented in the field and reviewed by analysts before use. They require that all six measurements are within 0.5 m of each other. When the difference between the maximum and minimum Secchi measurements at a site exceeds 0.5 m, the field crew repeats the entire set of measurements [24]. Data analysts again check Secchi data; values exceeding the maximum difference of 0.5 m among measurements at a site are reviewed and obvious transcription errors are corrected. Final values that do not meet the quality requirement are excluded from analysis. Table OS-3 cdn.intechopen.

[1] Due to space limitations, Tables OS-1 through OS-11 are provided as Online Supporting Information cdn.intechopen.com/public/259766_osi.zip.

com/public/259766_osi.zip illustrates the decisions made when reviewing Secchi data collected at 20 sites during the NCCA 2010 field season. For PAR, light sensors and data loggers are required to have been calibrated within 2 years prior to use and NCCA analysts conduct post measurement data checks to verify data quality. To ensure that the underwater light measurements decrease with depth (that is, light attenuation increases with depth), the PAR attenuation coefficient (Kd) is first calculated as the negative of the natural log of the ratio of underwater light to ambient light [−ln(UW/AMB)]. Kd is then plotted on the *Y* axis against the measurement depth on the *X* axis. If there is a negative slope of the resulting least squares regression line, or the coefficient of determination (R^2) ≤ 0.75[2], measurements are investigated further. When specific measurements are found to be incorrect, they are excluded from regression [30]. Figure OS-1 cdn.intechopen.com/public/259766_osi.zip illustrates an example of erroneous UW PAR measurements that were excluded from analysis at a site sampled during the 2010 NCCA field season.

3.3 Chemical

3.3.1 Algal toxins

Recent NARS, including the National Lakes Assessment (NLA 2017), National Rivers and Streams Assessment (NRSA 2018/2019), and the NCCA (2020), sampled assessment locations (sites) from across the US. Locations were selected using a probability-based approach to provide representative results to estimate conditions at broad spatial scales. For purposes of discussion in this section, we will focus on water grab samples that were collected from a subset of sites representing lakes, streams and rivers, and coastal areas for analysis of cyanobacteria-produced algal toxins (microcystins and cylindrospermopsin).

As part of the effort to meet programmatic data quality requirements [18, 20, 23], EPA designed a performance analysis to document the reliability and consistency with which analytical laboratories detected the presence and concentration of the algal toxins cylindrospermopsin and microcystins. With a focus on accuracy (percent recovery), the design provided performance test (PT) samples to state and national laboratories analyzing field samples for which the nominal concentrations were known to the NARS QC administrators. The objective of the PT analysis is to allow use of the results to evaluate the quality of the analytical procedures, specifically through use of enzyme-linked immunosorbent assay (ELISA) test kits, and potentially develop recommendations for improvement in sample handling, preparation, and analytical techniques.

Sets or "waves" of PT samples were prepared and delivered to the target laboratories during the same period that primary project samples were undergoing analysis. Two waves were analyzed for the NLA (2017), and three waves of PT samples each were analyzed for the NRSA (2018/2019) and the NCCA (2020). The procedures for analyzing microcystins and cylindrospermopsin included necessary cleanup steps for samples with salinity >3.5 parts per thousand, as well as dilution steps for samples with concentrations >upper detection limit (UDL) of the ELISA test kits. The PT

[2] The protocol in [30] calls for a minimum *R* of 0.95; the minimum *R* for the NCCA is relaxed to 0.75 to allow for variability in measurement due to factors such as differing sun angles throughout the day or underwater light reflection at shallower estuarine sites.

samples were subjected to multi-temperature stability studies before shipment, and then shipped on ice packs overnight to the laboratories analyzing NARS field samples.

PT samples were prepared to specified concentrations of cyanotoxins (Table OS-4 cdn.intechopen.com/public/259766_osi.zip) and distributed to the target laboratories. We used two performance measures in evaluating the PT results. First, percent recovery was used for accuracy, and RPD or relative standard deviation (RSD) [40, 48–50] for precision. Although all PT concentrations are shown (**Table 1**), for reasons of space limitations we have selected example results to illustrate results for one round of analyses for which the most accurate % recovery results were obtained and another for the least accurate from the most recent NARS, including NLA2017, NRSA2018/2019, and NCCA2020.

Both Lab A and Lab B met the % recovery goal of 70–130% [38] for the freshwater microcystins 2018/2019 NRSA Round 1 PT samples (Table OS-5 cdn.intechopen.com/public/259766_osi.zip). In comparison, Lab A did not meet the % recovery goal for the two of the freshwater microcystins 2017 NLA Round 1 PT samples. It should be noted that the results for sample M-7 were only slightly outside the % recovery range. In addition, although the results for M-10 were lower than 70% recovery, the PT sample concentration was much higher than the test kit range and required several dilutions for analysis.

Lab A met or nearly met the % recovery goal of 70–130% [38] for the estuarine microcystins 2020 NCCA Round 3 PT samples (Table OS-6 cdn.intechopen.com/public/259766_osi.zip). In contrast, Lab D did not meet the % recovery goal for 2 of the estuarine microcystins 2020 NCCA Wave 1 PT samples. The 2020 NCCA Wave 1 estuarine microcystins % recovery results ranged from 63.0 to 131.1, excluding the two non-detect results from Lab D. The 63.0% recovery value was a calculated PT sample concentration above the upper limit of detection (20MC-9) and the 131.1 % recovery value was calculated for the lowest microcystins concentration (20MC-8). The non-detect results reported by Lab D were for concentrations at the lower end of detection (20MC-8 and 20MC-10).

Lab A met the % recovery goal of 70–130% [39] for the freshwater cylindrospermopsin 2020 NCCA Wave 3 PT sample (Table OS-7 cdn.intechopen.com/public/259766_osi.zip). In comparison, Lab A did not meet the % recovery goal for four of the freshwater cylindrospermopsin 2017 NLA Wave 1 PT samples. It should be noted that of the 2017 NLA Wave 1 PT sample concentrations with % recovery value outside the % recovery goal, only C-4 had a concentration within the detection range of the test kit.

Lab A met the % recovery goal of 70–130% [39] for the estuarine cylindrospermopsin 2020 NCCA Wave 3 PT sample (Table OS-8 cdn.intechopen.com/public/259766_osi.zip). In contrast, Lab A did not meet the % recovery goal for all five of the estuarine cylindrospermopsin 2020 NCCA Wave 1 PT samples and Lab D did not meet the % recovery goal for one of the estuarine cylindrospermopsin 2020 NCCA Round 1 PT samples. The vendor laboratory noted that the salts used to prepare the estuarine PT samples might have caused the elevated % recovery values for the lower concentrations (<1 μg/L) due to background interference. The vendor laboratory indicated that the salts would not lead to false positive results if there were no cylindrospermopsin in the sample.

The analyses and comparisons of analytical results highlighted potential issues that allowed the QC coordinators to inquire for additional information. Although these particular instances did not result in anomalous results, the evaluations did help improve understanding of the sample handling and analysis process.

3.3.2 Per- and polyfluoroalkyl substances in residuals

Entities permitted to sell or distribute wastewater residuals for land application in Massachusetts were required by the Massachusetts Department of Environmental Protection (MDEP)[3] to collect and submit quarterly samples in 2020–2021 for analyses of 16 PFAS (Table OS-9 cdn.intechopen.com/public/259766_osi.zip). In 2020–2021, no EPA-approved methods were available for testing residuals for PFAS. Laboratories used "modified" EPA Method 533 (*Determination of Per- and Polyfluoroalkyl Substances in Drinking Water by Isotope Dilution Anion Exchange Solid Phase Extraction and Liquid Chromatography/Tandem Mass Spectrometry*) [35] to analyze samples. The laboratory SOP were reviewed and approved by the MassDEP before they were used to analyze the residuals samples. In addition, a standardized data quality evaluation checklist was developed and used to consistently perform reviews of the quality of results reported in laboratory data packages. Implementing these steps allowed for evaluation of whether the analytical results met the quality requirements outlined in EPA Method 533 "modified" [35], as well as the overall analytical quality requirements in 40 CFR Part 136.7 (*Guidelines Establishing Test Procedures for the Analysis of Pollutants, Quality Assurance and Quality Control*).

In 2021, an evaluation of the analytical results was performed for quarterly residuals samples collected during the last quarter of 2020 through the third quarter of 2021 using the standardized data evaluation checklists. The method quality objectives (e.g., holding times, minimum reporting limits, RPD for laboratory or field duplicates) presented (**Table 1**) were evaluated and documented for each sample using a standardized data quality evaluation checklist. Additional issues that the laboratories encountered during analysis were also documented in these checklists. Results from these standard evaluations were used to qualify the data to enable end users to interpret the quality of results. We provide a summary of the qualifiers used (and frequency of use) for each of the reported 16 analytes from a total of 164 samples (Table OS-10 cdn.intechopen.com/public/259766_osi.zip).

Elevated reporting limits (>1 ng/g) were the most frequently used qualifier (Table OS-10 cdn.intechopen.com/public/259766_osi.zip). The R qualifier was used for at least one analyte for 79% of the samples analyzed. These elevated reporting limits were less frequently observed in samples with low moisture content, with all samples with less than 28.3% solids having elevated reporting limits for at least one analyte. It should be noted that the remaining qualifiers used for the results were only applied when the results were greater than the detection limit. The J1- qualifier, indicating that the isotopically labeled analogue recovery was below the lower acceptance limit and that the residual result is estimated (could be biased low) for the corresponding target PFAS, was used for at least one analyte for 37% of samples analyzed. The J6+ qualifier, indicating that the ratio of the quantifier ion response to qualifier ion response (i.e., primary mass transition) falls outside of the laboratory established criteria (i.e., outside ratio limits) and that results are estimated maximum concentrations, was used for at least one analyte for 37% of the samples analyzed. The J5+/− qualifier was used for at least one analyte for 34% of the samples, commonly indicating that the RPD for the field sample duplicate (or less commonly the MSD)

[3] 310 CMR 32.00: Land Application of Sludge and Septage, which states "any additional substance for which sampling and analysis is required by the Department, before or after the sludge or septage is approved by the Department pursuant to 310 CMR 32.11." Also, see URL: https://www.mass.gov/doc/required-laboratory-procedures-for-testing-pfas-in-residuals/download.

was above the upper acceptance limit or not analyzed with the residual extraction batch; this indicated that the residual PFAS results above the RL were estimated (could be biased high or low).

Results of the 2020–2021 QC evaluations were used to inform ongoing residual analyses in Massachusetts. MassDEP communicated results for individual data packages and for the overall analysis to the laboratories, contributing facilities, and their management to refine protocols and execution of the residual PFAS monitoring program. Additional analyses of the magnitudes of PFAS concentrations over time and of duplicate precision were used to recommend field sampling and duplication frequency and is a technical issue many states and other entities are beginning to address.

3.3.3 Tissue contaminants

As with biological monitoring and bioassessments, performance measures and MQOs are essential for both the field and laboratory aspects of tissue contaminant monitoring studies of aquatic biota (e.g., fish, mollusk, or crustacean tissue studies for human health or ecological risk management and communication). QA planning and implementation should focus on defining DQOs, designing a QC system to measure data quality, and assessing data quality to determine its suitability to support management decisions regarding future monitoring, risk assessment, or issuance of consumption advisories [31, 51].

Field QC procedures need to be detailed in SOPs and as noted previously, sampling practitioners need to be trained in those program-specific procedures. A primary QA concern for the field collection, handling, preservation, and shipping stages of tissue contaminant studies is the preservation of tissue sample integrity. The accuracy of analytical results depends in part on the immediate preservation (i.e., freezing) of tissues and the prevention of exposure to extraneous sources of contamination. Those sources need to be identified and avoided or eliminated. Field blanks, or rinsates of empty field sample containers have been used by some investigators to evaluate field sample packaging materials as sources of contamination, with a control limit of less than the MDL as determined for the particular analytical method or monitoring program [51]; however, immediate freezing of whole organisms in the field (and preparation of tissue in the laboratory) and the use of food-grade packaging materials reduces or even eliminates the need for field blanks. Some studies may require tissue resection in the field, but sample processing (including resections) conducted under controlled laboratory conditions reduces the potential for sample contamination. One means of evaluating the efficacy of tissue preparation cleaning and decontamination procedures is the preparation and analysis of processing blanks or rinsates of the equipment used for dissecting and homogenizing tissues. As with field contamination QC measures, the control limit for processing blanks would also be <MDL for the particular analytical method or monitoring program. Control limit exceedances require suspension of sample preparation and specific corrective action by the preparation laboratory before resection or homogenization may resume.

Overall completeness is the number of valid sample measurements relative to the number of samples planned for collection, and it may be impacted by a variety of circumstances, e.g., storm events, samples lost during shipment, etc. Completeness objectives vary by study administrators and can range from 80% to 99%, with levels <80% generally requiring corrective action such as resampling or reanalysis [33, 34, 51]. Sampling precision (or the degree of agreement among replicate measurements caused by random error) can be estimated by comparing field replicates using RSD; however, acceptable field replicate samples require the collection of target organisms of the same species and the same sizes collected from the same location which may not always be possible. Rather than establishing acceptance limits for sampling precision, some researchers have instead used field replicate results to aid in the evaluation of study results and characterize the variability of the sampled population [32, 34]. Variability arising from tissue preparation (e.g., homogenization, compositing, and aliquoting), shipping, and laboratory analysis processes can be estimated by having the sample preparation laboratory prepare duplicate tissue homogenate or processed composite sample pairs to be analyzed as blind duplicates. [32] applied a MQO specifying that the RPD for these duplicate tissue composite pairs should be <50% for values greater than 5× the minimum level of quantification (ML) for each target contaminant and <100% for values <5× the ML.

In addition to the use of duplicate homogenate or composite sample pairs, a standard suite of laboratory QC measures including initial precision and recovery (IPR) samples, and matrix spike and matrix spike duplicate samples provides information about the precision associated with various components of the analytical process. IPR samples are used to demonstrate that a laboratory can achieve precision and accuracy using a particular analytical method prior to the analysis of any tissue study samples. They consist of a reference matrix (i.e., one that matches the study tissue matrix) that is spiked to a known level with the target contaminant. Accuracy is measured by the average recovery of the target chemical in replicate IPR samples. Precision is assessed by calculating RSD of the measured concentrations of the target chemical in the IPR samples. Matrix spike samples are field sample tissue homogenates with known amounts of a target chemical spiked into the sample to assess the effect of matrix interferences on compound identification and quantitation (measured as percent recovery of the chemical). Duplicate matrix spike samples consist of additional aliquots of matrix spike samples that are analyzed to assess the effect of tissue matrix interferences and are routinely used to assess method precision. Summarizing measurement QC limits for tissue studies is not as straightforward as identifying measurement quality indicators. Analytical QC limits vary with target chemicals and analytical methods. [51] provides general control limit recommendations and associated corrective actions for fish and shellfish tissue studies.

3.4 Ecotoxicology

Ecotoxicology tests are used in many countries and environmental programs as one of several approaches to assess environmental condition of soils, sediments, and water, toxicity of chemicals (including pesticides), and compliance with environmental regulatory statutes (e.g., the Clean Water Act in the U.S.). Toxicity testing for these various programs is largely conducted in a controlled laboratory setting according to specific test method protocols, e.g., [46, 52, 53], although mesocosm and in situ toxicity testing is also used in some cases in aquatic testing of chemicals, for example, (e.g., [54, 55]). Toxicity test results consist of two types of information: biological measurements and statistical interpretation of the observed biological data. Biological measurements are the raw data recorded when conducting toxicity tests (e.g., survival, weight, number of young produced). The statistical interpretation of a toxicity test is derived from the observed biological data.

Like other types of methods that rely on biological data, results of a toxicity test depend on the method used. Ecotoxicological testing relies on several QC procedures and analyses to help document that the test method performs acceptably given program DQO [46, 52]. Two key QA procedures used in all ecotoxicology testing are: (1) results from testing with a reference toxicant and (2) meeting minimum test acceptability criteria.

In reference toxicant testing, test organisms are exposed to a range of concentrations of a known toxicant or positive control (e.g., a metal such as copper or a salt such as potassium chloride for aquatic testing, e.g., [56–58]). Organism response to that toxicant is compared against an acceptable range of response previously established by the laboratory for the test organism and test method. Control charts are developed based on several reference toxicant tests for a given test species and test method to document an acceptable range of response to the toxicant [46]. In practice, statistical point estimate endpoints rather than the raw data are used to document results of each test and establish an acceptable range of response for a test method and reference toxicant. Often, a series of performance measures is used with corresponding MQO to address a range of relevant concerns (Table OS-11 cdn.intechopen.com/public/259766_osi.zip). Examples of point estimate endpoints include the lethal concentration to 50% of the test organisms (LC50) and the concentration resulting in a 25% inhibition in response compared to the control organisms (IC25). Point estimate endpoints have the advantage of generating 95% CIs around the mean value so that within test variability as well as between test variability can be established. These endpoints can be compared across tests and laboratories for a given chemical because the endpoint is not dependent on the concentration series used.

The second key QA requirement is that each test method has minimum test acceptance criteria (TAC) for control organisms that should (must for some programs such as the NPDES program in the U.S.) be met in a test for the results to be considered of acceptable quality. Examples of TACs include metrics such as minimum acceptable percent survival for organisms in a clean control matrix, minimum growth, and minimum number of offspring per female that must be achieved in the controls in a test [46].

A key performance measure in ecotoxicological testing is within-test variability or precision, both in the controls alone and for the entire test. Laboratories track performance metrics for the control over time to assess within-test variability. This is accomplished by calculating the mean, standard deviation (SD), and coefficient of variation (CV) of the control replicate data for each test conducted by the laboratory for a given test method. A statistical metric that is used to calculate within-test variability for the test as a whole is percent minimum significant difference (PMSD) [59, 60], which is derived from an Analysis of Variance (ANOVA) and Dunnetts Multiple Comparison Analysis. The PMSD documents the percent effect that can be statistically distinguished as compared to the control in the test based on the within-test variability observed.

Allowable ranges of PMSD values were derived by EPA using multiple tests for a given test method [59]. Controlling both minimum as well as maximum intra-test variability in whole effluent toxicity (WET) tests is seen as an important test acceptance factor. Too much variability among control replicates reduces the ability to distinguish statistical difference in organism response among treatments. Too little variability among control replicates, on the other hand, can yield statistically significant differences among-test concentrations and the control that are *biologically meaningless*. Controlling within-test precision is key to achieving the optimal sensitivity possible using a particular test species and ecotoxicology test method.

4. Conclusions

It should be noted that there has been vigorous debate on the appropriateness of actions that should result from interpreting statistical deviation in terms of process or QC [6], including that practitioners should avoid over-interpretation. This includes suggestions that unnecessary adjustments in processes could actually increase frequency of anomalous results. The implication here is that someone interpreting and developing recommendations from QC analysis who is not knowledgeable about the field of practice or study risks having a program just working toward a number, rather than truly trying to improve a process or determine the quality of environmental data for use in assessing ecological outcomes.

Recognition of the causes, magnitude, and effects of variability and error is attained through consistent observation and measurement and can simultaneously provide direction on the need for and types of corrective actions. Appropriately developed and implemented MQO, as part of consistent and routine measurement and monitoring programs, not only function to keep them on-track, but in the long run can also lead to more cost- and time-efficient processes.

Acknowledgements

The authors here express gratitude to Dr. Richard Mitchell (US EPA/Office of Water) for participation in and contribution to discussions related to the structure and content of this chapter. We would also like to recognize Lealdon Langley, Susannah King, and Jennifer Wood of the Massachusetts Department of Environmental Protection for their contributions during reviews of the Massachusetts PFAS data packages.

Author details

James Stribling[1]*, Susan Lanberg[2], Blaine Snyder[1], Jerry Diamond[1], Benjamin Jessup[3] and Hugh Sullivan[4]

1 Tetra Tech, Inc., Center for Ecological Sciences, Owings Mills, Maryland, USA

2 Tetra Tech, Inc., Environmental Sciences, Fairfax, Virginia, USA

3 Tetra Tech, Inc., Center for Ecological Sciences, Montpelier, Vermont, USA

4 US Environmental Protection Agency, Office of Water, Washington, DC, USA

*Address all correspondence to: james.stribling@tetratech.com

References

[1] IOM (Institute of Medicine). Environmental Decisions in the Face of Uncertainty. Institute of Medicine. Washington, DC: The National Academies Press; 2013. p. 281. DOI: 10.17226/12568

[2] Bradshaw G and Borchers J. Uncertainty as information: Narrowing the science-policy gap. Conservation Ecology. 2000;**4**(1):7. Available from: https://www.ecologyandsociety.org/vol4/iss1/art7/

[3] Lindley D. Uncertainty Einstein, Heisenberg, Bohr, and the Struggle for the Soul of Science. New York, NY: Anchor Books, a Division of Random House; 2007. p. 257. ISBN: 978-1-4000-7996-4

[4] Shewhart W. Economic Control of Quality of Manufactured Product. 7th Printing. D. Van Nostrand and Company; 1923 (Note, contemporary references to this book use 1931 as the publication date; 1923 is the date at the bottom of the Preface)

[5] Shewhart W. Statistical Methods from the Viewpoint of Quality Control. Washington, D.C.: The Graduate School, U.S. Department of Agriculture; 1939, 1939. Republished 1986, with a new Foreword by W.E. Deming. Mineola, NY: Dover Publications, Inc. p. 105

[6] Woodall W. Controversies and contradictions in statistical process control. Journal of Quality Technology. 2000;**32**(4):341-350. DOI: 10.1080/00224065.2000.11980013

[7] Divorski S, Scheirer M. Improving data quality for performance measures: Results from a GAO study of verification and validation. Evaluation and Program Planning. 2001;**24**:83-94. DOI: 10.1016/S0149-7189(00)00049-5

[8] Copeland C, Simpson M. The Information Quality Act: OMB's Guidance and Initial Implementation. Congressional Research Bulletin Report for Congress. Order Code RL32532. Washington, D.C.: The Library of Congress, Congressional Research Service; 2004

[9] ANSI/ASQC. American National Standard Specifications and Guidelines for Quality Systems for Environmental Data Collection and Environmental Technology Programs. E4-1994. American National Standards Institute/American Society for Quality Control, Milwaukee, WI: ASQ Quality Press; 1995

[10] U.S. EPA. Overview of the EPA Quality System for Environmental Data and Technology. EPA/240/R-02/003. Washington, D.C.: U.S. Environmental Protection Agency, Office of Water; 2002a

[11] U.S. EPA. Guidance on Systematic Planning Using the Data Quality Objectives Process. EPA QA/G-4. EPA/240/B-06/001. Washington, D.C.: U.S. Environmental Protection Agency, Office of Environmental Information; 2006

[12] U.S. EPA. EPA Requirements for Quality Assurance Project Plans. EPA QA/R-5. EPA/240/B-01/003. 2001; Reissued May 2006. U.S. Environmental Protection Agency, Washington, D.C.: U.S. Environmental Protection Agency, Office of Environmental Information

[13] Barbour M, Gerritsen J, Snyder B, Stribling J. Revision to the Rapid Bioassessment Protocols for Streams and Wadeable Rivers: Periphyton, Benthic Macroinvertebrates and Fish. 2nd ed. Washington, D.C.: U.S. Environmental

Protection Agency, Office of Water; 1999 EPA/841-D-97-002

[14] Flotemersch J, Stribling J, Paul M. Concepts and Approaches for the Bioassessment of Non-Wadeable Streams and Rivers. EPA/600/R-06/127. Cincinnati, OH: U.S. Environmental Protection Agency, Office of Research and Development, National Exposure Research Laboratory; 2006

[15] U.S. EPA. National Lakes Assessment. Field Operations Manual. EPA 841-B-11-003. 2011. Washington, D.C.: U.S. Environmental Protection Agency, Office of Water; 2012

[16] U.S. EPA. National Lakes Assessment. Laboratory Operations Manual. Version 1.1. EPA-841-B-11-004. Washington, D.C.: U.S. Environmental Protection Agency, Office of Water; 2012

[17] U.S. EPA. National Lakes Assessment 2017 Field Operations Manual Version 1.1. EPA 841-B-16-002. Washington, D.C.: U.S. Environmental Protection Agency, Office of Water; 2017a

[18] U.S. EPA. National Lakes Assessment 2017 Laboratory Operations Manual Version 1.1. EPA 841-B-16-004. Washington, D.C.: U.S. Environmental Protection Agency, Office of Water; 2017b

[19] U.S. EPA. National Rivers and Streams Assessment 2018-19: Laboratory Operations Manual Version 1.1. EPA-841-B-17-004. Washington, D.C.: U.S. Environmental Protection Agency, Office of Water; 2018

[20] U.S. EPA. National Rivers and Streams Assessment 2018/19 Quality Assurance Project Plan, Version 1.2. EPA 841-B-17-001. Washington, D.C.: United States Environmental Protection Agency, Office of Water; 2019b

[21] U.S. EPA. National Rivers and Streams Assessment 2018/19: Field Operations Manual – NonWadeable. EPA-841-B-17-003b. Washington, D.C.: U.S. Environmental Protection Agency, Office of Water; 2019c

[22] U.S. EPA. National Rivers and Streams Assessment 2018/19 Field Operations Manual Wadeable Version 1.2. EPA-841-B-17-003a. Washington, D.C.: U.S. Environmental Protection Agency, Office of Water; 2019d

[23] U.S. EPA. National Coastal Condition Assessment 2020 Quality Assurance Project Plan, Version 1.2. EPA 841-F-19-003. Washington, D.C.: U.S. Environmental Protection Agency, Office of Water; 2021a

[24] U.S. EPA. National Coastal Condition Assessment 2020 Field Operations Manual. EPA 841-F-19-005. Washington, D.C.: U.S. Environmental Protection Agency, Office of Water; 2021b

[25] Stribling J. Partitioning error sources for quality control and comparability analysis in biological monitoring and assessment. Chapter 4. In: Eldin A, editor. Modern Approaches to Quality Control. London, UK: INTECH Open Access Publisher; 2011. pp. 59-84. ISBN 978-953-307-971-4. DOI: 10.5772/22388

[26] Stribling J, Jessup B, Feldman D. Precision of benthic macroinvertebrate indicators of stream condition in Montana. Journal of the North American Benthological Society. 2008a;**27**(1):58-67. DOI: 10.1899/07-037R.1

[27] Stribling J, Pavlik K, Holdsworth S, Leppo E. Data quality, performance, and uncertainty in taxonomic identification for biological assessments. Journal of the North American Benthological Society. 2008b;**27**(4):906-919. DOI: 10.1899/07-175.1

[28] Klemm D, Stober Q, Lazorchak J. Fish Field and Laboratory Methods for Evaluating the Biological Integrity of Surface Waters. EPA/600/R-92/111. Cincinnati, OH: U.S. Environmental Protection Agency, Office of Research and Development; 1993

[29] IDEM. Quality Assurance Project Plan for Biological Community and Habitat Measurements. B-003-OWQ-WAP-XXX-20-Q-R0. Indianapolis, IN: Indiana Department of Environmental Management, Office of Water Quality, Watershed Assessment and Planning Branch; 2020

[30] Kopp B, Neckles H. A Protocol for Monitoring Estuarine Nutrient Enrichment in Coastal Parks of the National Park Service Northeast Region. Natural Resource Report NPS/NCBN/NRR—2009/110, Augusta. ME: U.S. Geological Survey Patuxent Wildlife Research Center; 2009

[31] U.S. EPA. Quality Assurance Project Plan for Analytical Control and Assessment Activities in the National Study of Chemical Residues in Lake Fish Tissue. EPA 823-R-02-006. Washington, D.C.: U.S. Environmental Protection Agency, Office of Water; 2000

[32] U.S. EPA. The National Study of Chemical Residues in Lake Fish Tissue. EPA-823-R-09-006. Washington, D.C.: U.S. Environmental Protection Agency, Office of Water; 2009

[33] Neebling T. Standard Operating Procedure for Fish Tissue Contaminant Sampling – Field and Laboratory Methods, Casper. WY: Wyoming Game and Fish Department, Fish Division; 2012

[34] Seiders K, Sandvik P. Freshwater Fish Contaminant Monitoring Program. Publication 20-03-106, Olympia. WA: Washington State Department of Ecology; 2020

[35] U.S. EPA. Method 533, Determination of Per-and Polyfluoroalkyl Substances in Drinking Water by Isotope Dilution Anion Exchange Solid Phase Extraction and Liquid Chromatography/Tandem Mass Spectrometry. EPA 815-B-19-020. Cincinnati, OH: U.S. Environmental Protection Agency, Office of Ground Water and Drinking Water; 2019a

[36] U.S. EPA. National Lakes Assessment 2017 Quality Assurance Project Plan Version 1.1. EPA-841-B-16-003. Washington, D.C.: U.S. Environmental Protection Agency, Office of Water; 2017c

[37] U.S. EPA. National Coastal Condition Assessment 2020 Laboratory Operations Manual, Version 1.2. EPA 841-F-19-004. Washington, D.C.: U.S. Environmental Protection Agency, Office of Water; 2021c

[38] U.S. EPA. Method 546: Determination of Total Microcystins and Nodularins in Drinking Water and Ambient Water by Adda Enzyme-Linked Immunosorbent Assay. EPA 815-B-16-011. Cincinnati, OH: U.S. Environmental Protection Agency, Office of Ground Water and Drinking Water; 2016

[39] U.S. EPA. Method 545: Determination of Cylindrospermopsin and Anatoxin-a in Drinking Water by Liquid Chromatography Electrospray Ionization Tandem Mass Spectrometry (LC/ESI-MS/MS). EPA 815-R-15-009. Cincinnati, OH: U.S. Environmental Protection Agency, Office of Ground Water and Drinking Water; 2015

[40] Berger W, McCarty H, and Smith R. Environmental Laboratory Data Evaluation. Schenectady, NY: Genium Publishing Company. 1996. ISBN 1-931690-91-9

[41] U.S. EPA. Single Laboratory Validated Method for Determination of Cylindrospermopsin and Anatoxin-a in Ambient Freshwater by Liquid Chromatography Electrospray Ionization Tandem Mass Spectrometry (LC/ESI-MS/MS). EPA/600/R-17/130. 2017d. Cincinnati, OH: U.S. Environmental Protection Agency, Office of Research and Development;

[42] U.S. EPA. Single Laboratory Validated Method for Determination of Microcystins and Nodularin in Ambient Freshwater by Solid Phase Extraction and Liquid Chromatography/Tandem Mass Spectrometry (LC/MS/MS). EPA/600/R-17/344. Washington, D.C.: U.S. Environmental Protection Agency, Office of Research and Development; 2017e

[43] ASTM. Standard Test Method for Measuring the Toxicity of Sediment-Associated Contaminants with Freshwater Invertebrates. ASTM E1706-19, West Conshohocken. PA: American Society for Testing and Materials; 2019

[44] U.S. EPA. Methods for Measuring the Toxicity and Bioaccumulation of Sediment-associated Contaminants with Freshwater Invertebrates. EPA/600/R-99/064. Washington, D.C.: U.S. Environmental Protection Agency, Office of Research and Development and Office of Water; 2000b

[45] U.S. EPA. Methods for Measuring the Acute Toxicity of Effluents and Receiving Waters to Freshwater and Marine Organisms. EPA-821-R-02-012. 2002b. Washington, D.C.: U.S. Environmental Protection Agency, Office of Water;

[46] U.S. EPA. Short-term Methods for Estimating the Chronic Toxicity of Effluents and Receiving Waters to Freshwater Organisms. EPA-821-R-02-013. Washington, D.C.: U.S. Environmental Protection Agency, Office of Water; 2002c

[47] Kirk J. Light and Photosynthesis in Aquatic Ecosystems. Cambridge University Press; 1994. DOI: 10.1017/CBO9780511623370

[48] APHA. Standard Methods for the Examination of Water and Wastewater. 21st ed. Washington, D.C.: American Public Health Association, American Water Works Association, and Water Environment Federation; 2005

[49] Keith L. Environmental Sampling and Analysis. A Practical Guide. 1991. Chelsea, MI: Lewis Publishers. p. 160. ISBN: 9780873713818

[50] Smith R. Interpretation of Organic Data. Amsterdam, New York: Genium Publishing Corporation, Genium Group, Inc., 2000. ISBN: 1-890911-19-4

[51] EPA US. Guidance for Assessing Chemical Contaminant Data for Use in Fish Advisories, Volume 1: Fish Sampling and Analysis. 3rd ed. EPA-823-B-00-007. Washington, D.C.: U.S. Environmental Protection Agency, Office of Water; 2000

[52] ASTM. Standard Test Method for Measuring the Toxicity of Sediment-Associated Contaminants with Freshwater Invertebrates. ASTM E1706-19, Philadelphia. PA: American Society for Testing and Materials; 2016

[53] OECD (Organization for Economic Cooperation and Development). OECD Test Guidelines for Chemicals [Internet]. 2022. Available from: https://www.oecd.org/chemicalsafety/testing/oecdguidelinesforthetestingofchemicals.htm [Accessed: 08 February 2022]

[54] U.S. EPA. Series 850 - Ecological Effects Test Guidelines. Washington,

D.C.: U.S. Environmental Protection Agency, Office of Pesticides and Toxic Substances; 2022

[55] Pereira A, Soares A, Alves F, Ribeiro R. Test chambers and test procedures for in situ toxicity testing with zooplankton. Environmental Toxicology and Chemistry. 1999;**18**: 1956-1964. DOI: 10.1002/etc.5620180914

[56] ECCC. Guidance Document on Control of Toxicity Test Precision Using Reference Toxicants. Environmental Protection Series. [Reference methods] EPS 1/RM/12. Ottawa, Ontario: Environment and Climate Change Canada, Environment Canada; August 1990

[57] Lazorchak J, Smith M. Rainbow trout (*Oncorhynchus mykiss*) and brook trout (*Salvelinus fontinalis*) 7-day survival and growth test method. Archives of Environmental Contamination and Toxicology. 2007;**2007**(53):397-405. DOI: 10.1007/s00244-006-0227-8

[58] Environmental Health Standing Committee (enHealth). The Role of Toxicity Testing in Identifying Toxic Substances in Water: A Framework for Identification of Suspected Toxic Compounds in Water. Canberra: Environmental Health Standing Committee, Australian Health Protection Principal Committee; 2012

[59] EPA US. Understanding and Accounting for Method Variability in Whole Effluent Toxicity Applications Under the National Pollutant Discharge Elimination System Program. EPA 833-R-00-003. Washington, D.C.: U.S. Environmental Protection Agency, Office of Wastewater Management; 2000d

[60] Denton D, Fox J, Fulk F. Enhancing toxicity test performance by using a statistical criterion. Environmental Toxicology and Chemistry. 2003;**22**: 2323-2328. DOI: 10.1897/02-384

Chapter 2

Quality Impairments in Flexible Road Pavements

Samuel I. Egwunatum, Ovie I. Akpokodje and Andrew I. Awo-Osagie

Abstract

The purpose of this chapter is to present the reader with the physical processes of how flexible road pavements progressively fail and impair the quality of finished roads arising from non-adherence to roads construction quality outlines and requirements. This was achieved by investigating eight (8) roads from a sample of nineteen (19) roads based on purposive sampling. Using instruments of steel tapes, paints for failed sections, rolling rule and pictures, measurement of length, width and depth of various failed sections were taken for five (5) daily measurements at three (3) monthly visit intervals for Four Hundred and Thirty Five (435) days to show the rate of deterioration. Data obtained were analyzed for reliability of pavements using Weibull distribution statistics on ReliasoftWeibull^{++}to extrapolate pavement reliability from bathtub function. Findings showed that roads failed progressively within six (6) months after finished construction and deteriorated fast with increased failures on length, depth and width of pavements. The practical implications of this is that the process of construction did not conform with required/stipulated quality control metrics of flexible road construction especially in the areas of geomaterials compaction, temperature and density of materials laid. It was recommended that organization adhere to quality control guidelines and requirements to forestall quality impairment.

Keywords: flexible pavement, Weibull test, quality control, quality impairment, deterioration rate, reliability

1. Introduction

Issues of total quality management implementation in different construction industries around the world are well validated in studies to be at various levels of implementation between developing and developed countries [1]. Structural failures of roads before the designed lifetime are regular features especially amongst developing countries in the form of evidential failure of a small structural component, accelerative failure with visible weakness such as cracks and abrupt, sharp failures [2] etc.

Non-compliance to specification outlines of projects demand –amongst others- the use of non-standard materials, ineffective/unqualified team members of quality control rangers, fast track construction, poor detailed designs, etc. and may be regarded as are remote and immediate factors leading to failures [3]. Quality Control as a process in road construction ensures conformity of the finished road pavement with required standards [4]. Quality defects are obtainable from the difference in the coefficient of variation of required elasticity modulus (C_v^{req}) from the standard elasticity modulus (C_v^{std}) which is based on deflection patterns of geomaterials composition and dynamic load intensity on pavements. The wider the difference between such elasticity moduli, the higher the propensity to failure of pavements [5]. Owing to such probability outlook, their reliability estimate, takes the form of

$$R_{p_t} = 0.5 + F\left[\frac{E_{eq} - E_m}{\sqrt{\sigma_{eq}^2 + \sigma_m^2}}\right] \tag{1}$$

where E_{eq} is the equivalent modulus of elasticity, E_m is the maximum modulus of elasticity, σ_{eq} is the equivalent modulus of elasticity and σ_m is the mean square deviation of the maximum modulus of elasticity.

The process of quality control in road pavements follows the examination and test of composite materials towards meeting the correct specifications and required quality. Quality control in road projects follow stratification checks by separating roads composite materials and bringing them to specialized and accredited laboratories in order to conduct a series of tests on them. However, it is worth noting at this point that specific tests may also be done on site using checklists as the construction progresses.

Absence of quality control checks has often resulted to impaired quality outputs and poor workmanship [6]. Quality impairment in road construction processes shows that finishing road surfaces, construction process, labour workforce and materials used are in need of quality review and standardization for improvement [7]. The same applies to the workforce involved. Evidences of quality failures in constructed roads is revealed in their reliability values from their mean survival time to failure time, which are consequential fall-outs of quality management principles not being implemented. Quality in the context of road construction is when functionality is at equilibrium with a construction process output based on road utilization from effective road performance, durability, conformance, reliability, uniformity and serviceability [8]. Further to this, impaired quality of constructed roads are revealing in varying forms of cracks, potholes, bulges and surface depressions that often results in poor transportation systems, and delayed economic growth [9]. Quality impairment of roads indicates an increased level of reliability failures. The aim of this paper is to parametrically estimate their durability.

2. Road construction and quality practice

Road constructions are either flexible or rigid highway pavements with most or all of the following construction materials *viz.*, soil, aggregates, admixtures, Portland cement concrete, Bituminous materials, structural steel and pavement markers [10]. All of these materials are compositely layered together in a definite mix and proportion to output a quality road carriageway [11]. Determinants of high quality roads are subjects of quality tests on the various road materials enumerated above. Test on highway materials such as, Moisture Content Value (MCV), Los Angeles Abrasion Value, Dynamic Cone Penetrometer, Flakiness Index, Penetration Test on Bitumen,

California Bearing Ratio (CBR), Softening point test on Bitumen and Ground Penetrating Radar tests are various laboratory test prerequisites for quality road [12].

Table 1 presents road construction tests for quality assured output.

Flexible road pavements construction primarily consist of 70% asphalt bitumen content that provides binder mix with aggregate to produce asphalt concrete. This is laid on a bituminous base of a binder course. Stabilization of this process is followed by the application of tack coat of 0.75 kg per sq. metre [8]. Quality control standard as required in the preparation and placing of premix material is that bitumen is heated in the temperature range of (150–177^0) C within which aggregate temperature must not differ by 14°C from the binder temperature [13]. The hot mixed material of the bitumen and the aggregate together with the binder is then paved at a satisfactory temperature of not more than 163°C. This is followed for a smoother surface with a roller compaction at a speed not exceeding 5 km per hour. Preliminary or breakdown rolling uses 8 to 12 tonnes rollers and further pressurized or intermediate rolling is done using 15 to 30 tonnes fixed wheel pneumatic rollers.

During construction, the routine quality control checks carried-out to ensure quality output are often stipulated in the watch-out for resulting pavement mix, temperature at point of laying and pavement gauge or thickness. Other checks not necessarily routine but periodical are checks for aggregate grading, bitumen content grade, temperature of aggregate temperature of paving mix at mixing and compaction [14]. At every 100 tonnes of mix discharged by the hot mix plant samples are collected for the above tests. Another test for quality compliance is carried-out by implementing the Marshall test for every 100m^2 paved and compacted [15]. This is also followed by the field density check to see if 95% of laboratory density obtained shows congruency in the field. Tolerance of 6 mm per 5 m length of paved surface is allowed for variations in depth of pavements [16]. Variations from longitudinal undulations along the straight edge at every 3.0 m check must not exceed 8.00 mm and the number of undulations higher than 6.0 mm should not exceed 10 for every 300 m of road. Near absence of quality checks in road construction projects are traceable to road failures in the form of cracks, potholes, bulge and creter depressions. A typical quality controlled road pavement construction is shown in **Figure 1**.

Failed roads maybe regarded as evidences of quality neglects. Road failures are progressive in nature with monotonic properties of lebesgue measure theory with respect to progressive road component failures. A collection of road used in a similar traffic pressured fashion normally will show propensity to fail within predictable time measures [17]. Determination of such failings owing to quality neglect is provided for in Weibull reliability analysis under the scheme of plotting the percentage of road sections that have yielded to failure over a randomized time period measurable in cycle-starts, hours of run-times, miles driven, etc. [18]. Usually, classification of quality impairment is obtainable from Weibull reliability analysis with non-linear bathtub graph having to be approximated with line of best-fit, with β describing the classification in:

$\beta < 1.0 =>$ Infant mortality $=>$ Optimum quality impairment in construction.

$\beta = 1.0 =>$ Randomized failure $=>$ Progressive quality impairment during construction.

$\beta > 1.0 =>$ Wear-Out Failure $=>$ High quality impairment during construction.

Most decent and prudent statistical inferences in Weibull test are parametrized with Time-to-Failure component of the road. This is historically accounted for by B (F) with 'F' representing the percentage of road section that have failed, while some parametrize by lifetime L(F) and 'B' representing bearing time. In the Weibull

S/N	Test type	Purpose	Test methods	Quality criterion	Expected outcome
1.	CBR – Test for Subgrade	A penetration test for the determination of mechanical fitness strength of the natural ground, subgrade and base course underweight on the carriageway	• Load bearing capacity • Moisture content • Potential for shrinkage and/or swelling	• Ease of compaction • Strength retention • Low volume response to adverse weather condition and capillary movement of ground water • Inability to compress • Bearing capability for stability	• Ability to furnish and dispense support to the finished pavements in resistance to traffic loads • Must have enough stability under inclement weather and heavy stack situation
2.	Aggregate Testing	Load transfer potentials or capability of finished pavement	• Crushing test • Abrasion test • Impact test • Soundness test • Bituminious adhesion test • Specific gravity and water absorption test	• Enabling relative offer of resistance to gradual traffic load • Ability to show hardness property of aggregate material • Ability to offer resistance to impacts on aggregate obtained as a percentage of aggregate passing sieve • Showing potential to resisting actions of weathering on aggregate under conditions of varying temperatures in sulfate solutions of sodium and magnesium. Weight loss not exceeding 12% and 18% on these solutions • Offering propensity to resist water permeability in voids on road surfaces. • Ability to show adhesion of bitumen binding to aggregate free from moisture and has no permeable water inlet	Aggregates in finished pavements must show promise of load transfer potential and capability according to test pass.
3.	Penetration test	Hardness or softness of Bitumen	• Penetration depth under the action of standard loaded needle	• Able to show penetration resistance with reference to hardness or softness of bitumen when needle load is applied under conditions of pouring temperature, size of needle and loading weight on needle	A desirable penetration value of 150 – 200 mm within 5 seconds, for cold climates or lower for hot climates
4.	Ductility test	Envisaged Bitumen deformation or elongation	• Measurement of distance to which a standard field sample of Bitumen material will be elongated without breaking at 27°C and 90 minutes rapid cooling	• Output a minimum ductility value of 75 cm under stressed condition of pulling rate, test temperature and pouring temperature	Bitumen must show ability to slow gradual deformation or elongation even at quick optimized stress and strain.

S/N	Test type	Purpose	Test methods	Quality criterion	Expected outcome
5.	Softening Point Test	To show at what temperature bitumen attains a specific point of softening	Using ring and ball apparatus where a brass ring holding sample of bitumen is placed in water or glycerin at a given temperature. Then the steel ball is placed on a bitumen sample also in the liquid medium and heated to 50°C in one minute	Output a temperature for which the softened bitumen touches the metal plate at a designated distance	Higher softening point shows lower temperature propensity and rudimentary in hot weather regions.
6.	Specific Gravity test	Determination of Bitumen binder density variation with aggregate	Specific gravity test by pycnometer or using weight of samples in air and water at 27°C.	Ensuring that mineral impurities of aromatic types are separated in the chemical composition of bitumen to keep the density at normal. With such mineral impurities, specific gravity of bitumen may increase	Obtaining a specific gravity of bitumen within 0.97 to 1.02
7.	Water content test	Prevention of bitumen foaming on heating to boiling point of water	Water distillation from a known weight of Bitumen specimen in a pure petroleum distillate, free of water. On heating, the water content in the specimen is collected from condensation and expressed as a percentage of weight of original specimen.	Water distilled is aimed at determination of allowable water content in the bitumen which it must contain to prevent foaming	Expected water content in Bitumen must be within the range of 0.2% by weight
8.	Heating loss test	Determination of volatility loss	A sample of about 50gm of bitumen is weighed and heated to 161°C for 5 hours in a specified oven. The sample is weighed again after heating and loss expressed as percent of weight of original sample.	Loss in weight of bitumen after heating shows not exceed 1% so as to retain its volatility	Relationship between Bitumen penetration value and weight loss must be 150–200 to 2% loss in weight.

Table 1.
Quality test metrics in flexible road pavements.

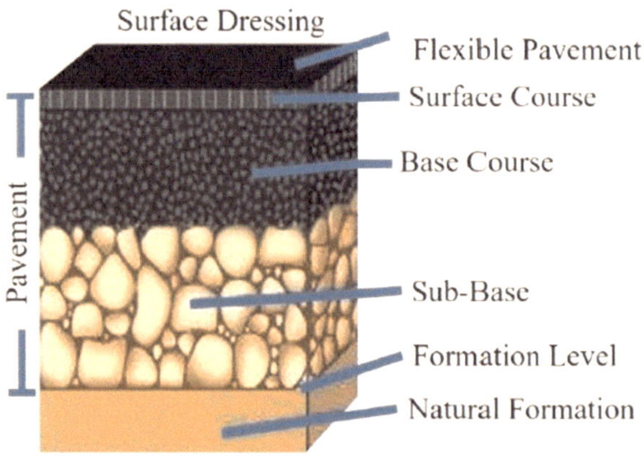

Figure 1.
Components of flexible pavements in Hassan and Sobhan [13].

statistics, the distribution shows the relationship between failed percentage with respect to time governed by constant shape factors 'β' and 'η' that determines shape and scale of distribution respectively by the function:

$$F(t) = 1 - e^{\left(\frac{-t}{\eta}\right)^{\beta}} \tag{2}$$

Summing the monotonic progressive failures over time to the point of measurement generates a probability density function (PDF) describing the frequency of failures over time estimates as:

$$f(t) = \left(\frac{\beta}{\eta}\right)\left(\frac{t}{\eta}\right)^{\beta-1} e^{-\left(\frac{t}{\eta}\right)^{\beta}} \tag{3}$$

Quality control checks on compacted geomaterials such as sub-base by a more precise measuring light weight deflectometer (LWD) device in the study of Duddu and Chennarapu [19] as against density and stiffness base methods with the aim of obtaining pavement deformation modulus (E_{LWD} have shown better predictive ability of deformations). For instance LWD tests on geomaterials such as soils, aggregates, and asphalt had output of 35/60 MP_a and 120/170 MP_a. As a quality control reference, LWD tests presents the user with information on longer life cycle pavement performance and predictive failure indicator time. Confirmation of such parametric evaluations follows the regressive test between LWD and other density/stiffness methods with better coefficient of determinations *(R^2)*. For instance, as outlined in the work of aforementioned researchers investigation of Sandy soil regression-correlation on California bearing ratio (CBR) and E_{LWD} by Dwivedi and Suman [4] gave R^2 values of 0.807 for unsoaked sand (US), 0.805 for soaked (S) sand and dry density 0.77 with the following relationship:

$$CBR_{(US)} = 0.0009E_{LWD}^{\;\;2}$$

$$CBR_{(S)} = 0.0001E_{LWD}^{\;\;2} \tag{4}$$

$$\gamma_d = 1 \times 10^{-5}E_{LWD}^{\;\;2}$$

Quality controlled output limits using their coefficient of determination (R^2) on lime based stabilized subgrade soil from correlative studies with E_{LWD} by in the literature of Bisht, Dhar and Hussain [20] for unconfined compressive strength (UCS) at $R^2 = 0.99$ and CBR at $R^2 = 0.93$ showed the following relation:

$$UCS = 4.9E_{LWD} \tag{5}$$

$$CBR = 0.15E_{LWD} \tag{6}$$

Studies by Nazzal, Abu-Farsakh, Alshibli and Mohammad [21] on crushed limestone and sandy soil geomaterials gave R^2 value of 0.83 between CBR and E_{LWD} with the following relation:

$$CBR = -14 + 0.66E_{LWD} \tag{7}$$

$$E_{v2} = (600 - 300)/(300 - E_{LWD-L_3}) \tag{8}$$

as a correlate between Static modulus of layer 2 (E_{v2}) and modulus of deformation measured by a Zorn LWD device with 300 mm diameter plate. Such stress/strain on flexible pavement layers often transfer elasticity modulus for determining pavement structural durability between layers. This is governed from the computation of road's elastic modulus (E_{gen}) based on 'g' the bearing capacity reserve of road bed and pavement in:

$$E_{gen} = \frac{E_1E_2\left[1 + \left(\frac{2h}{D}\right)^2\left(\frac{E_1}{E_2}\right)^{\frac{2}{3}}\right]^{\frac{1}{2}}}{E_1 - E_2\left\{1 - \left[1 + \left(\frac{2h}{D}\right)^2\left(\frac{E_1}{E_2}\right)^{\frac{2}{3}}\right]^{\frac{1}{2}}\right\}} \tag{9}$$

A similar correlation investigation on soil classification test between static modulus of pavement layer 1and deformation modulus using light weight deflectometer (LWD) by Alshibli, Abu-Farsakh and Seyman [22] showed a quality allowable R^2–value of 0.84. That of Rao, Shiva and Shankar [23] on subgrade geomaterials between CBR and E_{LWD} gave an R^2 value of 0.90 with the following regression result;

$$E_{v1} = 0.91E_{LWD-P_3} - 1.81 \tag{10}$$

$$CBR = -2.754 + 0.2867E_{LWD} \tag{11}$$

where E_{LWD-P_3} is the modulus of deformation measured by Prima 100 Cohesive and non-cohesive soils. Adam and Kopf [24] provided regression functions between static modulus of layer 1 (E_{vI}) and modulus of deformation from a Zorn LWD device with a 300 mm plate diameter. Deformation thresholds are predictable for quality control reasons for cohesive soils by the relationship:

$$E_{v1} = 0.833 \times E_{LWD-z3} \tag{12}$$

And for non-cohesive soils with the relation;

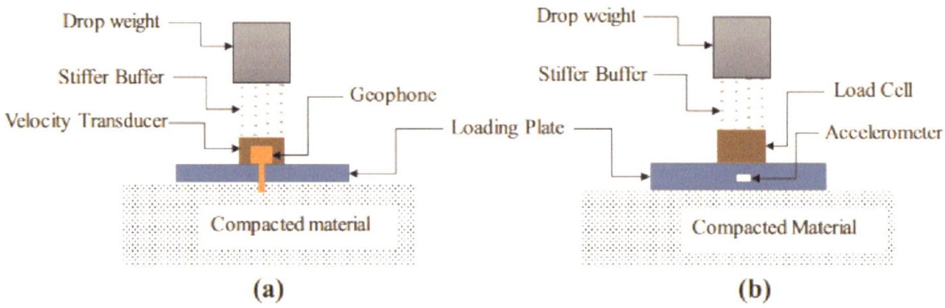

Figure 2.
*Schematic sketch of the location and type of transducer: **a** geophone measures velocity and is located on the compacted material, **b** accelerometer measures vibrations and is located in the plate. Photo credit: Duddu and Chennarapu [19].*

$$E_{v1} = 1.25 \times E_{LWD-z3} - 12.5(E_{LWD-z3})\vee_{range} at 10 - 90MPa \tag{13}$$

Quality control checks by light weight deflection (LWD) devices are conducted by velocity tracks using geophones or vibration tracks using accelerometer which is located on the test plate (see **Figure 2** from Duddu and Chennarapu [19]).

On the basis of limit state engineering designs, there are progressive failures at retail scales to yield a point of total failure beyond which roads become unserviceable to users before their expected lifetime span. Bazhanov and Saksonova [25] and Hassan and Sobhan [13] have shown that yield point in a quality impaired constructed road is attainable after a dynamic load is applied on pavements surface originating from a plastic deformation. Forms and types of road failures are shown in the accompanying **Table 2**.

According to Gupta [26], points of statistical references in reliability of pavement estimations are marked in the pavements failure rate (hazard rate) defined by:

$$r_F(k) = \frac{P(k)}{\sum_{i=k} P^{(i)},} \tag{14}$$

$$\frac{P(X=k)}{P(X \leq k)}, k = 0, 1, 2, \ldots$$

With $P(k) = P(X = k)$ being the mass function, cumulative distribution function $f(k) = P(X \leq k)$ and pavement survival function $\bar{F}(k) = 1 - f(k)$ respectively. The pavements' mean residual life, $(\mu_F(k)$is indicated by estimation bias as:

$$\mu_F(k) = E(X - k | X \geq k) = \frac{\sum_{x=k} \bar{F}(x)}{F(k-1)}, k = 0, 1, 2, \ldots.. \tag{15}$$

This estimation is premised on the deterioration force of decrement on the pavement lifespan which bears representation in plastic deformation in other to understand pavement tolerance [27]. Consequently, the pavement failure rates or hazard rates which are competing in risk value by a mortal force of decrement with mean residual life of pavement are relationally obtained by:

$$r_f(k) = \frac{1 + \mu_f(k+1) - \mu_f(k)}{1 + \mu_f(k+1)} \tag{16}$$

DOI: http://dx.doi.org/10.5772/intechopen.105697

Failure Type	Figure	Description
1. Cracks		
i. Alligator cracking		Originates from inadequate structural bends, poor drains. If poorly managed can deteriorate to potholes
ii. Transverse cracking		Originates from heavy traffic and poor materials mix design
iii. Block cracking		Originates from poor unstabilized base and defective ramming. This crack enables percolation of moisture and infiltration.

Failure Type	Figure	Description
iv. Longitudinal cracking		This form of crack is traceable to unstable base and defective construction. Moisture infiltration into the bed is prevalent.
v. Slippage Cracking		This form of crack failure originate from unstable wearing course and poor drainage
2. Depression	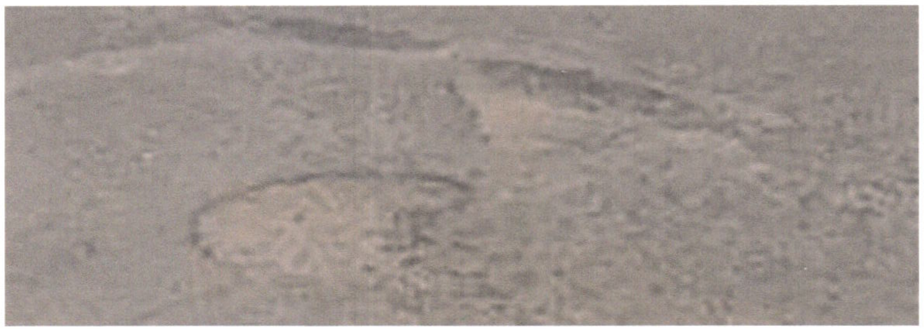	This form of road failure is very visible with creter form of depression. Its occurrence is traceable to heavy rainfall and improper side drainage

Failure Type	Figure	Description
3. Raveling		Failure is traceable to the inability of asphalt primers, binders and tack coats to hold aggregate in place, partial compaction and wearing off or weakening of asphalt binders.
4. Potholes		Failure is attributed to the exposure of road structural members to gradual wearing by heavy vehicular loads due to accumulation of surface rain water from cracks on pavements. It produces visible three (3) dimensional failures of depth, width and length.
5. Water Bleeding		Failure is occasioned by poor mix design of structural materials and with unsuitable binder. Failures of this nature lessens car skidding resistance.

Failure Type	Figure	Description
6. Corrugation and Shoving		Failure is often associated with roughness and elevated sections of the pavement. Poor materials mix design, high traffic loads and unstable binders are most the causes of this failure

Table 2.
Types of road failure.

$$1 - \frac{\mu_f(k)}{1 + \mu_f(k+1)}, k = 0, 1, 2, \dots..$$

The augmented pavements failure rate, mean residual life and its survival which are estimable consequences of an impaired engineering works is signified in a quality deficit index by relating the three statistical variables as:

$$\bar{F}(k) = \prod_{0 \le i \le k} [1 - r_f(i)]$$

$$\prod_{0 \le i \le k} \left[\frac{\mu_f(i)}{1 + \mu_f(i+1)}\right], \mu(0) = E(x) \tag{17}$$

Following the competing mortal forces of decrement on pavements with failure induced components yield from several real time traffic loadings, correspond to variations in the lifetime survival of pavement obtainable by:

$$\sigma_F^2(k) = Var(x - k \vee x \ge k)$$

$$k^2 + \frac{\sum_{i=k}^{\infty}(2i+1)\bar{F}(i)}{\bar{F}(k-1)} - \left(\frac{\sum_{i=k}^{\infty}\bar{F}(i)}{\bar{F}(k-1)} + k\right)^2$$

$$2\frac{\sum_{i=k}^{\infty}\bar{F}(i)}{\bar{F}(k-1)} - (2k-1)\mu_F(k) - \mu_F^2(k) \tag{18}$$

In order to idealize how quality is impaired by statistical reliability variables, the pavement's failure rate, mean residual life and variance residual life functions have causal aggregation and estimated by:

$$\sigma_F^2(k+1) - \sigma_F^2(k) = r_F(k)$$

Consequently, decreasing pavement variance residual life is X if X

$$\sigma_F^2(k+1) \le \mu_F(k)[1 + \mu_F(k+1)].$$

and it is an increasing variance residual life if

$$\sigma_F^2(k+1) \ge \mu_F(k)[1 + \mu_F(k+1)]$$

These statistical narrations in their numerical values are indicators of progressive failures with monotonicity properties for quality impairments assessment. In recent times, researches into deterioration rates of road pavements particularly in Riveros and Arredondo [28] and Al-Zahrani and Stoyanov [29] with transition probabilities indicated changes from one state to another (owing to deterioration). This illustrates precision predictability by Weibull distribution estimation. The probability density function are parametrized by $\propto$- and β- for which $(\propto > 0, \beta > 0)$ and given as:

$$F_{(t)} = \int_{\frac{\propto}{\beta}}^{o} \left(\frac{t}{\beta}\right)^{\propto - 1} exp\left[-\left(\frac{t}{\beta}\right)^{\propto}\right] \begin{matrix} fort < 0 \\ fort \ge 0 \end{matrix} \tag{19}$$

And its distribution function as:

$$F(x) = \int_1^0 exp\left[-\left(\frac{x}{\beta}\right)^{\propto}\right] \begin{matrix} forx < 0 \\ forx \geq o \end{matrix} \quad (20)$$

Under the Weibull test for pavement deterioration, expected values and variance are estimated by:

$$\mu = \beta^{-1}\left(1+\frac{1}{\propto}\right), \sigma^2 = \beta^2\left[\left(1+\frac{2}{\propto}\right) - {}^{2}\left(1+\frac{1}{\propto}\right)\right] \quad (21)$$

In this case, rather than Laplacian integral, the Weibull distribution is predicted on the gamma function

with: $((x))$

$$(x) = \int_o^{\infty} t^{x-1}e^{-t}dtforx > 0 \quad (22)$$

This chapter deployed the use of Weibull test to obtaining the deterioration rates of selected Benin city roads in cluster from generating their deterioration model by linear regression having deterioration state as a dependent variable and pavement as an independent variable.

3. Methodology

In this research study, quality impairment in road construction was assessed by field investigation of eight (8) failing roads from a purposive sampling from 19 failed roads in the Benin city metropolis of Nigeria. Obtaining life right censored data through measurements of component failed depth, width and length with a start and end observation times were obtained. In achieving this, a seven (7) days growth rate study of failed portions in five (5) different field visitations at an interval of three (3) months for each visit was conducted to enable the capture of variation in growth rate between visits. The research team also engaged four daily undergraduate students to support obtaining measurements and controlling traffic. From the historical data gathered, mixed Weibull distribution software (Reliasoft Weibull^{++}) was used to analyze data goodness-of-fit test. Their reliability function was also tested in terms of their failure rate function and mean life function by estimating the parameters that makes the reliability function most closely fit the life data set. A review of the statistical criterion reference analytically for model fitness, shape parameters, assumed *βs* and graphically for fit to line, S-shape and minimum life was done from the reliability bathtub curve plot while computing their statistical function at 90% confidence bounds. **Tables 3–5** depict life data measurement of failed roads.

4. Results and discussion

Figures 3–12 and **Tables 3–5** are discussed in this section.

Figure 3.
Showing how water aids road failure.

Figure 4.
Showing how water aids road failure.

Figure 5.
Showing how failed portion of roads affects or increase journey time.

Figure 6.
Failed portion in Luckyway Road.

Figure 7.
Failed portion in Mission Road.

Figure 8.
Failed portion in New Benin Road.

5. Weibull reliability test for the deterioration of the width of the roads in Lucky Way, Mission Road, New Benin Road, Ogida Road, Ring Road, Sapele Road, Technical College Road and Textile Mill Road

The shape parameter β and the 95% confidence interval of β for the data regarding the width of the roads in Lucky Way, Mission Road, New Benin Road, Ogida Road, Ring Road, Sapele Road, Technical College Road and Textile Mill Road are given in

Figure 9.
Failed portion in Technical College Road.

Figure 10.
Failed portion in Textile Mill Road.

Figure 11.
Failed portion in Ogida/Upper Siluko Road.

Table 6. The road is reliable and there is not enough evidence for the deterioration of the road if the shape parameter β is close to 1 and the 95% confidence interval of β contains 1.

There is enough evidence though for the deterioration of the width of the roads in Ikpoba Hill, Lucky Way, Mission, New Benin, Ogida, Ring, Sapele and Technical College since the value of $\hat{\beta} = 22.29$ from **Table 6** is far from 1, and the approximately 95% confidence interval for $\hat{\beta}[11.45, 43.40]$ does not contain 1. The plot of the failure

Figure 12.
Failed portion in Upper Sapele Road.

rate or hazard function, which describes the likelihood of deterioration in width during the next time increment is given in **Figure 13**.

There is a steeper increase after 225 days in the hazard function in **Figure 13**. This shows that the tendency of the width of the road to deteriorate increases after the 225th day. This is due to the value of $\hat{\beta} = 22.29$ being far from 1, and the approximately 95% confidence interval for $\hat{\beta}[11.45, 43.40]$ does not contain 1. The plot of the reliability test that shows the trend of reliability (the probability of the width of the road not deteriorating at time t) with time is given in **Figure 14**. The deterioration started after the 225th day. There was a sharp rate of deterioration (decrease in the reliability status) of the width of the roads in Lucky Way, Mission Road, New Benin Road, Ogida Road, Ring Road, Sapele Road, Technical College Road and Textile Mill Road after the 225th day.

6. Weibull reliability test for the deterioration of the depth of the roads in lucky way, Mission road, new Benin road, Ogida road, ring road, Sapele road, technical college road and textile mill road

The shape parameter β and the 95% confidence interval of β for the data pertaining to depth of the roads in Lucky Way, Mission Road, New Benin Road, Ogida Road, Ring Road, Sapele Road, Technical College Road and Textile Mill Road are given in **Table 7**. It may be stated the road is reliable and there is not enough evidence for the deterioration of the road if the shape parameter β is close to 1 and the 95% confidence interval of β contains 1.

There is enough evidence for the deterioration of the depth of the roads in Lucky Way, Mission Road, New Benin Road, Ogida Road, Ring Road, Sapele Road, Technical College Road and Textile Mill Road since the value of $\hat{\beta} = 22.50$ from **Table 7** is far from 1, and the approximately 95% confidence interval for $\hat{\beta}[10.02, 50.53]$ does not contain 1. The plot of the failure rate or hazard function, which describes the likelihood of deterioration in depth during the next time increment is given in **Figure 15**.

DOI: http://dx.doi.org/10.5772/intechopen.105697

Routes	Serial No	1st Visit			2nd Visit			3rd Visit			4th Visit			5th Visit		
		Length (M)	Width (M)	Depth (M)	Length (M)	Width (M)	Depth (M)	Length (M)	Width (M)	Depth (M)	Length (M)	Width (M)	Depth (M)	Length (M)	Width (M)	Depth (M)
Lucky Way	1	18	11	0.15	18.6	11	0.16	20	11	0.17	20	11	0.17	19.7	11	0.16
	2	6	2	0.23	6.8	2.2	0.23	7	4	0.24	7.3	4.2	0.24	7	3	0.24
	3	3	4	0.2	3.2	4.1	0.21	4	5	0.22	4.1	5.3	0.24	4	4.5	0.21
	4	135	11	0.1	137	11	0.11	141	11	0.12	141	11	0.12	139	11	0.11
Mission Road	1	2	3	0.1	2.3	3.7	0.11	2.5	4	0.12	2.5	4	0.12	2.4	3.9	0.11
	2	3	1	0.15	3.6	1.8	0.18	3.7	2	0.18	3.7	2	0.18	3.6	2	0.18
	3	5	2	0.089	5.3	2.3	0.1	6.2	3	0.11	6.2	3	0.11	6	2.5	0.1
	4	1	1	0.1	1.3	1.5	0.12	1.8	2	0.12	1.8	2	0.12	1.5	1.8	0.12
	5	1	1	0.076	2	1.7	0.097	2.4	2	0.1	2.4	2	0.1	2.2	1.9	0.097
	6	3	3	0.18	4	3.6	0.2	4.5	4	0.21	4.5	4	0.21	4.1	3.6	0.2
New Benin	1	2	3	0.051	2.2	3.8	0.61	2.7	3.9	0.071	2.7	3.9	0.071	2.5	3.9	0.066
	2	1	1	0.076	1.2	1.5	0.079	1.5	3	0.091	1.5	3	0.091	1.3	2.8	0.089
Ogida (Upper Siloku)	1	473	11	0.076	474	11	0.084	483	11	0.094	483	11	0.094	479	11	0.089
	2	204	8	0.051	205	8.2	0.061	208	8.5	0.066	208	8.5	0.066	208	8.3	0.061
	3	6	1	0.063	6.5	1.5	0.066	7	2	0.074	7	2	0.074	6.8	1.8	0.07
	4	5	2	0.058	5.1	2.1	0.061	6	2.5	0.074	6	2.5	0.074	5.3	2.3	0.07
	5	23	1.6	0.076	24	1.7	0.081	25	2	0.091	25	2	0.091	25	2	0.084
	6	4.7	1.5	0.089	5	2.5	0.1	5.2	3	0.1	5.2	3	0.1	5	2.8	0.1
	7	14	1.3	0.053	14.2	1.8	0.074	14.8	2	0.089	14.8	2	0.089	14.5	2	0.079

Routes	Serial No	1st Visit			2nd Visit			3rd Visit			4th Visit			5th Visit		
		Length (M)	Width (M)	Depth (M)	Length (M)	Width (M)	Depth (M)	Length (M)	Width (M)	Depth (M)	Length (M)	Width (M)	Depth (M)	Length (M)	Width (M)	Depth (M)
Ring Road	1	1	1	0.025	1.7	2.1	0.033	2	2.5	0.041	2	2.5	0.041	2	2.2	0.038
	2	2	4	0.13	2.3	4.6	0.14	2.4	4.9	0.13	2.4	4.9	0.13	2.3	4.7	0.13
Sapele Road	1	1	1	0.051	2.5	3.6	0.069	3	4	0.076	3	4	0.076	2.8	4	0.069
	2	2	3	0.13	5.6	6.2	0.13	6.9	7	0.13	6.9	7	0.13	6	6.7	0.13
Technical College Road	1	2	5	0.051	2.6	5.3	0.056	3	5.5	0.066	3	5.5	0.066	2.9	5.4	0.058
	2	1	1	0.1	2	1.7	0.1	2.2	1.9	0.12	2.2	1.9	0.12	2	1.9	0.11
Textile Mill Road	1	1005	11	0.22	1008	11	0.23	1010	11	0.23	1010	11	0.23	1008	11	0.23
	2	258	11	0.22	259	11	0.23	262	11	0.24	262	11	0.24	261	11	0.23
	3	50	9	0.22	50	9	0.23	53	9.3	0.24	53	9.3	0.24	50	9.2	0.24

Table 3.
Life data measurement of failed road sections.

Routes	2nd and 1st Visits			3rd and 2nd Visits			4th and 3rd Visits			5th and 4th Visits		
	Length (M)	Width (M)	Depth (M)	Length (M)	Width (M)	Depth (M)	Length (M)	Width (M)	Depth (M)	Length (M)	Width (M)	Depth (M)
Lucky Way	0.6	0	0.01	1.1	0	0	0.3	0	0.01	0.3	0	0.01
	0.8	0.2	0	0.2	0.8	0.01	0	1	0	0	1	0
	0.2	0.1	0.01	0.8	0.4	0	0	0.5	0.01	0	0.5	0.01
	2	0	0.01	2	0	0	2		0.01	2		0.01
Mission Road	0.3	0.7	0.01	0.2	0.2	0	0.1	0.1	0.01	0.1	0.1	0.01
	0.6	0.8	0.03	0	0.2	0	0.1	0	0	0.1	0	0
	0.3	0.3	0.011	0.7	0.2	0	0.2	0.5	0.01	0.2	0.5	0.01
	0.3	0.5	0.02	0.2	0.3	0	0.3	0.2	0	0.3	0.2	0
	1	0.7	0.021	0.2	0.2	0	0.2	0.1	0.003	0.2	0.1	0.003
	1	0.6	0.02	0.1	0	0	0.4	0.4	0.01	0.4	0.4	0.01
New Benin Lagos Rd	0.2	0.8	0.01	0.3	0.1	0.005	0.2	0	0.005	0.2	0	0.005
	0.2	0.5	0	0.2	1.3	0.01	0.2	0.2	0.002	0.2	0.2	0.002
Ogida	1	0	0.008	5	0	0.005	4	0	0.005	4	0	0.005
	1	0.2	0.01	3	0.1	0.001	0	0.2	0.006	0	0.2	0.006
	0.5	0.5	0.003	0.3	0.3	0.004	0.2	0.2	0.004	0.2	0.2	0.004
	0.1	0.1	0.003	0.2	0.2	0.009	0.7	0.2	0.001	0.7	0.2	0.001
	1	0.1	0.005	1	0.3	0.003	0	0	0.007	0	0	0.007
	0.3	1	0.011	0	0.3	0	0.2	0.2	0	0.2	0.2	0
	0.2	0.5	0.021	0.3	0.2	0.005	0.3	0	0.01	0.3	0	0.01
Ring Road	0.7	1.1	0.008	0.3	0.1	0.005	0	0.3	0.003	0	0.3	0.003
	0.3	1.4	0	0	0.1	0	0.1	0.2	0	0.1	0.2	0

Routes	2nd and 1st Visits			3rd and 2nd Visits			4th and 3rd Visits			5th and 4th Visits		
Sapele	1.5	2.6	0.018	0.3	0.4	0	0.2	0	0.007	0.2	0	0.007
	3.6	3.2	0	0.4	0.5	0	0	0.3	0	0	0.3	0
Technical College Rd	0.6	0.3	0.005	0.3	0.1	0.002	0.1	0.1	0.008	0.1	0.1	0.008
	1	0.7	0	0	0.2	0.01	0.2	0	0.01	0.2	0	0.01
Textile Mill Road	3	0	0.01	0	0	0	2	0	0	2	0	0
	1	0	0.01	2	0	0	1	0	0.01	1	0	0.01
	0	0	0.01	0	0.2	0.01	3	0.1	0	3	0.1	0

Table 4.
Alternate visit comparison showing deterioration rate.

Routes	Ltime	Wtime	Dtime	Lstatus	Wstatus	Dstatus
Lucky Way	81	384	27	0	1	0
	39	42	165	0	0	0
	48	35	18	0	0	0
	62	380	54	0	1	0
Mission Road	68	36	58	0	0	0
	78	45	36	0	0	0
	49	85	45	0	0	0
	41	39	48	0	0	0
	40	31	75	0	0	0
	50	47	80	0	0	0
New Benin	65	49	65	0	0	0
	68	54	68	0	0	0
Ogida (Upper Siloku)	74	370	76	0	1	0
	72	62	16	0	0	0
	49	92	35	0	0	0
	68	88	65	0	0	0
	60	77	48	0	0	0
	38	74	49	0	0	0
	74	73	52	0	0	0
Ring Road	72	45	50	0	0	0
	69	65	385	0	0	1
Sapele Road	21	81	69	0	0	0
	92	70	378	0	0	1
Technical College Road	43	60	74	0	0	0
	25	42	132	0	0	0
Textile Mill Road	92	364	64	0	1	0
	88	388	63	0	1	0
	394	149	58	1	0	0

Table 5.
Road deterioration status change.

There is a steeper increase after 225 days in the hazard function in **Figure 15**. This shows that the tendency of the depth of the road to deteriorate increases after the 225th day. This is due to the value of $\hat{\beta} = 22.50$ being far from 1, and the approximately 95% confidence interval for $\hat{\beta}[10.02, 50.53]$ does not contain 1. The plot of the reliability test that shows the trend of reliability (the probability of the depth of the road not deteriorating at time t) with time is given in **Figure 16**. The deterioration started after the 225th day. There was a sharp rate of deterioration (decrease in the reliability status) of the depth of the roads Lucky Way, Mission Road, New Benin

β	LCL	UCL
22.29	11.45	43.40

Table 6.
The shape parameter β and the 95% confidence interval of β.

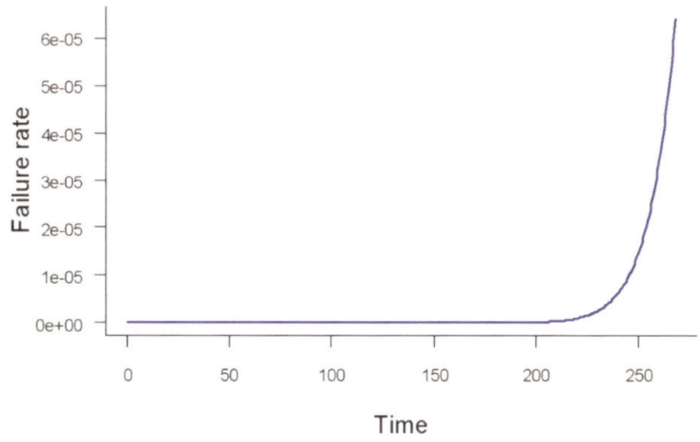

Figure 13.
The failure rate or hazard function plot for the width of the roads.

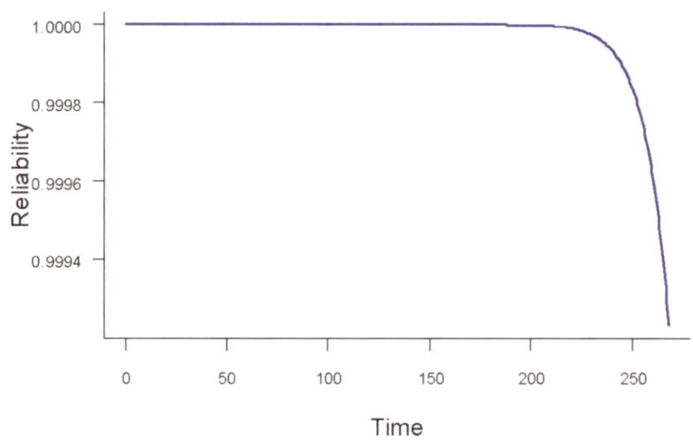

Figure 14.
The reliability test plot for the width of the roads.

β	LCL	UCL
22.50	10.02	50.53

Table 7.
The shape parameter β and the 95% confidence interval of β.

Road, Ogida Road, Ring Road, Sapele Road, Technical College Road and Textile Mill Road after the 225th day.

Weibull Reliability Test for the Deterioration of the Length of the Roads in Lucky Way, Mission Road, New Benin Road, Ogida Road, Ring Road, Sapele Road, Technical College Road and Textile Mill Road.

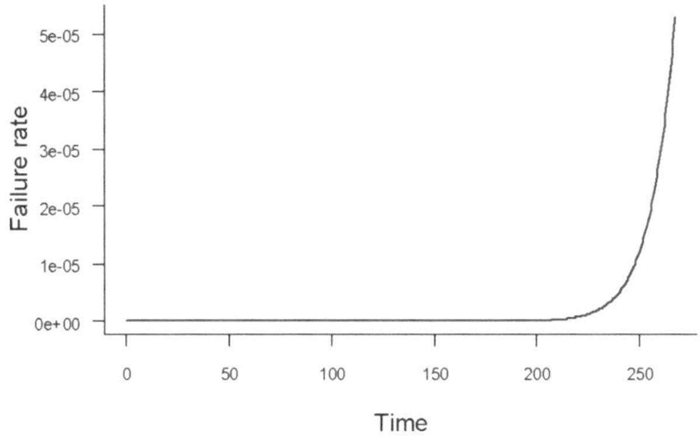

Figure 15.
The failure rate or hazard function plot for the depth of the roads.

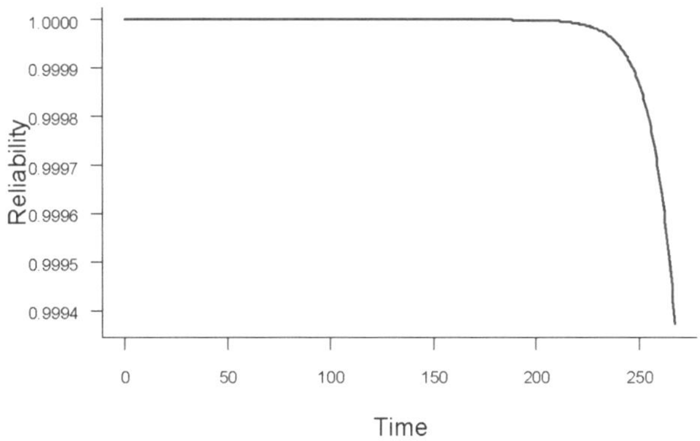

Figure 16.
The reliability test plot for the depth of the roads.

β	LCL	UCL
6.71	2.50	17.99

Table 8.
The shape parameter β and the 95% confidence interval of β.

The shape parameter β and the 95% confidence interval of β for the data on length of the roads in Lucky Way, Mission Road, New Benin Road, Ogida Road, Ring Road, Sapele Road, Technical College Road and Textile Mill Road are given in **Table 8**. The road is reliable and there is not enough evidence for the deterioration of the road if the shape parameter β is close to 1 and the 95% confidence interval of β contains 1.

There is sufficient evidence for the deterioration of the length of the roads in Lucky Way, Mission Road, New Benin Road, Ogida Road, Ring Road, Sapele Road, Technical College Road and Textile Mill Road since the value of $\hat{\beta} = 6.71$ from **Table 8** is far from 1, and the approximately 95% confidence interval for $\hat{\beta}[2.50, 17.99]$ does not contain 1. The plot of the failure rate or hazard function, which describes the likelihood of deterioration in length during the next time increment is given in **Figure 17**.

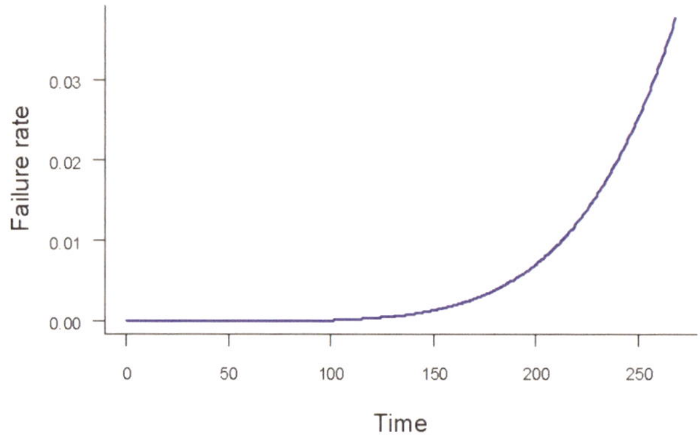

Figure 17.
The failure rate or hazard function plot for the length of the roads.

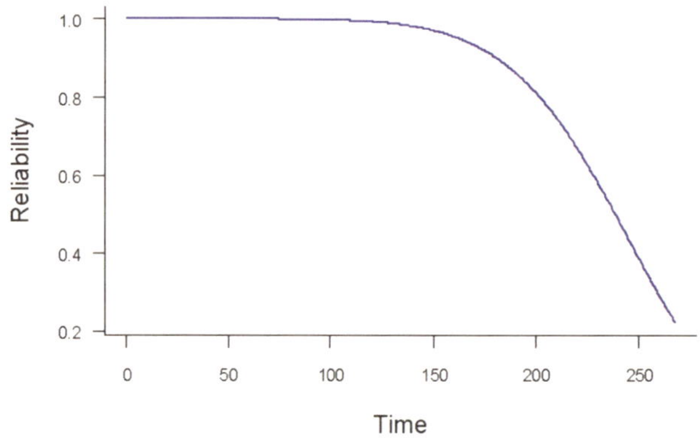

Figure 18.
The reliability test plot for the length of the roads.

There is a steeper increase after 140 days in the hazard function in **Figure 17**. This shows that the tendency of the length of the road to deteriorate increases after the 140th day. This is due to the value of $\hat{\beta} = 6.71$ being far from 1, and the approximately 95% confidence interval for $\hat{\beta}[2.50, 17.99]$ does not contain 1. The plot of the reliability test that shows the trend of reliability (the probability of the length of the road not deteriorating at time t) with time is given in **Figure 18**. The deterioration started after the 100th day. The rate of deterioration was minimal between the 100th to 150thday; after the 150thday, there was a sharp decline in the reliability status (increase in the deterioration rate) of the length of the roads in Lucky Way, Mission Road, New Benin Road, Ogida Road, Ring Road, Sapele Road, Technical College Road and Textile Mill Road.

7. Conclusion

In this study, eight (8) roads from Nineteen (19) mapped failing roads by means of purposive sampling the Benin city area of Edo state in Nigeria were assessed. Data collection follows five different routes in five (5) different field visitations at intervals of three (3) months for each visit. During visits variational failure growth in terms of length, width and depth was obtained by tape rule measurements, rolling measuring rule, pegs, paints and photographs. Data collected were scrutinized and subjected to Weibull analysis to obtain road failure-specific sequences in reliability terms to validate and underscore quality impairments in constructed road pavements. Quality impairments originating from road component materials failure were evident in roads failing as early as six (6) months after construction. Progressive failure was noticed to be aided by further deterioration owing to lack of maintenance according to the types of road failures photographed in this paper. This is further augmented by relation to monotonicity failure theory elucidated in Gupta [26] with steady state progressive deterioration shown in the bathtub log-convexity property of the Weibull measurement count. A validation of quality impairment was deduced from a degenerating reliability Weibull analysis as corroborated in the literature of Efimenko and Moisejenko [2] and Bazhanov [5]. By undermining quality control process at construction, steep failures from a deteriorating pavement aided by the stress/strain mortal force of decrement prevailed early enough in the lifetime of the pavement to cause road failure.

8. Recommendation

Arising from the study and the observations and analysis conducted, the following recommendations are made:

1. It is recommended that quality control should be acculturated in organizations specializing in road construction and set-up procedures and instruments for quality control at points of raw materials storage including blending, mixing and placing of asphalts.

2. Deploy a quality control metrics for checking the level of quality-specific work output during the entire construction process.

3. Evolve a quality and maintenance sequence of road construction for every and any activity in the construction process.

4. Overhauling and maintaining construction equipment as required in the ISO 9002 Quality Assurance Framework. This will enable equipment work optimally especially laboratory equipment so as measurements and investigations can be precise

5. Setting site and organization's agenda around QA/AC needs at project review phase and staff meetings. This aspect of quality management encourages team building and common purpose focus towards quality objective.

6. Go round site supervision with Engineers Instruction cards to direct the rework of construction defects. This is with the aim of ensuring that site instructions for defective works are carried out without omission.

7. Allow proper physical and chemical properties limits of materials to be attained before use, in order to avoid materials failure. It will also help in quality enhancement of finished output.

Author details

Samuel I. Egwunatum[1*], Ovie I. Akpokodje[2] and Andrew I. Awo-Osagie[3]

1 Department of Quantity Surveying, Federal University of Technology, Owerri, Nigeria

2 Department of Civil Engineering, Delta State University of Science and Technology, Ozoro, Nigeria

3 Department of Quantity Surveying, Delta State University of Science and Technology, Ozoro, Nigeria

*Address all correspondence to: samuelegwunatum@gmail.com

References

[1] Egwunatum SI, Anumudu AC, Eze EC, Awodele IA. Total quality management (TQM) implementation in the Nigerian construction industry. Engineering Construction and Architectural Management. 2021;**29**(1): 354-382. DOI: 10.1108/ECAM-08-2020-0639

[2] Efimenko V, Moisejenko R. Calculation of road reliability based on mathematical theory of reliability. MATEC Web of Conference. 2018;**143**: 01004

[3] Wada SA. Bituminous Pavement failures. International Journal of Engineering Research and Applications. 2016;**6**(2):94-100

[4] Dwivedi S, Suman SK. Quality assessment of road shoulders using light weight deflectometer and geoguage. International Journal of Recent Technology and Engineering. 2019; **8**(2):1-9. DOI: 10.35940/ijrte.B2649.07821

[5] Bazhanov AP. General issues of assessing the reliability of roads. IOP Conference series: Materials Science and Engineering. 2019;**632**(2019):012012. DOI: 10.1088/1757-899X/632/1/012012

[6] Mahmood QY, Mohammed W, Misnan MS, Yusof ZM, Bakri A. Development of Quality Culture in the Construction Industry. Beijing: *ICCI 2006*; 2006. pp. 1-11

[7] Magdi ME. Zumrawi investigating causes of pavement deterioration in Khartoum State *Sudan Digital Open Science Index*. *International* Journal of Civil Environment, Structure, Construction and Architectural Engineering. 2015;**9**(11):1450-1455

[8] Imran MA et al. Assessment on the road pavement failures and maintenance of Rajshahi City. International Conference on Recent Innovation in Civil Engineering for Sustainable Development. 2016:787-791

[9] Imran MA, Rabbany HM, Islam MT, Sharon MH. Assessment on the road pavement failure and maintenance of Rajshahi City. In: International Conference on Recent Innovation in Civil Engineering for Sustainable Development. Gazipur, Bangladesh: ICSD Department of Civil Engineering DUET; 2015

[10] MoRTH. Specifications for Road and Bridge Works of Ministry of Road Transport & Highways (MoRTH), Indian Roads Congress and Standard. 5th ed. New Delhi: Indian Road Congress; 2013

[11] Wee SY, Chan RB, Teo HW. Potential modeling of pavement deterioration rate due to cracking. UNIMASE-Journal of Civil Engineering. 2009;**1**(1):1-6

[12] Petersen DL, Siekmeier JA, Nelson CR, Peterson RL. Intelligent soil compaction – Technology results and a roadmap toward widespread use. Transportation Research Record. 2006; **1975**(1):81-88. DOI:10.1177/0361198106197500109

[13] Mehedi Hassan, Abdus Sobhan, Highway failure and their maintenance: A study of Rajshahi Metropolitan City, Rajshashi, Bangladesh. International Journal of Sustainable Transportation Technology. 2020;**3**(2 October):45-50

[14] Ola SA. Estimating road pavement failure susceptibility using a modified TDRAMS model in South-Western

Nigeria. International Journal of Engineering Research In Africa. 2018;**40**: 63-67

[15] Agrawal AK, Kawaguchi A, Qian G. Bridge Deterioration Rates. Albany, N. Y: Transportation Infrastructure Research Consortium, New York State Department of Transportation; 2008

[16] Hnin E, Khaing E, Htwe TT. Study on failures and maintenance of flexible pavement. International Journal of Scientific Engineering and Technology Research (IJSETR). 2014;**2**(4):2984-2990

[17] Madanat S, Mishalani R, Ibrahim WHW. Estimation of infrastructure transition probabilities from condition rating data. Journal of Infrastructure Systems. 1995;**1**(2): 120-125

[18] Hughes M. How the Weibull Distribution is used in Reliability Engineering. Technical Article. 2019. Available from: https://www.allaboutcircuits.com/technical. [Accessed: January 14, 2022]

[19] Duddu SR, Chennarapu H. Quality control of compaction with lightweight deflectometer (LWD) device: A state-of-art. International Journal of Geo-Engineering. 2022;**13**:6. DOI: 10.1186/s40703-021-00171-2

[20] Bisht S, Dhar S, Hussain M. Performance evaluation of lime stabilized sub-grade soil using light weight deflectometer. In: Indian Geotechnical Conference. IIT Guwahati, India: December GeoNEst; 2017. pp. 14-16

[21] Nazzal M, Abu-Farsakh M, Alshibli E, Mohammad L. Evaluating light falling-weight deflectometer device for in-situ measurement of elastic modulus of pavement layers. Transportation Research Record. 2007; **2016**(1):13-23. DOI: 10.3141/2016-2

[22] Alshibli KA, Abu-Farsakh M, Seyman E. Laboratory evaluation of the geoguage and light falling weight deflectometer as construction control tools. Journals of Materials in Civil Engineering. 2005;**17**(5):560-569. DOI: 10.1061/(ASCE)0899-1561(2005)17:5(560)

[23] Rao C, Shiva GV, Shankar R. PFWD, CBR and DCP evaluation of lateritic subgrades of Dakshina Kannada, India. In: 12th International Conference of International Association for Computer Methods and Advances in Geo-Technics (IACMAG). Mangalore, India, Goa, India: National Institute of Technology Karnataka; 2008. pp. 4417-4423

[24] Adam D, Kopf F. Metrological and Theoretical Investigations as A Basis for The Further Development and Normative Application of the Dynamic Load Plate. Report No. 68, Lecture. Germany: Fairs in the Geotechnik, Technical University of Braunschweig; 2002

[25] Bazhanov AP, Saksonova ES. The method of carrying out heuristic procedures in the selection and justification of the composition of the determining indicators of the quality of pavements that affects the reliability of roads. Preparing a heuristic procedure and interviewing experts. Transport Facilities. 2018;**5**(1):18

[26] Gupta PL. Properties of reliability functions of discrete distributions. Communications in Statistics–Theory and Methods. 2015;**44**(19):4114-4131. DOI: 10.1080/03610926.2014.982827

[27] Gupta PL, Gupta RC, Ong SH, Srivastava HM. Reliability and non-reliability studies of Poisson variables in

series and parallel systems. Applied Mathematics and Computation. 2012; **218**:5112-5120

[28] Riveros GA, Arredondo E. Guide to the Development of a Deterioration Rate Curve Using Condition State Inspection Data. ERDC/CHL CHETN-IX-25. United States: US Army Corps of Engineers; 2010

[29] Al-Zahrani N, Stoyanov J. On some properties of life distributions with increasing elasticity and log-concavity. Applied Mathematical Sciences. 2008; **2**:2349-2361

Chapter 3

Practicing Hypothesis Tests in Textile Engineering: Spinning Mill Exercise

Nefise Gönül Şengöz

Abstract

A novel statistical approach for multiple-stream processes is proposed in this manuscript. As important as quality control in manufacturing is, hypothesis tests are an important part of it if utilized and constructed the most logically to evaluate and decide on a special matter in a production line or a production machine. The proposed statistical approach is explained in detail in a spinning mill having 20 spinning frames. The spinning frames are adjusted according to customers' orders and to the technology of spinning frames first. Then, the result of that adjustment is controlled statistically by means of hypothesis testing, χ^2, t-test, and F statistics are used. Later, they are pooled one by one, and at the end, all 20 spinning frames are considered as one machine producing the same yarn, the same variance of yarn count, and the same yarn count. Performed literature review claims that control charts are appropriate for multiple-stream processes. But, the application of this proposed statistical approach guarantees that production starts with correct adjustments on machines, and control charts become more sensitive to the assignable causes. The application area of this proposed statistical approach is wide, leading to higher quality in products, a requirement that is in demand more every day.

Keywords: textile engineering, hypothesis tests, spinning mill, spinning frame, chi-square statistics, t-tests, F statistics, pooled estimator of sigma, pooled estimate of standard deviation, pooled estimator of variance, pooled t-Test, distribution, multiple-stream processes

1. Introduction

Quality is demanded by every customer in the products they purchase in this era of science and technology, claiming for better products and services alike. This demand produces pressure on the manufacturers to conform to customers' wishes by offering products and/or services incorporating increased quality levels, applying quality control methods, practicing statistical quality control, etc. Manufacturers intensely control and improve the quality of their products in order to make them better while also aiming at establishing a competitive edge.

Textiles are regarded as fundamental items in everyday life. They are indeed used in every field of daily life like apparel, home textiles, technical textiles (automotive, aerospace, geological, agriculture, civil, medical, sport, packaging, protective, military, art, etc.). Similar to every engineering branch, quality is the main requirement in textile engineering. The only way to achieve this is the application of quality control tools which are mostly applied in every step of textile production in order to fulfill the demands of consumers.

The main steps in textile manufacturing are yarn production, weaving, knitting, and ready-wear; besides nonwovens, texturing, finishing, dyeing, printing, etc. Yarn production is the primary step among these because if a good yarn is produced at the beginning the rest of the steps will probably end up likewise good. Good yarn provides the properties required for the next step, and for any succeeding step thereafter until the end product is reached, namely the one used in daily life. In a reverse pattern, first, the usage area of that special textile product to be manufactured has to be decided on as well as determining the requirements of properties in that special unit. Then one step backwards, weaving or knitting-specific quality properties are determined, followed by the properties of yarn to meet the properties of fabric. Finally, the latter are determined together with the fibers to be used and thus, production starts. It is very important to keep the quality properties of yarn correct and stable in order for the rest of the steps to be good. This is why quality control tools have to be applied in yarn production. Besides, technology in machinery is another grand field where huge improvements are achieved so as to manufacture products with the aimed properties. Textile machinery is an area where many technological improvements are successfully applied, yielding production of yarn with better properties.

Textile manufacturing is a multiple-stream process where one operation is usually done by more than one machine. The product of every machine is mixed into one lot. In literature, it is stated that in processes consisting of several machines producing the same material which pool their output into a common stream, control charts are appropriate to use in order to keep quality under control. In this case, machines producing the same material form a rational subgroup. Separate control charts are advised for each rational subgroup, each individual machine, or sometimes even for the different heads on the same machine. Therefore, the proper selection of samples is very important within the rational subgroup concept; the process is to be consistent and careful by extracting as much useful information as possible from the evaluation of the control charts. Even more, simultaneous monitoring of all streams is impractical when the streams are large in number, identical, and independent. Also, control charts are sensitive to assignable causes that affect the uniformity across the streams and between-stream variability [1–7].

The main concept of control charts is: Sampling the material of which the property/properties to be investigated, testing the property/properties, obtaining the results, plotting the values on the control charts, and interpreting the charts. Production is under control while the plot falls between the upper and lower control limits. If not, then the precautions needed are taken and adjustments to the machines are done. Not only one machine produces the same product but there may be more than one machine producing the same material which will be mixed and shipped into one lot, and every machine producing the same material will have to do so. The customer does not need to know which machine produced which constitutes the lot; it is the responsibility of the factory to ship a lot containing the same properties in every piece [8].

In this manuscript, it is worth noting at this point that before constructing the control charts for rational subgroups, the adjustments on the subgroups have to be controlled statistically first. The subgroups are machines in this case. Control charts may keep the control limits after the correct adjustments on the machines are successfully done. It is thought that controlling the adjustments of the machines to produce the right material is different than keeping it under control with control charts. If the adjustments of the machines are correct at the beginning, then the purpose of the control charts will only be sensitive to assignable causes. Otherwise, it may be as if it is expected too much from the control charts; however, in this proposed novel statistical approach the purposes are separated and may help to understand processes better and keep quality under control. When quality will be set at the beginning and tested statistically then control charts will help to carry it forward in a stable manner. In this study, a different approach will be presented which is applying hypothesis tests to the adjustment of the multiple-stream machines prior to them starting production. A novel method for this kind of statistical control is proposed and explained in detail based on an example of a textile engineering spinning mill.

Hypothesis testing is a process of drawing conclusions on the collected data of statistical testing and is a specific approach for testing means or averages of that data. The purpose of statistical inference is to draw conclusions about a population on the basis of data obtained from a sample of that population. Hypothesis testing evaluates the strength of evidence from the sample and gives the basis to determine the relation to the population. Hypothesis testing equally indicates the chance about how reliably the observed results in a sample can be extrapolated to the larger population of collected samples. A specific hypothesis is formulated, the data from the sample is evaluated and if they support the specific hypothesis a statistical inference about the population is reached. Hypothesis testing is a dominant approach for data analysis in many fields of science [9].

In literature, it is discussed that there is a close connection between hypothesis testing and control charts. It is considered that if the obtained value of $\bar{x}$ is plotted and values fall in-between the control limits then it is expressed that the process mean is in control, and it is equal to a value μ_0. If $\bar{x}$ falls out of the upper or the lower control limits then it will have a value other than μ_0, it is concluded that the control chart is a kind of hypothesis testing and shows that the process is under statistical control. If the plots are in-between the control limits, this means the hypothesis is not rejected; if they are out of the control limits, this means the hypothesis is rejected [10].

On the other hand, there are some differences between hypothesis tests and control charts. The validity of assumptions, like the form of the distribution, independence, etc., are tested in hypothesis testing but not in control charts. Instead, the departures from $\bar{x}$ are seen in control charts so that the process variability may be reduced. There may be assignable causes in production and they result in different types of shifts in the process parameters. An assignable cause can result in an increase or a decrease to a new value but return quickly. It can have ups and downs in-between the control limits, and can shift to a new value but remain there; this is called a sustained shift. It is recognized in literature that only the sustained shift fits the statistical hypothesis testing model.

This chapter suggests that adjustment of the machines in a multiple-stream should be done with hypothesis testing at the beginning and then continuing production should be observed with control charts so that the quality will be under control at the beginning and will be kept stable during production. This proposed method will be done just at the beginning of production for once in order to confirm that the adjustments to produce the same lot are the same all throughout the lot, as well as considering that every centimeter of yarn will exactly be the same in the tons of guaranteed yarn production. Then, while the production continues the control charts will monitor that quality is kept stable. This novel approach of a statistical control method will be explained in detail given in an example of a textile engineering spinning mill. In this case, the type of the yarn, the properties of the yarn, the type of fibers used to produce the yarn will not be considered except for yarn count. Yarn count property will be mentioned in the proposed hypothesis testing method. One may bear in mind that the same application can be done for every property of yarn like twist, breaking strength, breaking elongation, elasticity, abrasion resistance, hairiness, unevenness, imperfection (thick place, thin place, neps), etc.

2. Spinning mill

When yarn production is considered, regardless of the type of fiber processed, yarn production generally consists of blowroom/blending, carding, drawing, roving, and spinning steps seen in **Figure 1**, whereas an example of a spinning mill is given in **Figure 2** [11]. The same concept mentioned above is applied in yarn production. In order to produce the yarn with the aimed properties at the end, the needed adjustments have to be done starting from the very beginning of the stream until the endpoint where the yarn is obtained. In every production step there is usually more than one machine producing the same product and pouring into a common stream.

Yarn is produced on spinning frames that are ring spinning machines. Rovings come from the top to the spindles, on the way they are drafted and twisted, and the yarn forms (**Figure 3**) [12, 13]. The yarn properties, which are yarn count and yarn twist, are adjusted on the frame, but the rest of the properties listed above are the result of

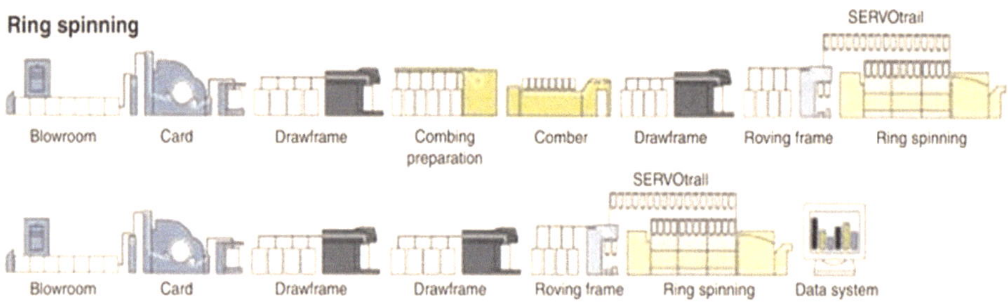

Figure 1.
Yarn production steps [11].

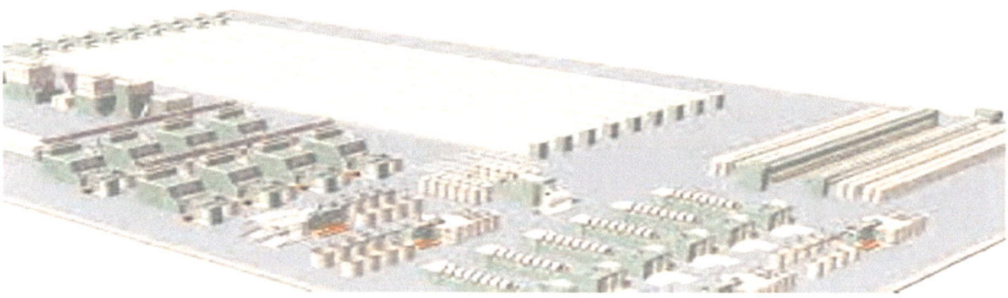

Figure 2.
An example of a spinning mill [11].

Figure 3.
Rovings, spindles, yarn [12, 13].

pressure between rollers, machine production speed, roller surfaces, delivery angles, climate, cleanness, human factor, gauge, etc. Since the yarn count is one of the adjustments done on the spinning frame, this will be considered in the rest of this chapter.

3. Reference statistical methodology in quality control

Hypothesis testing is one of the useful tools of statistical methodology in quality control and improvement. In hypothesis testing, there are the null hypothesis (H_0) and the alternative hypothesis (H_1). While the null hypothesis H_0 indicates a certain point of view of the research question, the alternative hypothesis H_1 indicates the opposite of that point of view. The opposite can be stated as not equal, greater than, or less than. Not equal is a two-sided alternative hypothesis, and the latter two are one-sided alternative hypotheses. Therefore, the determination of the parameter values in a hypothesis to be tested is very important, as they may either come from past information, a theory or model, or conformity. A statistical inference is reached with correct determination.

When working with test results, it is assumed that the obtained test results are normally distributed. If the underlying distribution of the obtained results deviate moderately from normal distribution, t-tests perform reasonably well because of the robustness of the test. If the underlying distribution of the obtained results deviates substantially from normal distribution, when the sample size is large, because of the central limit theorem (CLT), they approximate normal distribution [14]. Especially in textile manufacturing, it is considered that the test results of properties of a product exhibit normal distribution.

In statistical inference, there may be errors, especially in hypothesis testing, wherein two kinds of errors exist. The first one is the null hypothesis is rejected even if it is true, which is the wrong decision. This is Type I Error and is symbolized by α which is also called the level of significance. In this case, the null hypothesis is unable to be rejected by $1-\alpha$ probability, or which is the right decision. The second kind of error is the null hypothesis is unable to be rejected even if it is false, which is the wrong decision. This is Type II Error and is symbolized by β. In this case, the null hypothesis is rejected by $1-\beta$ probability, or which is the right decision. Hypothesis testing errors are shown in **Table 1**. The level of significance α would take values like 0.1, 0.05, 0.01, 0.001, etc.

By designing a test procedure in hypothesis testing, a value of the probability of Type I Error α is specified so that a small value of the probability of Type II Error β is obtained. The α risk can directly be controlled or chosen; the β risk can indirectly be controlled because it is the function of sample size; consequently, the larger the sample size, the smaller it is. In textiles production, Type I Error α is sufficient. The nature of textiles production for daily usage like apparel, home textiles (rugs, curtains, bedsheets, carpets, towels, etc.), Type I Error α is satisfactory, there is no requirement for Type II Error β in such cases. The important thing is to produce yarn, fabric, ready-wear, etc. with level of significance $\alpha = 0.05$, which is usually used and is deemed enough. On the other hand, Type II Error β is strongly reasonable for technical textiles like medical, aerotextiles, geotextiles, etc.; even there are cases where 6σ is applied (such as in vivo medical textiles, aerotextiles). These special cases will not be studied in this manuscript; for the rest, only Type I Error α will be considered.

A hypothesis test can be conducted by different test statistics like the z test, t-test, χ^2 test, the appropriate one is selected in accordance with the purpose of the hypothesis test. The set of values of the test statistic which lead to the rejection of H_0 is named as the critical region or rejection region for the test.

Therefore, the procedures for a hypothesis test can be listed as:

1. To determine the null hypothesis (H_0) and the alternative hypothesis (H_1),

2. To determine the level of significance (α),

3. To determine the appropriate test statistic,

4. To determine the test statistic limit(s) leading to rejection of the null hypothesis (critical region or rejection region),

5. To calculate,

6. To conclude if the null hypothesis is rejected or it is unable to be rejected,

7. To write the conclusion sentence.

	H_0	
Decision	**True**	**False**
H_0 Unable to reject	Right decision $1-\alpha$	Wrong decision Type II Error β
H_0 Reject	Wrong decision Type I Error α	Right decision $1-\beta$

Table 1.
Hypothesis testing errors.

Sampling is very important in hypothesis testing because an inference will be reached through the parameter information the samples contain and that conclusion will be applied to all of the rest of the population.

In a hypothesis testing, if x is a random variable with unknown mean μ and known variance σ^2, then the hypothesis testing is that the mean is equal to a chosen value, μ_0. The null hypothesis (H_0) and the alternative hypothesis (H_1) are stated as:

$$\begin{aligned} H_0 &: \mu = \mu_0 \\ H_1 &: \mu \neq \mu_0 \end{aligned} \tag{1}$$

Level of significance α is determined. n samples are taken from the random variable x and the z statistic is calculated:

$$Z_0 = \frac{\overline{x} - \mu_0}{\sigma/\sqrt{n}} \tag{2}$$

If $|Z_0| > Z_{\alpha/2}$ then H_0 is rejected, $Z_{\alpha/2}$ is the upper $\alpha/2$ percentage point of the standard normal distribution at a fixed significance level two-sided.

If x is a random variable with unknown mean μ and unknown variance σ^2, then the hypothesis testing is that the mean is equal to a chosen value, μ_0. The hypothesis is stated as:

$$\begin{aligned} H_0 &: \mu = \mu_0 \\ H_1 &: \mu \neq \mu_0 \end{aligned} \tag{3}$$

Since the variance is unknown, it is assumed that the x random variable has a normal distribution and deviations from normality will not affect the results much. Also, σ^2 is unknown and it is estimated by s^2. The level of significance α is determined. n samples are taken from the random variable x and the test statistic becomes a t-test:

$$t_0 = \frac{\overline{x} - \mu_0}{s/\sqrt{n}} \tag{4}$$

where instead of a normal distribution it becomes a t distribution with $n-1$ degrees of freedom.

If $|(t_0)| > t_{\alpha/2,n-1}$ then H_0 is rejected, $t_{\alpha/2,n-1}$ is the upper $\alpha/2$ percentage point of the t distribution with $n-1$ degrees of freedom at a fixed significance level two-sided.

Statistical tests on means are very little sensitive to normality assumptions but the tests on variances are sensitive. To test the variance of a normal distribution is equal to a chosen variance, σ_0^2, then the hypothesis is stated as:

$$H_0 : \sigma^2 = \sigma_0^2$$
$$H_1 : \sigma^2 \neq \sigma_0^2 \tag{5}$$

and the test statistic becomes a χ^2 test:

$$\chi_0^2 = \frac{(n-1)s^2}{\sigma_0^2} \tag{6}$$

where s^2 is the sample variance of n repeats. The level of significance α is determined. If $\chi_0^2 > \chi_{\alpha/2,n-1}^2$ or if $\chi_0^2 < \chi_{1-\alpha/2,n-1}^2$ then the null hypothesis H_0 is rejected for a fixed significance level, $\chi_{\alpha/2,n-1}^2$ is the upper $\alpha/2$ upper percentage point of the chi-square distribution with $n-1$ degrees of freedom and $\chi_{1-\alpha/2,n-1}^2$ is the lower $1-(\alpha/2)$ percentage. If a one-sided alternative is specified, then the hypothesis is:

$$H_0 : \sigma^2 = \sigma_0^2$$
$$H_1 : \sigma^2 < \sigma_0^2 \tag{7}$$

and the null hypothesis is rejected if $\chi_0^2 < \chi_{1-\alpha,n-1}^2$. For the other one-sided alternative, the hypothesis is:

$$H_0 : \sigma^2 = \sigma_0^2$$
$$H_1 : \sigma^2 > \sigma_0^2 \tag{8}$$

and the null hypothesis is rejected if $\chi_0^2 > \chi_{\alpha,n-1}^2$.

Chi-square testing is applied a lot in quality improvement by monitoring and control procedures. There may be a normal random variable with mean μ and variance σ^2. If $\sigma^2 \leq \sigma_0^2$, σ_0^2 being a chosen value, then the natural inherent variability of the process will be within the requirements of the design and the production will mostly be within the specification limits. But if $\sigma^2 > \sigma_0^2$, this means that the natural variability in the process is exceeding the specification limits. This case increases the percentage of non-conforming production items.

If there are two independent populations, as shown in **Figure 4**, then it will statistically be tested for the difference in means $\mu_1 - \mu_2$. It is assumed that μ_1, $\overline{x}_1$, σ_1^2, and n_1 are known and belonging to Population 1; whereas μ_2, $\overline{x}_2$, σ_2^2, and n_2 are known and belonging to Population 2. Both samples of the populations are random, and both populations are normally distributed; if they are not normal, the conditions of the CLT applies.

The point estimator of $\mu_1 - \mu_2$ is the difference in sample means $\overline{x}_1 - \overline{x}_2$ and from the properties of expected values:

$$E(\overline{x}_1 - \overline{x}_2) = E(\overline{x}_1) - E(\overline{x}_2) = \mu_1 - \mu_2 \tag{9}$$

is obtained and the variance of $\overline{x}_1 - \overline{x}_2$ is:

$$V(\overline{x}_1 - \overline{x}_2) = V(\overline{x}_1) + V(\overline{x}_2) = \frac{\sigma_1^2}{n_1} + \frac{\sigma_2^2}{n_2} \tag{10}$$

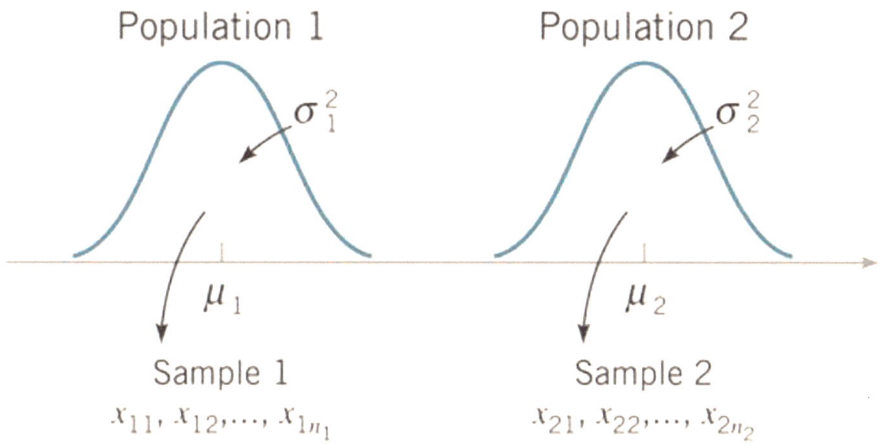

Figure 4.
Symbolization of two independent populations [1].

From the assumptions and the preceding results, the quantity Z with N(0,1) distribution can be stated as:

$$Z = \frac{\overline{x}_1 - \overline{x}_2 - (\mu_1 - \mu_2)}{\sqrt{\frac{\sigma_1^2}{n_1} + \frac{\sigma_2^2}{n_2}}} \tag{11}$$

If it is tested that the difference in means $\mu_1 - \mu_2$ is zero, that they are equal, the hypothesis is:

$$\begin{aligned} H_0 &: \mu_1 - \mu_2 = 0 \\ H_1 &: \mu_1 - \mu_2 \neq 0 \end{aligned} \tag{12}$$

Substituting 0 for $\mu_1 - \mu_2$, becomes:

$$Z_0 = \frac{\overline{x}_1 - \overline{x}_2}{\sqrt{\frac{\sigma_1^2}{n_1} + \frac{\sigma_2^2}{n_2}}} \tag{13}$$

If $|Z_0| > Z_{\alpha/2}$ then H_0 is rejected, $Z_{\alpha/2}$ is the upper $\alpha/2$ percentage point of the standard normal distribution at a fixed significance level two-sided.

If there are two independent populations, then the difference in means $\mu_1 - \mu_2$ will statistically be tested. It is assumed that μ_1, $\overline{x}_1$, and n_1 are known belonging to Population 1; μ_2, $\overline{x}_2$, and n_2 are known belonging to Population 2, but σ_1^2 and σ_2^2 are unknown. Both samples of the populations are random, and both populations are normally distributed; if they are not normal, the conditions of the CLT applies. The two σ_1^2 and σ_2^2 may be equal or not. In this manuscript, the condition that they are equal will be considered, becoming $\sigma_1^2 = \sigma_2^2 = \sigma^2$. Since σ_1^2 and σ_2^2 are unknown, t-statistic will be used and sample variances of the two populations would be s_1^2, and s_2^2, respectively.

The expected value of the difference in sample means $\overline{x}_1 - \overline{x}_2$ which is an unbiased estimator of the difference in means is:

$$E(\overline{x}_1 - \overline{x}_2) = \mu_1 - \mu_2 \tag{14}$$

The variance of $\overline{x}_1 - \overline{x}_2$ is:

$$V(\overline{x}_1 - \overline{x}_2) = \frac{\sigma^2}{n_1} + \frac{\sigma^2}{n_2} = \sigma^2\left(\frac{1}{n_1} + \frac{1}{n_2}\right) \tag{15}$$

Estimator of σ^2 is the combination of s_1^2 and s_2^2 it is the pooled estimator of σ^2, denoted by s_p^2, which is:

$$s_p^2 = \frac{(n_1 - 1)s_1^2 + (n_2 - 1)s_2^2}{n_1 + n_2 - 2} \tag{16}$$

s_p^2 is the weighted average of the two sample variances s_1^2 and s_2^2.

The z test statistic for unknown σ is:

$$z = \frac{\overline{x}_1 - \overline{x}_2 - (\mu_1 - \mu_2)}{\sigma\sqrt{\frac{1}{n_1} + \frac{1}{n_2}}} \tag{17}$$

and then, for t-statistic σ is replaced by s_p.

$$t = \frac{\overline{x}_1 - \overline{x}_2 - (\mu_1 - \mu_2)}{s_p\sqrt{\frac{1}{n_1} + \frac{1}{n_2}}} \tag{18}$$

a t distribution with $n_1 + n_2 - 2$ degrees of freedom, also called the pooled t-test.

If it is tested that the difference in means $\mu_1 - \mu_2$ is zero - meaning they are equal- the hypothesis is:

$$\begin{aligned} H_0 &: \mu_1 - \mu_2 = 0 \\ H_1 &: \mu_1 - \mu_2 \neq 0 \end{aligned} \tag{19}$$

Substituting 0 for $\mu_1 - \mu_2$, it becomes:

$$t_0 = \frac{\overline{x}_1 - \overline{x}_2}{s_p\sqrt{\frac{1}{n_1} + \frac{1}{n_2}}} \tag{20}$$

If $|t_0| > t_{\alpha/2, n_1+n_2-2}$ then H_0 is rejected, $t_{\alpha/2}$ is the upper $\alpha/2$ percentage point of the t-distribution with $n_1 + n_2 - 2$ degrees of freedom at a fixed significance level two-sided.

If the variances of two independent normal distributions are tested if they are equal, σ_1^2, s_1^2 and n_1 for Population 1, and σ_2^2, s_2^2 and n_2 for Population 2, then the hypothesis is:

$$\begin{aligned} H_0 &: \sigma_1^2 = \sigma_2^2 \\ H_1 &: \sigma_1^2 \neq \sigma_2^2 \end{aligned} \tag{21}$$

F statistics is the ratio of the two sample variances:

$$F_0 = \frac{s_1^2}{s_2^2} \tag{22}$$

H_0 is rejected if $F_0 > F_{\alpha/2, n_1-1, n_2-1}$ or $F_0 < F_{1-(\alpha/2), n_1-1, n_2-1}$, which denote the upper $\alpha/2$ and lower $1-(\alpha/2)$ percentage points of the F distribution with degrees of freedom $n_1 - 1$ and $n_2 - 1$, respectively, at a fixed significance level two-sided.

4. Proposed statistical approach

In hypothesis testing, sampling is very important because an inference is reached from the values in the sample about the values in the population. Therefore sampling has to be done very carefully and samples should represent the population. Sampling is a wide subject in textile engineering. Regular sampling during production and acceptance sampling from a static lot are two grand different subjects. This broad topic of sampling in textile engineering can well be covered in a separate manuscript, so the details of sampling will not be dealt herein. Instead, the number of samples, which are repeats, will be indicated as n_i.

In ring spinning yarn production the number of spindles per spinning frame is the determining factor for sampling. As a general application, at least five bobbins per 500 spindles per spinning frame are taken randomly for the tests of yarn properties. Different frame brands have different spindle numbers such as 576, 1008, 1296, and 1824. depending on the model of the frame. For example, at least 10 bobbins have to be chosen randomly for 576 spindles per frame, at least 15 bobbins have to be chosen randomly for 1008 spindles per frame, at least 15 bobbins again for 1296 spindles per frame, or at least 20 bobbins for 1824 spindles per frame.

When sampling for hypothesis testing in this chapter, the yarn lot is the population, and statistical inference and decisions will be made about the yarn lot from the samples selected from it. In order to conclude that the machine is adjusted correctly or to make a decision about its status, samples are randomly selected as different bobbins from independent, identical, and with equal probability of being chosen spindles on a spinning frame which are adjusted to produce a special yarn. Test results of the samples will give information about the yarn population. **Figure 5** shows the relationship between a population and a sample. In textile engineering, it is assumed that the property values of a textile material have a normal distribution, consequently in yarn spinning, yarn properties also exhibit normal distribution for property values.

The constant of variation (CV%) is a frequently used value in textile engineering. Starting from fibers to the end product, say apparel, fiber (fineness, length, breaking strength – breaking elongation, etc.), yarn (count, twist, irregularity, etc.), fabric (warp and weft density, fabric thickness, etc.), and apparel (measurements, weight, etc.) properties are all tested and the results are statistically analyzed; and the mean x, standard deviation s, and CV% $\left(\frac{s}{\bar{x}} \times 100\right)$ are calculated. The value of CV% indicates much information about the property it was calculated from. Furthermore, comparisons and evaluations are done making it possible to have a comprehensible understanding of how production is continuing. The constant of variation of yarn count can be expressed as $CV\%_{YarnCount}$. The value $CV\%_{YarnCount}$ has a close relationship with the technology of textile machinery. Technology of textile machinery developed considerably a lot when compared to the 1970s and 1980s. Textile machinery producers incorporate broad R&D departments and one of the main topics of their researches is

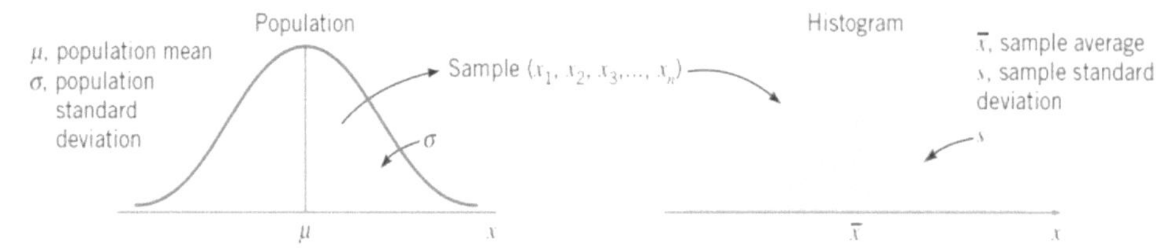

Figure 5.
Relationship between a population and a sample [1].

on $CV\%_{YarnCount}$. As a general consideration, $CV\%_{YarnCount}$ 5% was acceptable until the late 80's, whereas the $CV\%_{YarnCount}$ decreased to 3% in the mid 90's, then to 1–1.5% in the late 90s, and since 2000s this value is acceptable if it is less than 1%. In order for the $CV\%_{YarnCount}$ to be less than 1%, the variance of yarn count s has to be < 1 also. Within the context of this paper, it will be considered that the spinning frames were produced after the year 2000; therefore, the proposed statistical approach will be explained by considering s as < 1 in accordance with textile industry.

Yarn count is adjusted on the machine according to what the customer ordered. Yarn count will be indicated as μ_0 in this chapter.

The main aspect in both sampling, variation of yarn count, and yarn count is that every machine has to be adjusted to produce the yarn the customer ordered. The lot will be shipped as one and it does not make any difference for the customer which machine produced which yarn. The customer ordered the yarn lot and will regard it all the same at every single centimeter of yarn produced.

Suppose now a special yarn count of μ_0 in tex unit will be produced in twenty spinning frames in a spinning mill (**Figure 6**).

In this proposed statistical approach, the procedure starts with adjusting the Spinning Frame 1 (SF 1). The necessary adjustments to produce μ_0 tex yarn is done on the SF 1, the frame will run for a few minutes, the yarn will be produced a little bit, and n_1 bobbins from spindles are chosen as samples randomly from this normal distribution. The first thing is to test if the adjustments are correct and confront them with what the customer ordered. Since all the frames were produced after the year 2000, of the same brand, the same model, and the same technical specifications, the variance of yarn count has to be less than 1, with the latter being thus, the specified value for these hypothesis tests. In this manuscript it is argued that if the variance of yarn count is less than 1 it has to be tested before the yarn count. Then, the level of significance α is determined which is equal to 0.05 for ordinary textiles. n_1 bobbins from spindles of SF 1 are taken for yarn count tests done in the laboratory. The one-sided hypothesis is:

$$\begin{aligned} H_0 &: \sigma_1^2 = 1 \\ H_1 &: \sigma_1^2 < 1 \end{aligned} \quad (23)$$

and the χ^2 test statistic is:

$$\chi_1^2 = \frac{(n_1 - 1)s_1^2}{1} \quad (24)$$

s_1^2 is the sample variance of n_1 repeats from SF 1. The null hypothesis of variance of yarn count is rejected if $\chi_1^2 < \chi_{1-\alpha,n_1-1}^2$. If it is unable to be rejected, then the procedure

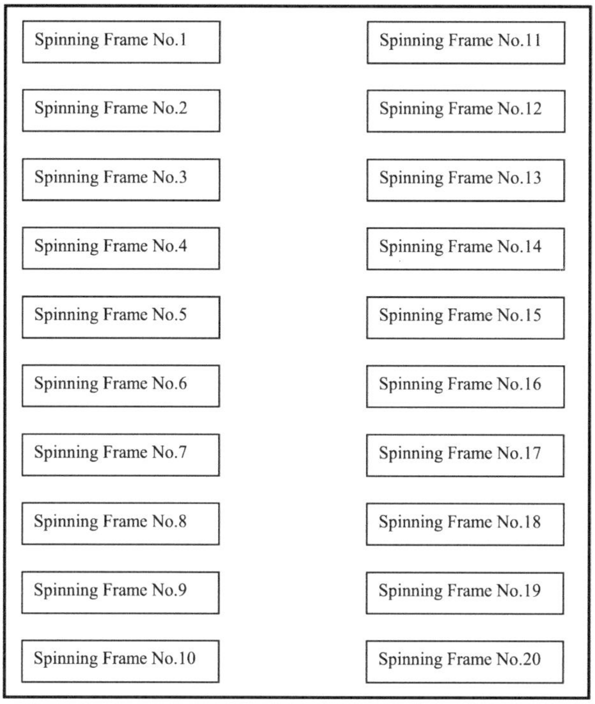

Figure 6.
Representation of spinning frames in a spinning mill.

continues by going back to the SF 1 and doing some more adjustments on the frame and repeating this test until the null hypothesis of variance of yarn count is rejected.

When it is guaranteed that the variance of yarn count is less than 1, then comes the yarn count statistics tests. The average of yarn count of SF 1 would be μ_1 and the two-sided hypothesis of yarn count is:

$$\begin{aligned} H_0 &: \mu_1 = \mu_0 \\ H_1 &: \mu_1 \neq \mu_0 \end{aligned} \tag{25}$$

Variance is estimated by s_1^2, $\overline{x}_1$ is the average of the n_1 repeats of yarn count from SF 1, the t-test statistic is:

$$t_1 = \frac{\overline{x}_1 - \mu_0}{s_1/\sqrt{n_1}} \tag{26}$$

where instead of a normal distribution it is a t distribution with $n_1 - 1$ degrees of freedom. If $|(t_1)| > t_{\alpha/2, n_1 - 1}$ then H_0 is rejected, $t_{\alpha/2, n_1 - 1}$ is the upper $\alpha/2$ percentage point of the t distribution with $n_1 - 1$ degrees of freedom at a fixed significance level two-sided. If the null hypothesis of yarn count is rejected, then the procedure continues by going back to the SF 1 and doing some more adjustments on the frame and repeating these tests until the null hypothesis of yarn count is unable to be rejected.

Now the SF 1 is ready to produce what the customer ordered, so the procedure will continue with the statistics to make the SF 2 to produce what the customer ordered and also the same as SF 1. The necessary adjustments to produce μ_0 tex yarn is done on the SF 2 and n_2 bobbins from spindles of SF 2 are chosen randomly. To test if the variance of yarn count of SF 2 is less than 1, the one-sided hypothesis is:

$$\begin{aligned} H_0 &: \sigma_2^2 = 1 \\ H_1 &: \sigma_2^2 < 1 \end{aligned} \tag{27}$$

and the χ^2 test statistic is:

$$\chi_2^2 = \frac{(n_2 - 1)s_2^2}{1} \tag{28}$$

s_2^2 is the sample variance of n_2 repeats from SF 2. The null hypothesis of variance is rejected if $\chi_2^2 < \chi_{1-\alpha,n_2-1}^2$. If it is unable to be rejected, then the procedure continues by going back to the SF 2 and doing some more adjustments on the frame and repeating this test until the null hypothesis of variance of yarn count is rejected.

Both of the variances of yarn counts of SF 1 and SF 2 may be less than 1 but their equality has to be tested also. This is justified because they will all mix into one lot and it is not important from the view of point of customer which frame produced which yarn. To test their equality, the hypothesis is:

$$\begin{aligned} H_0 &: \sigma_1^2 = \sigma_2^2 \\ H_1 &: \sigma_1^2 \neq \sigma_2^2 \end{aligned} \tag{29}$$

and the F statistics is:

$$F_{(1,2)} = \frac{s_1^2}{s_2^2} \tag{30}$$

H_0 is rejected if $F_{(1,2)} > F_{\alpha/2,n_1-1,n_2-1}$ or $F_{(1,2)} < F_{1-(\alpha/2),n_1-1,n_2-1}$ which denote the upper $\alpha/2$ and lower $1-(\alpha/2)$ percentage points of the F distribution with degrees of freedom $n_1 - 1$ and $n_2 - 1$, respectively. If the null hypothesis is rejected, then the procedure continues by going back to the SF 2 and doing some more adjustments on the frame and repeating these tests until the null hypothesis of equality of variances of yarn counts is unable to be rejected.

When it is guaranteed that both the variance of yarn count is less than 1 for SF 2 and the two frames' variances are equal, then comes the yarn count statistics tests for SF 2. The average of yarn count of n_2 samples from SF 2 would be μ_2 and the two-sided hypothesis of yarn count is:

$$\begin{aligned} H_0 &: \mu_2 = \mu_0 \\ H_1 &: \mu_2 \neq \mu_0 \end{aligned} \tag{31}$$

Variance is estimated by s_2^2, $\overline{x}_2$ is the average of the n_2 repeats of yarn count from SF 2, the t-test statistic is:

$$t_2 = \frac{\overline{x}_2 - \mu_0}{s_2/\sqrt{n_2}} \tag{32}$$

where instead of a normal distribution it is a t distribution with $n_2 - 1$ degrees of freedom. If $|(t_2)| > t_{\alpha/2, n_2 - 1}$ then H_0 is rejected, $t_{\alpha/2, n_2 - 1}$ is the upper $\alpha/2$ percentage point of the a t distribution with $n_2 - 1$ degrees of freedom at a fixed significance level two-sided. If the null hypothesis of yarn count is rejected, then the procedure continues by going back to the SF 2 and doing some more adjustments on the frame and repeating these tests until the null hypothesis of yarn count is unable to be rejected.

Both of the yarn counts of SF 1 and SF 2 may be equal to what the customer ordered but their equality with each other has to be tested also because they will all mix into one lot and it is not important from the view of point of customer which frame produced which yarn. To test the yarn count equality of SF 1 and SF 2, even there is only one μ_0, and for ease of reference, the hypothesis is:

$$\begin{aligned} H_0 &: \mu_1 - \mu_2 = 0 \\ H_1 &: \mu_1 - \mu_2 \neq 0 \end{aligned} \tag{33}$$

and the pooled t-test statistic is:

$$t_{(1,2)} = \frac{\overline{x}_1 - \overline{x}_2}{s_{p(1,2)} \sqrt{\frac{1}{n_1} + \frac{1}{n_2}}} \tag{34}$$

$$s^2_{p(1,2)} = \frac{(n_1 - 1)s_1^2 + (n_2 - 1)s_2^2}{n_1 + n_2 - 2} \tag{35}$$

$s^2_{p(1,2)}$ is the pooled estimator of variance of SFs 1 and 2 with $(n_1 + n_2 - 2)$ degrees of freedom.

If $|t_{(1,2)}| > t_{\alpha/2, n_1 + n_2 - 2}$ then H_0 is rejected, $t_{\alpha/2}$ is the upper $\alpha/2$ percentage point of the t-distribution with $n_1 + n_2 - 2$ degrees of freedom at a fixed significance level two-sided. If the H_0 of yarn count is rejected, then the procedure continues by going back to the SF 2 and doing some more adjustments on the frame and repeating this and the above tests until the H_0 of yarn count is unable to be rejected. The operation steps can be summarized as below:

Step 1) Yarn count adjustment of SF 1.
Go to Step 1 and repeat until H_0 is unable to be rejected

Step 2) Testing the variance of yarn count of SF 1 to be less than 1.
Go to Step 1 and repeat until H_0 is unable to be rejected

Step 3) Testing the yarn count of SF 1 with μ_0.
Go to Steps 1 and 2, and repeat until H_0 is unable to be rejected

Step 4) Yarn count adjustment of SF 2.
Go to Step 4 and repeat until H_0 is unable to be rejected

Step 5) Testing the variance of yarn count of SF 2 to be less than 1.
Go to Step 4 and repeat until H_0 is unable to be rejected

Step 6) Testing the equality of variances of SF 1 and SF 2.
Go to Steps 4 and 5, and repeat until H_0 is unable to be rejected

Step 7) Testing the yarn count of SF 2 with μ_0.
Go to Steps 4, 5, and 6, and repeat until H_0 is unable to be rejected.

Step 8) Testing the equality of yarn counts of SF 1 and SF 2.
Go to Steps 4, 5, 6, and 7, and repeat until H_0 is unable to be rejected.

The same will be repeated for the rest of the spinning frames until SF 20.

Now the SFs 1 and 2 are producing the same yarn having the same yarn count and same variance of yarn count. The SFs 1 and 2 can be considered as one machine producing the same product. A representation is given in **Figure 7**.

The procedure will continue with the statistics to make the SF 3 to produce what the customer ordered and also the same as SFs 1 and 2. The necessary adjustments to produce μ_0 tex yarn is done on the SF 3 and n_3 bobbins are chosen randomly, having s_3^2 variance and $\bar{x}_3$ average yarn count. To test if the variance of yarn count of SF 3 is less than 1, the one-sided hypothesis is:

$$H_0 : \sigma_3^2 = 1$$
$$H_1 : \sigma_3^2 < 1 \quad (36)$$

and the χ^2 test statistic is:

$$\chi_3^2 = \frac{(n_3 - 1)s_3^2}{1} \quad (37)$$

s_3^2 is the sample variance of yarn count of n_3 repeats from SF 3. The null hypothesis of variance of yarn count is rejected if $\chi_3^2 < \chi_{1-\alpha,n_3-1}^2$. If it is unable to be rejected, then the procedure continues by going back to the SF 3 and doing some more adjustments on the frame and repeating this test until the H_0 of variance of yarn count is rejected.

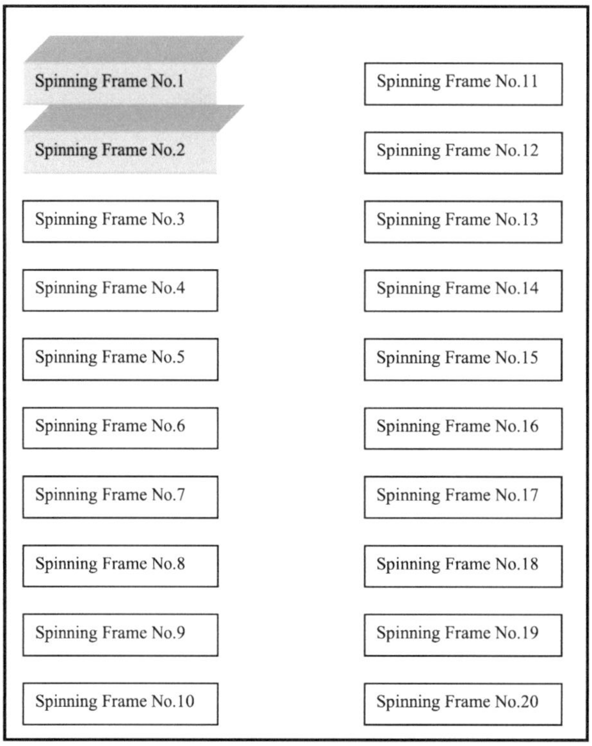

Figure 7.
Representation of SFs 1 and .2 producing the same yarn.

Both of the variances of yarn count of (SFs 1 and 2) and SF 3 may be less than 1 but their equality has to be tested also because they will all mix into one lot. To test their equality, the hypothesis is:

$$\begin{aligned} H_0 &: \sigma^2_{(1,2)} = \sigma^2_3 \\ H_1 &: \sigma^2_{(1,2)} \neq \sigma^2_3 \end{aligned} \tag{38}$$

and the F statistics is:

$$F_{(1-3)} = \frac{s^2_{p(1,2)}}{s^2_3} \tag{39}$$

H_0 is rejected if $F_{(1-3)} > F_{\alpha/2, n_1+n_2-2, n_3-1}$ or $F_{(1-3)} < F_{1-(\alpha/2), n_1+n_2-2, n_3-1}$, which denote the upper $\alpha/2$ and lower $1-(\alpha/2)$ percentage points of the F distribution with degrees of freedom n_1+n_2-2 and n_3-1, respectively. If the H_0 of variance of yarn count is rejected, then the procedure continues by going back to the SF 3 and doing some more adjustments on the frame and repeating these tests until the H_0 of variance of yarn count is unable to be rejected.

When it is guaranteed that both the variance of yarn count is less than 1 for SF 3 and it is equal with the first two frames' pooled variance of yarn count, then come the yarn count statistics tests for SF 3. The average of yarn count of SF 3 would be μ_3 and the two-sided hypothesis:

$$\begin{aligned} H_0 &: \mu_3 = \mu_0 \\ H_1 &: \mu_3 \neq \mu_0 \end{aligned} \tag{40}$$

Variance is estimated by s^2_3, $\overline{x}_3$ is the average of the n_3 repeats of yarn count from SF 3, the t-test statistic is:

$$t_3 = \frac{\overline{x}_3 - \mu_0}{s_3/\sqrt{n_3}} \tag{41}$$

where instead of a normal distribution it is a t distribution with $n_3 - 1$ degrees of freedom. If $|(t_3)| > t_{\alpha/2, n_3-1}$ then H_0 is rejected, $t_{\alpha/2, n_3-1}$ is the upper $\alpha/2$ percentage point of the a t distribution with $n_3 - 1$ degrees of freedom at a fixed significance level two-sided. If the H_0 of yarn count is rejected, then the procedure continues by going back to the SF 3 and doing some more adjustments on the frame and repeating these tests until the H_0 of yarn count is unable to be rejected.

Both of the yarn counts of (SFs 1 and 2) and SF 3 may be equal to what the customer ordered but their equality with each other also has to be tested because they will all mix into one lot. To test the yarn count equality of (SFs 1 and 2) and SF 3, there is only one μ_0, and for ease of reference, the hypothesis is:

$$\begin{aligned} H_0 &: \mu_{(1,2)} - \mu_3 = 0 \\ H_1 &: \mu_{(1,2)} - \mu_3 \neq 0 \end{aligned} \tag{42}$$

and the pooled t-test statistic is:

The average of yarn counts of SF 1 and SF 2 is:

$$\frac{\overline{x}_1 + \overline{x}_2}{2} = \overline{x}_{(1,2)} \tag{43}$$

then,

$$t_{(1-3)} = \frac{\overline{x}_{(1,2)} - \overline{x}_3}{s_{p(1-3)}\sqrt{\frac{1}{n_1+n_2-2} + \frac{1}{n_3}}} \tag{44}$$

$$s^2_{p(1-3)} = \frac{(n_1 + n_2 - 2)s^2_{p(1,2)} + (n_3 - 1)s^2_3}{n_1 + n_2 + n_3 - 3} \tag{45}$$

$s^2_{p(1-3)}$ is the pooled estimator of variance of SFs 1–3 with $(n_1 + n_2 - 3)$ degrees of freedom.

If $|t_{(1-3)}| \succ t_{\alpha/2, n_1+n_2+n_3-3}$ then H_0 is rejected, $t_{\alpha/2}$ is the upper $\alpha/2$ percentage point of the t-distribution with $n_1 + n_2 + n_3 - 3$ degrees of freedom at a fixed significance level two-sided. If the H_0 of yarn count is rejected, then the procedure continues by going back to the SF 3 and doing some more adjustments on the frame and repeating this and the above tests until the H_0 of yarn count is unable to be rejected. The operation steps can be summarized as below:

Step 1) Yarn count adjustment of SF 1.
Go to Step 1 and repeat until H_0 is unable to be rejected
Step 2) Testing the variance of yarn count of SF 1 to be less than 1.
Go to Step 1 and repeat until H_0 is unable to be rejected
Step 3) Testing the yarn count of SF 1 with μ_0.
Go to Steps 1 and 2, and repeat until H_0 is unable to be rejected
Step 4) Yarn count adjustment of SF 2.
Go to Step 4 and repeat until H_0 is unable to be rejected
Step 5) Testing the variance of yarn count of SF 2 to be less than 1.
Go to Step 4 and repeat until H_0 is unable to be rejected
Step 6) Testing the equality of variances of SF 1 and SF 2.
Go to Steps 4 and 5, and repeat until H_0 is unable to be rejected
Step 7) Testing the yarn count of SF 2 with μ_0.
Go to Steps 4, 5, and 6, and repeat until H_0 is unable to be rejected
Step 8) Testing the equality of yarn counts of SF 1 and SF 2.
Go to Steps 4, 5, 6, and 7, and repeat until H_0 is unable to be rejected.
Step 9) Yarn count adjustment of SF 3.
Go to Step 9 and repeat until H_0 is unable to be rejected
Step 10) Testing the variance of yarn count of SF 3 to be less than 1.
Go to Step 9 and repeat until H_0 is unable to be rejected
Step 11) Testing the equality of variances of (SFs 1 and 2) and SF 3.
Go to Steps 9 and 10, and repeat until H_0 is unable to be rejected
Step 12) Testing the yarn count of SF 3 with μ_0.
Go to Steps 9, 10, and 11, and repeat until H_0 is unable to be rejected
Step 13) Testing the equality of yarn counts of (SFs 1 and 2) and SF 3.
Go to Steps 9, 10, 11, and 12, and repeat until H_0 is unable to be rejected

The same will be done in a repeating pattern for the rest of the spinning frames until SF 20.

Now the SFs 1–3 are producing the same yarn having the same yarn count and the same variance of yarn count. The SFs 1–3 can be considered as one machine producing the same product. A representation is given in **Figure 8**.

The procedure will continue with the statistics to make the SF 4 to produce what the customer ordered and also the same as (SFs 1–3). The necessary adjustments to

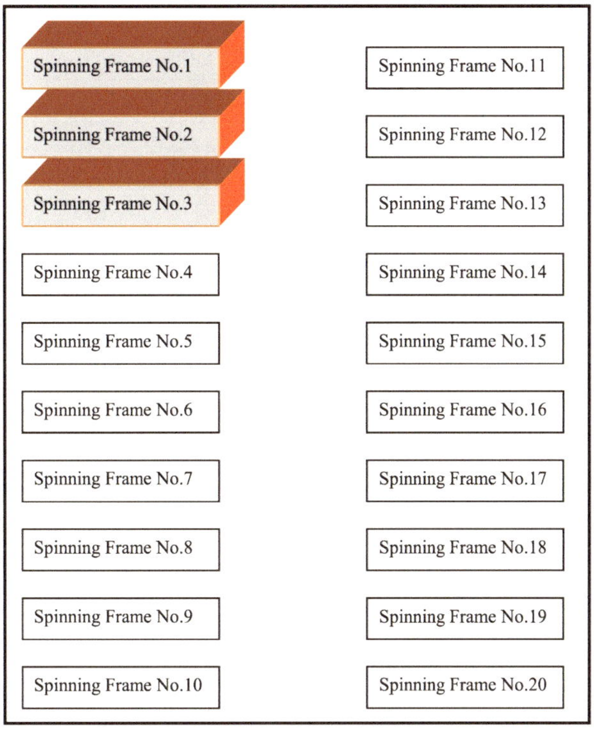

Figure 8.
Representation of SFs 1–3 producing the same yarn.

produce μ_0 tex yarn is done on the SF 4 and n_4 bobbins from spindles are chosen randomly, having s_4^2 variance of yarn count and $\overline{x}_4$ average yarn count. To test if the variance of yarn count of SF 4 is less than 1, the one-sided hypothesis is:

$$\begin{aligned} H_0 &: \sigma_4^2 = 1 \\ H_1 &: \sigma_4^2 < 1 \end{aligned} \tag{46}$$

and the χ^2 test statistic is:

$$\chi_4^2 = \frac{(n_4 - 1)s_4^2}{1} \tag{47}$$

The H_0 of variance of yarn count is rejected if $\chi_4^2 < \chi^2_{1-\alpha, n_4-1}$. If it is unable to be rejected, then the procedure continues by going back to the SF 4 and doing some more adjustments on the frame and repeating this test until the H_0 of variance of yarn count is rejected.

Both of the variances of yarn count of (SFs 1–3) and SF 4 may be less than 1 but their equality has to be tested also because they will all mix into one lot. To test their equality, the hypothesis is:

$$\begin{aligned} H_0 &: \sigma^2_{(1-3)} = \sigma_4^2 \\ H_1 &: \sigma^2_{(1-3)} \neq \sigma_4^2 \end{aligned} \tag{48}$$

and the F statistics is:

$$F_{(1-4)} = \frac{s^2_{p(1-3)}}{s^2_4} \tag{49}$$

H_0 is rejected if $F_{(1-4)} > F_{\alpha/2, n_1+n_2+n_3-3, n_4-1}$ or $F_{(1-4)} < F_{1-(\alpha/2), n_1+n_2+n_3-3, n_4-1}$, which denote the upper $\alpha/2$ and lower $1-(\alpha/2)$ percentage points of the F distribution with degrees of freedom $n_1+n_2+n_3-3$ and n_4-1, respectively. If the H_0 is rejected, then the procedure continues by going back to the SF 4 and doing some more adjustments on the frame and repeating these tests until the H_0 is unable to be rejected.

When it is guaranteed that both the variance of yarn count is less than 1 for SF 4 and it is equal to the first three frames' pooled variance of yarn count, then come the yarn count statistics tests. The average of yarn count of SF 4 would be μ_4 and the two-sided hypothesis is:

$$\begin{aligned} H_0 &: \mu_4 = \mu_0 \\ H_1 &: \mu_4 \neq \mu_0 \end{aligned} \tag{50}$$

Variance is estimated by s^2_4, $\overline{x}_4$ is the average of the n_4 repeats of yarn count from SF 4, the t-test statistic is:

$$t_4 = \frac{\overline{x}_4 - \mu_0}{s_4/\sqrt{n_4}} \tag{51}$$

where instead of a normal distribution it is a t distribution with n_4-1 degrees of freedom. If $|(t_4)| > t_{\alpha/2, n_4-1}$ then H_0 is rejected, $t_{\alpha/2, n_4-1}$ is the upper $\alpha/2$ percentage point of the t distribution with n_4-1 degrees of freedom at a fixed significance level two-sided. If the H_0 of yarn count is rejected, then the procedure continues by going back to the SF 4 and doing some more adjustments on the frame and repeating these tests until the H_0 of yarn count is unable to be rejected.

Both of the yarn counts of (SFs 1–3) and SF 4 may be equal to what the customer ordered but their equality with each other has to be tested also because they will all mix into one lot. To test the yarn count equality of (SFs 1–3) and SF 4, there is only one μ_0, the hypothesis is:

$$\begin{aligned} H_0 &: \mu_{(1-3)} - \mu_4 = 0 \\ H_1 &: \mu_{(1-3)} - \mu_4 \neq 0 \end{aligned} \tag{52}$$

and the pooled t-test statistic is:

The average of yarn counts of (SFs 1 and 2) and SF 3 is:

$$\frac{\overline{x}_{(1,2)} + \overline{x}_3}{2} = \overline{x}_{(1-3)} \tag{53}$$

then,

$$t_{(1-4)} = \frac{\overline{x}_{(1-3)} - \overline{x}_4}{s_{p(1-4)}\sqrt{\frac{1}{n_1+n_2+n_3-3} + \frac{1}{n_4}}} \tag{54}$$

$$s^2_{p(1-4)} = \frac{(n_1+n_2+n_3-3)s^2_{p(1-3)} + (n_4-1)s^2_4}{n_1 + \ldots + n_4 - 4} \tag{55}$$

$s^2_{p(1-4)}$ is the pooled estimator of variance of SFs 1–4 with (n_1+n_2-4) degrees of freedom.

If $\left|t_{(1-4)}\right| > t_{\alpha/2, n_1+\ldots+n_4-4}$ then H_0 is rejected, $t_{\alpha/2}$ is the upper $\alpha/2$ percentage point of the t-distribution with $n_1 + \ldots + n_4 - 4$ degrees of freedom at a fixed significance level two-sided. If the H_0 of yarn count is rejected, then the procedure continues by going back to the SF 4 and doing some more adjustments on the frame and repeating this and the above tests until the H_0 of yarn count is unable to be rejected. The operation steps can be summarized as below:

(Continued)

Step 14) Yarn count adjustment of SF 4.
Go to Step 14 and repeat until H_0 is unable to be rejected

Step 15) Testing the variance of yarn count of SF 3 to be less than 1.
Go to Step 14 and repeat until H_0 is unable to be rejected

Step 16) Testing the equality of variances of (SFs 1–3) and SF 4.
Go to Steps 14 and 15, and repeat until H_0 is unable to be rejected

Step 17) Testing the yarn count of SF 3 with μ_0.
Go to Steps 14, 15, and 16, and repeat until H_0 is unable to be rejected

Step 18) Testing the equality of yarn counts of (SFs 1–3) and SF 4.
Go to Steps 14, 15, 16, and 17, and repeat until H_0 is unable to be rejected

The same will be repeated for the rest of the spinning frames until SF 20.

Now the SFs 1-4 are producing the same yarn having the same yarn count and same variance of yarn count. The SFs 1–4 can be considered as one machine producing the same product. A representation is given in **Figure 9**.

Suppose the same procedure is repeated for SFs 5, 6, and 7, and now the procedure will continue with the statistics to make the SF 8 to produce what the customer

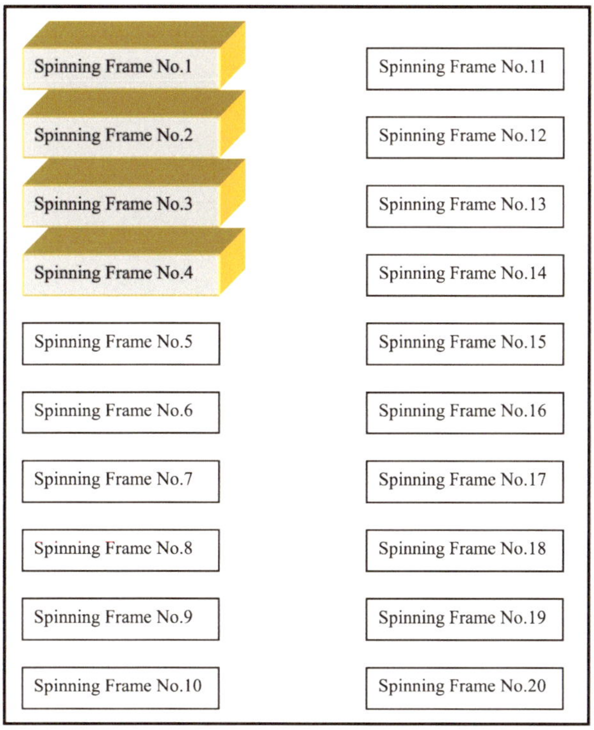

Figure 9.
Representation of SFs 1–4 producing the same yarn.

ordered and also the same as (SFs 1–7). The necessary adjustments to produce μ_0 tex yarn is done on the SF 8 and n_8 bobbins from spindles are taken randomly, having s_8^2 variance of yarn count and $\overline{x}_8$ average yarn count. To test if the variance of yarn count of SF 8 is less than 1, the one-sided hypothesis is:

$$H_0 : \sigma_8^2 = 1$$
$$H_1 : \sigma_8^2 < 1 \tag{56}$$

and the χ^2 test statistic is:

$$\chi_8^2 = \frac{(n_8 - 1)s_8^2}{1} \tag{57}$$

s_8^2 is the variance of yarn count of n_8 repeats from SF 8. The H_0 of variance of yarn count is rejected if $\chi_8^2 < \chi_{1-\alpha,n_8-1}^2$. If it is unable to be rejected, then the procedure continues by going back to the SF 8 and doing some more adjustments on the frame and repeating this test until the H_0 of variance of yarn count is rejected.

Both of the variances of yarn count of (SFs 1–7) and SF 8 may be less than 1 but their equality has to be tested also because they will all mix into one lot. To test their equality, the hypothesis is:

$$H_0 : \sigma_{(1-7)}^2 = \sigma_8^2$$
$$H_1 : \sigma_{(1-7)}^2 \neq \sigma_8^2 \tag{58}$$

and the F statistics is:

$$F_{(1-8)} = \frac{s_{p(1-7)}^2}{s_8^2} \tag{59}$$

H_0 is rejected if $F_{(1-8)} > F_{\alpha/2,n_1+\ldots+n_7-7,n_8-1}$ or $F_{(1-8)} < F_{1-(\alpha/2),n_1+\ldots+n_7-7,n_8-1}$ which denote the upper $\alpha/2$ and lower $1-(\alpha/2)$ percentage points of the F distribution with degrees of freedom $n_1 + \ldots + n_7 - 7$ and $n_8 - 1$, respectively. If the H_0 of variance of yarn count is rejected, then the procedure continues by going back to the SF 8 and doing some more adjustments on the frame and repeating these tests until the H_0 of variance of yarn count is unable to be rejected.

When it is guaranteed that both the variance of yarn count is less than 1 for SF 8 and it is equal to the first seven frames' pooled variance, then come the yarn count statistics tests for SF 8. The average of yarn count of SF 8 would be μ_8 and the two-sided hypothesis is:

$$H_0 : \mu_8 = \mu_0$$
$$H_1 : \mu_8 \neq \mu_0 \tag{60}$$

Variance is estimated by s_8^2, $\overline{x}_8$ is the average of the n_8 repeats of yarn count from SF 8, the t-test statistic is:

$$t_8 = \frac{\overline{x}_8 - \mu_0}{s_8/\sqrt{n_8}} \tag{61}$$

where instead of a normal distribution it is a t distribution with $n_8 - 1$ degrees of freedom. If $|(t_8)| > t_{\alpha/2, n_8 - 1}$ then H_0 is rejected, $t_{\alpha/2, n_8 - 1}$ is the upper $\alpha/2$ percentage point of the a t distribution with $n_8 - 1$ degrees of freedom at a fixed significance level two-sided. If the H_0 of yarn count is rejected, then the procedure continues by going back to the SF 8 and doing some more adjustments on the frame and repeating these tests until the H_0 of yarn count is unable to be rejected.

Both of the yarn counts of (SFs 1–7) and SF 8 may be equal to what the customer ordered but their equality with each other also has to be tested because they will all mix into one lot. To test the yarn count equality of (SFs 1–7) and SF 8, there is only one μ_0, the hypothesis is:

$$\begin{aligned} H_0 &: \mu_{(1-7)} - \mu_8 = 0 \\ H_1 &: \mu_{(1-7)} - \mu_8 \neq 0 \end{aligned} \tag{62}$$

and the pooled t-test statistic is:

The average of yarn counts of (SFs 1–6) and SF 7 is:

$$\frac{\overline{x}_{(1-6)} + \overline{x}_7}{2} = \overline{x}_{(1-7)} \tag{63}$$

then,

$$t_{(1-8)} = \frac{\overline{x}_{(1-7)} - \overline{x}_8}{s_{p(1-8)}\sqrt{\frac{1}{n_1 + \dots + n_7 - 7} + \frac{1}{n_8}}} \tag{64}$$

$$s^2_{p(1-8)} = \frac{(n_1 + \dots + n_7 - 7)s^2_{p(1-7)} + (n_8 - 1)s_8^2}{n_1 + \dots + n_8 - 8} \tag{65}$$

$s^2_{p(1-8)}$ is the pooled estimator of variance of SFs 1–8 with $(n_1 + n_2 - 8)$ degrees of freedom.

If $\left|t_{(1-8)}\right| > t_{\alpha/2, n_1 + \dots + n_8 - 8}$ then H_0 is rejected, $t_{\alpha/2}$ is the upper $\alpha/2$ percentage point of the t-distribution with $n_1 + \dots + n_8 - 8$ degrees of freedom at a fixed significance level two-sided. If the H_0 of yarn count is rejected, then the procedure continues by going back to the SF 8 and doing some more adjustments on the frame and repeating this and the above tests until H_0 of yarn count is unable to be rejected. The operation steps can be summarized as below:

(Continued)

Step 34) Yarn count adjustment of SF 8.
    Go to Step 34 and repeat until H_0 is unable to be rejected
Step 35) Testing the variance of yarn count of SF 8 to be less than 1.
    Go to Step 34 and repeat until H_0 is unable to be rejected
Step 36) Testing the equality of variances of (SFs 1–7) and SF 8.
    Go to Steps 34 and 35, and repeat until H_0 is unable to be rejected
Step 37) Testing the yarn count of SF 8 with μ_0.
    Go to Steps 34, 35, and 36, and repeat until H_0 is unable to be rejected
Step 38) Testing the equality of yarn counts of (SFs 1–7) and SF 8.
    Go to Steps 34, 35, 36, and 37, and repeat until H_0 is unable to be rejected

The same will be done in a repeating manner for the rest of the spinning frames until SF 20.

Now the SFs 1–8 are producing the same yarn having the same yarn count and same variance of yarn count. The SFs (1–8) can be considered as a one machine producing the same product. A representation is given in **Figure 10**.

Suppose the same procedure is repeated for SFs 9, 10, 11, 12, 13, 14, 15, 16, 17, 18, and 19, and now the last spinning frame is the 20th one, the procedure will continue with the statistics to make the SF 20 to produce what the customer ordered and also the same as (SFs 1–19). The necessary adjustments to produce μ_0 tex yarn is done on the SF 20 and n_{20} bobbins from spindles are chosen randomly, having s_{20}^2 variance of yarn count and $\overline{x}_{20}$ average yarn count. To test if the variance of yarn count of SF 20 is less than 1, the one-sided hypothesis is:

$$\begin{aligned} H_0 &: \sigma_{20}^2 = 1 \\ H_1 &: \sigma_{20}^2 < 1 \end{aligned} \tag{66}$$

and the χ^2 test statistic is:

$$\chi_{20}^2 = \frac{(n_{20} - 1)s_{20}^2}{1} \tag{67}$$

The H_0 of variance of yarn count is rejected if $\chi_{20}^2 < \chi_{1-\alpha,n_{20}-1}^2$. If it is unable to be rejected, then the procedure continues by going back to the SF 20 and doing some more adjustments on the frame and repeating this test until the H_0 of variance of yarn count is rejected.

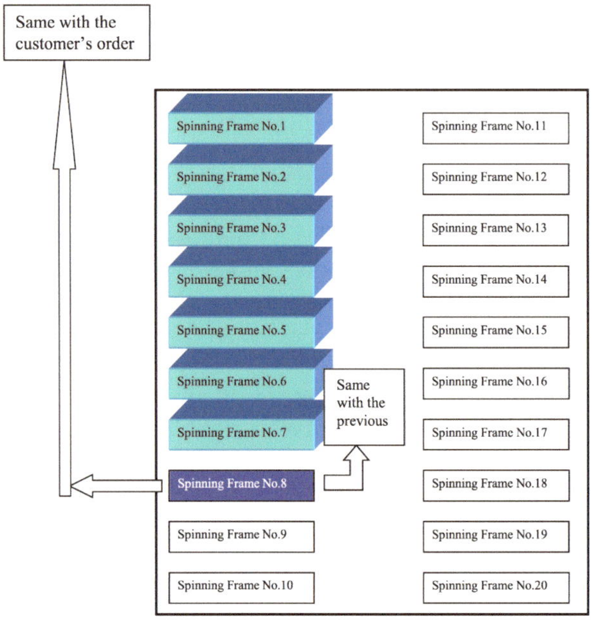

Figure 10.
Representation of SFs 1–8 producing the same yarn.

Both of the variances of yarn count of SFs (1–19) and SF 20 may be less than 1 but their equality has to be tested also because they will all mix into one lot. To test their equality, the hypothesis is:

$$\begin{aligned} H_0 &: \sigma^2_{(1-19)} = \sigma^2_{20} \\ H_1 &: \sigma^2_{(1-19)} \neq \sigma^2_{20} \end{aligned} \tag{68}$$

and the F statistics is:

$$F_{(1-20)} = \frac{s^2_{p(1-19)}}{s^2_{20}} \tag{69}$$

H_0 is rejected if $F_{20} > F_{\alpha/2, n_1 + \dots + n_{19} - 19, n_{20} - 1}$ or $F_{20} < F_{1-(\alpha/2), n_1 + \dots + n_{19} - 19, n_{20} - 1}$ which denote the upper $\alpha/2$ and lower $1 - (\alpha/2)$ percentage points of the F distribution with degrees of freedom $n_1 + \dots + n_{19} - 19$ and $n_{20} - 1$, respectively. If the H_0 is rejected, then the procedure continues by going back to the SF 20 and doing some more adjustments on the frame and repeating these tests until the H_0 is unable to be rejected.

When it is guaranteed that both the variance of yarn count is less than 1 for SF 20 and it is equal to the other nineteen frames' pooled variance of yarn count, then come the yarn count statistics tests. The average of yarn count of SF 20 would be μ_{20} and the two-sided hypothesis is:

$$\begin{aligned} H_0 &: \mu_{20} = \mu_0 \\ H_1 &: \mu_{20} \neq \mu_0 \end{aligned} \tag{70}$$

Variance of yarn count is estimated by s^2_{20}, $\overline{x}_{20}$ is the average of the n_{20} repeats of yarn count from SF 20, the t-test statistic is:

$$t_{20} = \frac{\overline{x}_{20} - \mu_0}{s_{20}/\sqrt{n_{20}}} \tag{71}$$

where instead of a normal distribution it is a t distribution with $n_{20} - 1$ degrees of freedom. If $|(t_{20})| > t_{\alpha/2, n_{20}-1}$ then H_0 is rejected, $t_{\alpha/2, n_{20}-1}$ is the upper $\alpha/2$ percentage point of the a t distribution with $n_{20} - 1$ degrees of freedom at a fixed significance level two-sided. If the H_0 of yarn count is rejected, then the procedure continues by going back to the SF 20 and doing some more adjustments on the frame and repeating this test until the H_0 of yarn count is unable to be rejected.

Both of the yarn counts of (SFs 1–19) and SF 20 may be equal to what the customer ordered but their equality with each other also has to be tested because they will all mix into one lot. To test the yarn count equality of SFs (1–19) and SF 20, there is only one μ_0, the hypothesis is:

$$\begin{aligned} H_0 &: \mu_{(1-19)} - \mu_{20} = 0 \\ H_1 &: \mu_{(1-19)} - \mu_{20} \neq 0 \end{aligned} \tag{72}$$

and the pooled t-test statistic is:

The average of yarn counts of (SFs 1–18) and SF 19 is:

$$\frac{\overline{x}_{(1-18)} + \overline{x}_{19}}{2} = \overline{x}_{(1-19)} \tag{73}$$

then,

$$t_{(1-20)} = \frac{\overline{x}_{(1-19)} - \overline{x}_{20}}{s_{p(1-20)}\sqrt{\frac{1}{n_1 + \ldots + n_{19} - 19} + \frac{1}{n_{20}}}} \tag{74}$$

$$s^2_{p(1-20)} = \frac{(n_1 + \ldots + n_{19} - 19)s^2_{p(1-19)} + (n_{20} - 1)s^2_{20}}{n_1 + \ldots + n_{20} - 20} \tag{75}$$

$s^2_{p(1-20)}$ is the pooled estimator of variance of SFs 1–20 with $(n_1 + n_2 - 20)$ degrees of freedom.

If $|t_{(1-20)}| > t_{\alpha/2, n_1 + \ldots + n_{20} - 20}$ then H_0 is rejected, $t_{\alpha/2}$ is the upper $\alpha/2$ percentage point of the t-distribution with $n_1 + \ldots + n_{20} - 20$ degrees of freedom at a fixed significance level two-sided. If the H_0 is rejected, then the procedure continues by going back to the SF 20 and doing some more adjustments on the frame and repeating this and the above tests until the H_0 of yarn count is unable to be rejected. The operation steps can be summarized as below:

(Continued)

Step 94) Yarn count adjustment of SF20.
Go to Step 94 and repeat until H_0 is unable to be rejected

Step 95) Testing the variance of yarn count of SF 20 to be less than 1.
Go to Step 94 and repeat until H_0 is unable to be rejected

Step 96) Testing the equality of variances of SFs (1–19) and SF 20.
Go to Steps 94 and 95, and repeat until H_0 is unable to be rejected

Step 97) Testing the yarn count of SF 20 with μ_0.
Go to Steps 94, 95 and 96, and repeat until H_0 is unable to be rejected

Step 98) Testing the equality of yarn counts of SFs (1–19) and SF 20.
Go to Steps 94, 95, 96 and 97, and repeat until H_0 is unable to be rejected.

Now the SFs 1–20 are producing the same yarn having the same yarn count and same variance of yarn count. The SFs (1–20) can be considered as one machine producing the same product, no difference between the yarns of twenty different spinning frames. A representation is given in **Figure 11**.

The logic in this proposed statistical approach is in a spinning mill having twenty spinning frames to adjust the first spinning frame according to what the customer ordered and to the technology of the spinning frame; take samples, statistically test them and if rejected, correct the adjustments, do the statistic tests again, and if unable to be rejected, adjust the second spinning frame according to what the customer ordered and to the technology of the spinning frame, take samples, statistically test them and if rejected, correct the adjustments, do the statistic tests again, pool the output of the first and second frames, if rejected, repeat, and if unable to be rejected, go on to the third frame, and so on until the twentieth frame. This approach pools the output of all the spinning frames in multiple-stream process of ring spinning. This will guarantee that the production starts correct and is pooled, producing yarn as per customers' order by incorporating the necessary technology and reducing variability. The whole lot will have the same yarn property at the beginning of production. During production, control charts will be performed and assignable causes will be seen if they occur, and will be

DOI: http://dx.doi.org/10.5772/intechopen.105643

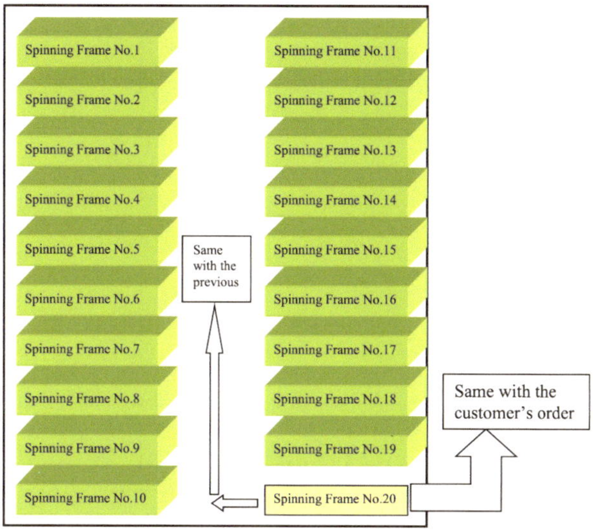

Figure 11.
Representation of SFs 1–20 producing the same yarn.

taken care of. Control charts will give much valuable information during production because it is assured that the production started correctly and all the frames are pooled. Additionally, instead of preparing separate control charts for each rational subgroup, even only one control chart for the whole lot would be enough, saving hence, time, cost, manpower, etc. This robust statistical approach can be incorporated in a statistics computer program, yielding a number of benefits for the enterprises.

On the other hand there is no restriction to employ boxplots, ANOVA, residual plots, etc. during production. These statistical methods will all add positive inferences on the data collected and support production and management. Claiming for better products and services alike will lead to new perspectives, ideas, point of views, etc.

Besides, spinning frames are not the only application area of this logic. Starting from the beginning of the stream, it can be applied to every machine in production, same two or more machines doing the same production, and so on. The first one will be adjusted at the beginning according to this logic, starting will be correct and will be pooled one by one, and continuing production will be controlled with the other statistical methods. Moreover, yarn count property is not the only application area falling under this logic. Yarn twist is also a property adjusted on the spinning frame. Other properties of textile materials adjusted on the machines can all be well worked with this proposed statistical approach.

A summary of the statistical procedures followed in this proposed statistical approach is given in **Table 2**.

5. Conclusion

This paper proposed a novel statistical approach for multiple-stream processes. Performed literature review suggests that control charts are used in multiple-stream processes but in this proposed statistical method, the expectations from the control

	SF 1		SF 2	SF 3	SF 4	… … … SF 8	… … … SF 20
Adjustment of variance of yarn count	$H_0 : \sigma_1^2 = 1$ $H_1 : \sigma_1^2 < 1$	Adjustment of variance of yarn count	$H_0 : \sigma_2^2 = 1$ $H_1 : \sigma_2^2 < 1$	$H_0 : \sigma_3^2 = 1$ $H_1 : \sigma_3^2 < 1$	$H_0 : \sigma_4^2 = 1$ $H_1 : \sigma_4^2 < 1$	$H_0 : \sigma_8^2 = 1$ $H_1 : \sigma_8^2 < 1$	$H_0 : \sigma_{20}^2 = 1$ $H_1 : \sigma_{20}^2 < 1$
	$\chi_1^2 = \frac{(n_1-1)s_1^2}{1}$		$\chi_2^2 = \frac{(n_2-1)s_2^2}{1}$	$\chi_3^2 = \frac{(n_3-1)s_3^2}{1}$	$\chi_4^2 = \frac{(n_4-1)s_4^2}{1}$	$\chi_8^2 = \frac{(n_8-1)s_8^2}{1}$	$\chi_{20}^2 = \frac{(n_{20}-1)s_{20}^2}{1}$
Rejection criteria	$\chi_1^2 < \chi^2_{1-\alpha,n_1-1}$	**Rejection criteria**	$\chi_2^2 < \chi^2_{1-\alpha,n_2-1}$	$\chi_3^2 < \chi^2_{1-\alpha,n_3-1}$	$\chi_4^2 < \chi^2_{1-\alpha,n_4-1}$	$\chi_8^2 < \chi^2_{1-\alpha,n_8-1}$	$\chi_{20}^2 < \chi^2_{1-\alpha,n_{20}-1}$
	———	Equalization of variance of yarn count with previous	$H_0 : \sigma_1^2 = \sigma_2^2$ $H_1 : \sigma_1^2 \neq \sigma_2^2$	$H_0 : \sigma^2_{(1,2)} = \sigma_3^2$ $H_1 : \sigma^2_{(1,2)} \neq \sigma_3^2$	$H_0 : \sigma^2_{(1-3)} = \sigma_4^2$ $H_1 : \sigma^2_{(1-3)} \neq \sigma_4^2$	$H_0 : \sigma^2_{(1-7)} = \sigma_8^2$ $H_1 : \sigma^2_{(1-7)} \neq \sigma_8^2$	$H_0 : \sigma^2_{(1-19)} = \sigma_{20}^2$ $H_1 : \sigma^2_{(1-19)} \neq \sigma_{20}^2$
			$F_{(1,2)} = \frac{s_1^2}{s_2^2}$	$F_{(1-3)} = \frac{s^2_{p(1,2)}}{s_3^2}$	$F_{(1-4)} = \frac{s^2_{p(1-3)}}{s_4^2}$	$F_{(1-8)} = \frac{s^2_{p(1-7)}}{s_8^2}$	$F_{(1-20)} = \frac{s^2_{p(1-19)}}{s_{20}^2}$
		Rejection criteria	$F_{(1,2)} > F_{\alpha/2,n_1-1,n_2-1}$ or $F_{(1,2)} < F_{1-(\alpha/2),n_1-1,n_2-1}$	$F_{(1-3)} > F_{\alpha/2,n_1+n_2-2,n_3-1}$ or $F_{(1-3)} < F_{1-(\alpha/2),n_1+n_2-2,n_3-1}$	$F_{(1-4)} > F_{\alpha/2,n_1+n_2+n_3-3,n_4-1}$ or $F_{(1-4)} < F_{1-(\alpha/2),n_1+n_2+n_3-3,n_4-1}$	$F_{(1-8)} > F_{\alpha/2,n_1+\ldots+n_7-7,n_8-1}$ or $F_{(1-8)} < F_{1-(\alpha/2),n_1+\ldots+n_7-7,n_8-1}$	$F_{20} > F_{\alpha/2,n_1+\ldots+n_{19}-19,n_{20}-1}$ or $F_{20} < F_{1-(\alpha/2),n_1+\ldots+n_{19}-19,n_{20}-1}$
Adjustment of yarn count	$H_0 : \mu_1 = \mu_0$ $H_1 : \mu_1 \neq \mu_0$	**Adjustment of yarn count**	$H_0 : \mu_2 = \mu_0$ $H_1 : \mu_2 \neq \mu_0$	$H_0 : \mu_3 = \mu_0$ $H_1 : \mu_3 \neq \mu_0$	$H_0 : \mu_4 = \mu_0$ $H_1 : \mu_4 \neq \mu_0$	$H_0 : \mu_8 = \mu_0$ $H_1 : \mu_8 \neq \mu_0$	$H_0 : \mu_{20} = \mu_0$ $H_1 : \mu_{20} \neq \mu_0$
	$t_1 = \frac{\bar{x}_1-\mu_0}{s_1/\sqrt{n_1}}$		$t_2 = \frac{\bar{x}_2-\mu_0}{s_2/\sqrt{n_2}}$	$t_3 = \frac{\bar{x}_3-\mu_0}{s_3/\sqrt{n_3}}$	$t_4 = \frac{\bar{x}_4-\mu_0}{s_4/\sqrt{n_4}}$	$t_8 = \frac{\bar{x}_8-\mu_0}{s_8/\sqrt{n_8}}$	$t_{20} = \frac{\bar{x}_{20}-\mu_0}{s_{20}/\sqrt{n_{20}}}$
Rejection criteria	$\lvert(t_1)\rvert > t_{\alpha/2,n_1-1}$	**Rejection criteria**	$\lvert(t_2)\rvert > t_{\alpha/2,n_2-1}$	$\lvert(t_3)\rvert > t_{\alpha/2,n_3-1}$	$\lvert(t_4)\rvert > t_{\alpha/2,n_4-1}$	$\lvert(t_8)\rvert > t_{\alpha/2,n_8-1}$	$\lvert(t_{20})\rvert > t_{\alpha/2,n_{20}-1}$
	———	Pooling of yarn count with previous	$H_0 : \mu_1 - \mu_2 = 0$ $H_1 : \mu_1 - \mu_2 \neq 0$	$H_0 : \mu_{(1,2)} - \mu_3 = 0$ $H_1 : \mu_{(1,2)} - \mu_3 \neq 0$	$H_0 : \mu_{(1-3)} - \mu_4 = 0$ $H_1 : \mu_{(1-3)} - \mu_4 \neq 0$	$H_0 : \mu_{(1-7)} - \mu_8 = 0$ $H_1 : \mu_{(1-7)} - \mu_8 \neq 0$	$H_0 : \mu_{(1-19)} - \mu_{20} = 0$ $H_1 : \mu_{(1-19)} - \mu_{20} \neq 0$
			———	$\frac{\bar{x}_1+\bar{x}_2}{2} = \bar{x}_{(1,2)}$	$\frac{\bar{x}_{(1,2)}+\bar{x}_3}{2} = \bar{x}_{(1-3)}$	$\frac{\bar{x}_{(1-6)}+\bar{x}_7}{2} = \bar{x}_{(1-7)}$	$\frac{\bar{x}_{(1-18)}+\bar{x}_{19}}{2} = \bar{x}_{(1-19)}$
	———		$t_{(1,2)} = \frac{\bar{x}_1-\bar{x}_2}{s_{p(1,2)}\sqrt{\frac{1}{n_1}+\frac{1}{n_2}}}$	$t_{(1-3)} = \frac{\bar{x}_{(1,2)}-\bar{x}_3}{s_{p(1-3)}\sqrt{\frac{1}{n_1+n_2-2}+\frac{1}{n_3}}}$	$t_{(1-4)} = \frac{\bar{x}_{(1-3)}-\bar{x}_4}{s_{p(1-4)}\sqrt{\frac{1}{n_1+n_2+n_3-3}+\frac{1}{n_4}}}$	$t_{(1-8)} = \frac{\bar{x}_{(1-7)}-\bar{x}_8}{s_{p(1-8)}\sqrt{\frac{1}{n_1+\ldots+n_7-7}+\frac{1}{n_8}}}$	$t_{(1-20)} = \frac{\bar{x}_{(1-19)}-\bar{x}_{20}}{s_{p(1-20)}\sqrt{\frac{1}{n_1+\ldots+n_{19}-19}+\frac{1}{n_{20}}}}$
			$s^2_{p(1,2)} = \frac{(n_1-1)s_1^2+(n_2-1)s_2^2}{n_1+n_2-2}$	$s^2_{p(1-3)} = \frac{(n_1+n_2-2)s^2_{p(1,2)}+(n_3-1)s_3^2}{n_1+n_2+n_3-3}$	$s^2_{p(1-4)} = \frac{(n_1+n_2+n_3-3)s^2_{p(1-3)}+(n_4-1)s_4^2}{n_1+\ldots+n_4-4}$	$s^2_{p(1-8)} = \frac{(n_1+\ldots+n_7-7)s^2_{p(1-7)}+(n_8-1)s_8^2}{n_1+\ldots+n_8-8}$	$s^2_{p(1-20)} = \frac{(n_1+\ldots+n_{19}-19)s^2_{p(1-19)}+(n_{20}-1)s_{20}^2}{n_1+\ldots+n_{20}-20}$
	———	**Rejection criteria**	$\lvert t_{(1,2)}\rvert > t_{\alpha/2,n_1+n_2-2}$	$\lvert t_{(1-3)}\rvert > t_{\alpha/2,n_1+n_2+n_3-3}$	$\lvert t_{(1-4)}\rvert > t_{\alpha/2,n_1+\ldots+n_4-4}$	$\lvert t_{(1-8)}\rvert > t_{\alpha/2,n_1+\ldots+n_8-8}$	$\lvert t_{(1-20)}\rvert > t_{\alpha/2,n_1+\ldots+n_{20}-20}$

Table 2.
Summary of the statistical procedures followed in this proposed statistical approach.

charts are divided into two: First adjust the machines correctly and pool production, then use control charts for assignable causes.

In this chapter, the proposed statistical approach is explained in detail being based on a spinning mill having twenty spinning frames. When the first spinning frame is adjusted according to what the customer ordered and to the technology of the spinning frame, the results of that adjustment is controlled statistically, by means of hypothesis testing. It is the yarn count property, being μ_0, the examples are given. Yarn wrap on bobbins on the spindles, from rovings coming from the top, are drafted, and twisted to produce the yarn. n_i samples are taken from independent, identical, and with equal probability of being chosen spindles, and yarn count property have a normal distribution, as the other properties of textile materials. The adjustments on the first spinning frame are done and the variance of yarn count is hypothesis tested with less than one because of the production year of the frame. The χ^2 test statistic is applied. If rejected, the adjustments are corrected, and the same test is repeated. If unable to be rejected, then yarn count is hypothesis tested with what the customer ordered μ_0, the t-test statistic is applied; if rejected, the adjustments are corrected, and the same tests are repeated. If unable to be rejected, the second spinning frame is adjusted, the variance of yarn count is hypothesis tested with a χ^2 test statistic, and the equality of variances of yarn count of the two spinning frames is hypothesis tested with an F statistic. If rejected, the tests are repeated, if unable to be rejected, the yarn count hypothesis is tested with the t-test statistic. If rejected, adjustments on the frames are done and the tests are repeated, if both are unable to be rejected, then the yarn count of the two spinning machines are pooled. Now, the two frames are considered as one machine producing the same yarn, same variance of yarn count and same yarn count property, variability reduced the most. This statistical approach continues until the twentieth spinning frame and one by one, all the frames are considered as one machine producing the same yarn, same variance of yarn count and same yarn count property at the end.

This novel statistical approach guarantees that production starts with correct adjustments of the machines. In the performed literature review, this however has not been come across. By applying this statistical approach at the beginning of production, the correct starting will be assured and the machines will all be pooled one by one. On the other hand, during production, control charts will be applied to see the assignable causes and quick care ought to be taken. Additionally, instead of preparing separate control charts for each rational subgroup, even only one control chart for the whole lot would be enough, saving time, cost, manpower etc. This robust statistical approach can be incorporated in a statistics computer program, ending up with many benefits for the companies. Other statistical methods like boxplots, ANOVA, residual plots will all provide additional information about how the production proceeds. In addition to the above, this novel statistical approach can be applied to machines starting from the beginning of the multiple-stream like blowroom, carding, drawing, roving, examples for a spinning mill, more than one machine producing the same material. Besides, it can equally be applied to the other properties of textile materials, both adjusted directly on the machines or which result indirectly with machine settings like pressure, speed, etc. Raw materials, products, efficiency, yield, waste reduction, shift management of workers, faults, machine breakdowns, spare parts, electricity, economics, and much other application areas would emerge in due time.

Author details

Nefise Gönül Şengöz
Uşak University, Uşak, Turkey

*Address all correspondence to: nefisegonul.sengoz@usak.edu.tr

References

[1] Montgomery DC. Introduction to Statistical Quality Control. 4th ed. New York: Wiley; 2001. p. 796

[2] Woodall WH. Controversies and contradictions in statistical process control. Journal of Quality Technology. 2000;**20**(4):515-521

[3] Püskülcü H, İkiz F. İstatistiğe Giriş, İzmir: E.Ü. Mühendislik Fakültesi Ders Kitapları Yayın No: 1; 1983. p. 166

[4] Boyd DF. Applying the group control chart for –*x* and *R*. Industrial Quality Control. 1950;**7**(3):22-25

[5] Mortell RR, Runger GC. Statistical process control of multiple stream processes. Journal of Quality Technology. 1995;**27**(1):1-12

[6] Lanning JW, Montgomery DC, Runger GC. Monitoring a multiple-stream filling operation using fractional samples. Quality Engineering. 2002–2003;**15**(2): 183-195. DOI: 10.1081/QEN-120015851

[7] Woodall WH, Montgomery DC. Research issues and ideas in statistical process control. Journal of Quality Technology. 1999;**31**(4):376-386. DOI: 10.1080/00224065. 1999.119 79944

[8] Şengöz NG. Control charts to enhance quality. In: Leo DK, editor. Quality management systems – A selective presentation of case-studies showcasing its evolution. InTech.; 2018. pp. 153-194. Print ISBN 978-953-51-3919- 5; Online ISBN 978-953-51-3920-1. DOI: 10.5772/ intechopen.73237

[9] Runger GC, Alt FB, Montgomery DC. Controlling multiple stream processes with principal components. International Journal of Production Research. 1996;**34**(11):2991-2999

[10] Ghosh A, Saha, B, Mal P. Statistical Techniques, Design of Experiments and Stochastic Modeling. Boca Raton FL: CRC Press; 2022. p. 478

[11] Available from: https://saurer.com/ en/systems/zinser-ring-spinning-system/zinser-systems-ring-spinning. [Accessed: 2022-03-21]

[12] Available from: Rieter-g32-brochure-2991-v2-89907-tr.pdf. [Accessed: 2022-03-16]

[13] Available from: Rieter-g38-ring-spinning-machine-brochure-3236-v3-en. pdf. [Accessed: 2022-03-27]

[14] Available from: https://www.aha journals.org/doi/10.1161/circulationaha. 105.586461. [Accessed: 2022-01-18]

Chapter 4

Significance of "Quality Control" in Leather Goods and Garment Production

Abduletif Hebo

Abstract

The leather industry is one of the priority sectors that contribute to export income and economic development in the majority of African countries, in terms of creating job opportunities. Leather products need high care during manufacturing because their quality should never be compromised. Quality is a universal term used to evaluate the performance of a product or a service and the acceptance by the customer(s) in terms of customer satisfaction. As such, understanding quality concepts such as quality control (QC), quality standards, procedures, and documents related to leather goods and garment production in accordance with manufacturing company's policy is deemed useful within the context of this paper. Supervisors, quality controllers, and operators in leather products manufacturing firms need to know required quality parameters and associated control mechanisms so that defect-free products will reach the end users. In order to achieve this, quality-influencing parameters such as performance, reliability, durability, serviceability, esthetics, features, and conformance are measured so as to verify set quality levels. Furthermore, factors that affect the quality of leather goods and garments as well as methods of identifying and isolating common defects and faulty pieces especially in the finishing activities of leather production are included herein. Hence, this paper covers quality control aspects on leather goods observed within the garment manufacturing subsector.

Keywords: quality, quality control, leather goods and garment, defects

1. Introduction

The leather sector is regarded in most African countries as a major economy driver that highly contributes to a country's economic growth by means of employment opportunities and foreign cash inflow. The leather sector includes tannery, footwear, leather gloves, and leather goods and garment subsectors. Leather by itself requires high care during the different phases of production, storage, and transportation. Hence, the issue of quality in leather manufacturing process is of paramount importance as most defects in leather and leather products are irreversible. Rework or correction of incurred damages during production of leather, leather goods, and garment leads to higher labor costs and sometimes to rejection of the products.

Therefore, the implementation of quality control (QC) concepts in every production step is associated with a valuable impact on the finished products so that defect-free products reach end users.

Most of the Ethiopian companies specializing in the leather sector prefer visual inspections and simpler physical testing methods to control product quality. This preference may work for the local market as the latter may not draw too much focus on quality aspects due to it being less aware for various quality dimensions. However, for penetrating and competing on an international level, special privileges such as African Growth Opportunity Act (AGOA) from importing countries or producing goods with the required quality and at a competitive price are required. Nonetheless, consistent with descriptive statistics results, econometric findings also reveal that exporting firms were found rather less efficient compared to those which are either powerless or have totally given up looking for the international market with respect to income and market sustainability. Once basic international standards are met and market access is established through various mechanisms including participation in trade fairs, the use of the internet, and buyer contacts, exporting companies have continued to benefit from the market due to the natural superiority of Ethiopian leather in terms of fineness, thickness, flexibility, strength, and compactness of texture, according to UNCTAD (2000) [1].

But this is not true for export markets. All leather and leather products-related quality standards need to be implemented which in turn includes, but may not be limited to, physical and/or chemical testing and inspections. In a perfectly competitive market setup consisting of a high number of buyers and sellers (also referred to as a thick market), price signals would reward high quality, and hence, producers and traders of substandard quality would either be driven out of the market or would be relegated to a distinct low-quality-oriented market [2].

This chapter provides a background on quality control aspects for the production of leather products that directly reach end users. It covers quality control aspects applicable to leather products and goods and the garment manufacturing subsector.

1.1 Aims and objectives

The objective of this chapter is to provide a background on quality control aspects required in the production of leather products. In doing so, this research work aims to address the significance of quality control and quality aspects in the leather products manufacturing subsector.

2. Literature review

Quality is an absolute term. Concepts of quality and quality control with regard to the manufacture of leather goods and garments need to be viewed in accordance with the policies of the relevant specialty industries. The outcome thereof in conjunction with the application of quality control concepts listed herein will serve as the basis for supervisors, team leaders, and even operators in those companies with advanced know-how to the parameters, check points, and control mechanisms so that defect-free products will reach end users.

Most Ethiopian leather and leather products manufacturing firms had implemented various quality-related improvement tools and systems including, ISO 9001:2018, Environmental Management Systems (EMS ISO 14001:2018, Occupational

Health and Safety Management Systems ISO 45001:2018), and a plethora of other quality management systems (QMSs) in order to enhance their local and global competitions. For instance, (ELICO) Ethiopian Leather Industries Company PLC, Pittards Glove Manufacturing Factory PLC, Modern Zege leather products and footwear Industry PLC had implemented these systems [3].

Therefore, they will be able to acquire and maintain quality concepts, agreed quality standards and procedures, and introduce quality control/quality assurance (QA) to organizational staff/personnel. Furthermore, they will apply these parameters in leather goods and garment production, identify accompanied issues, and provide related documents to employees in accordance with the organization policy.

To implement quality standards, the basic conditions of the customer are (a) the purpose and (b) the selling price of the product or service.

These basic conditions can be resolved in to the following 10 detailed conditions:

- specifications of dimensions,
- operating characteristics,
- life and reliability objectives,
- safety requirements,
- relevant standard,
- engineering,
- manufacturing and quality costs,
- production conditions,
- field installation,
- maintenance and service objectives,
- energy utilization and material conservation factors,
- environmental and other side effects, and
- cost of operation or use

2.1 Concept of quality

Quality is the totality of features and characteristics of a product or service that affect its ability to satisfy the specified or implied needs of a customer. Quality consistency requires from users to concentrate on the process rather than on the product alone. Quality gurus define quality as "conformance to requirement" and "fitness for use" [4]. Good quality will automatically result in productivity improvement. It is the author's view that the best policy should be to do the things right first time.

Quality helps determine a firm's success in a number of ways:

- customer loyalty: satisfied customers return, make repeat purchases, and recommend the product or service to others,
- strong brand reputation for quality: retailers want to stock the product; improved quality leads to fewer returns and replacements which in turn lead to reduced costs attracting thus and retaining good staff.

The term "Quality" can be measured aspects such as failure or reject rates, level of product returns, customer complaints, customer satisfaction, customer loyalty, evident from repeat purchases, or renewal rates and employee health and well-being.

2.2 Quality parameters

Quality is measured in a relative manner. It depends on how the user perceives or the way he/she get satisfied with that product/service. Once a product/service is accepted to customers, it can pull more new customers and may be produced/delivered in greater numbers, affecting in turn costs that are reduced and sales which will be increased. But, as quality has no universal meaning, the way users perceive it varies. Some users may like the performance or the reliability, while others may be happy with esthetic features and so on. What is reliable for a user may not be true for another. Hence, quality is an important factor which customers look for in a product or service in order to be rewarded with total satisfaction. Some of the important quality factors/parameters that customer considers in a product or services as stated by some quality gurus are listed as follows.

2.2.1 Dimensions (parameters) of quality

Performance: it evaluates if the product does the intended (planed or proposed) job or if the service delivered meets intended objective. Potential costumers usually evaluate a product to determine if it will perform certain specific functions and how well it will do them. For example, the production of a document holder or a leather bag with multifunction pockets would fall within this category.

Reliability: it indicates a product's failure rate. Different products may need repair over their service life. The leather machineries should be also reliable so as to increase productivity, i.e. when leather garments are produced, greater attention ought to be placed during, e.g. the stitching procedure. As the needle is typically of a cutter edge type, sometimes it stitches the component by cutting the part. So, if proper stitching is not done, the product is either repaired or rejected.

Durability: it shows the duration that the product is expected to last for. This is the effective service life of the product that customer wants over a long period of time, e.g. a customer that orders a leather jacket may expect this to last for at least 5 years.

Serviceability: this parameter stands for how easy the product may be repaired. There are many industries where the customer's view of quality is directly influenced by how quickly and economically a repair or routine maintenance activity can be accomplished' in this case-study, dyeing or changing color of the leather jacket after a number of uses can be an example for this.

Esthetics: this dimension shows what the product looks like externally. This is the visual appeal of the product, often taking into account factors such as style, color, shape, packaging alternatives, and other sensory features.

Features: it means what features the product possesses. Usually, customers associate high quality with products that have added features (such as special color, design, handles, and decorations), which go beyond the basic performance of the competition.

Conformance: it is used to evaluate if the product or service conforms to the specification. This means, if it is developed based on a performance specification; will it actually perform as specified? If it is developed based on a design specification, does it possess all of the features defined?

Perceived quality: The product or service may possess adequate or even superior dimensions of quality but still fall victim to negative customer or public perceptions. As an example, a high-quality product may get the reputation for being low quality based on poor service by installation or field technicians. If the product is not installed or maintained properly, and fails as a result, the failure is often associated with the product's quality rather than the quality of the service it receives.

2.3 Quality control and quality assurance

Quality control (QC) is a procedure or a set of procedures intended to ensure that a manufactured product or performed service adheres to a defined set of quality criteria or meets the requirements of the client or customer. While quality assurance (QA) is defined as a procedure or set of procedures projected to ensure that a product or service under development (before the work is complete, as opposed to afterward) meets specified requirements. QA is sometimes expressed together with QC as a single expression. There is plenty of quality control types. The following are used in the leather-related production controls [5].

Quality control of incoming material:

- Ensuring the right materials are available in the right quantity at the right time.
- Based on quality requirements, the purchase information such as specification, packing instruction, and transportation instruction should be clearly identified.
- For example, in the garment industry for finished leather, the parameters such as color fastness, light fastness, tensile strength and softness need to be checked.
- Physical characteristics such as lining, tensile strength and color fastness need to be checked. After finalizing the parameters to be assessed for each incoming material, the standards need to be met for each parameter of each incoming material.

Process control:

Process control (PC) can be defined as any activity that adds value to the product to be supplied or the service to be rendered. The term "process" in the leather garments manufacturing industry may include – but not be limited to – unit processes such as cutting, assembling and stitching, and finishing. According to the process control steps, the parameters for each process need to be identified first. For example, in assembling and stitching, the needle to be used, i.e. the needle number and needle point have an influence on the final product. Further, the thread used in bobbin (lower thread) and the sewing machine (top thread) also affects the quality of the final product.

Process control is carried out by the following steps:

- Identification of process control parameters,
- Establishing the standards for each parameter of each process (internal process control standard/working standard for process control),
- Fixing the limits for each parameter.

Figure 1 shows the intended process control.

- **Product control:**

Product refers to the physical output produced by supplying in the inputs or raw materials and carrying out any production process. The final product is what is dispatched to the customer or the end users. Apart from this, there are components sometimes referred to as intermittent products. For a leather garment manufacturer, these are prepared sleeves, pockets, collars, etc. Therefore, the output after each operation or process is an intermediate product. Product control generally refers to the control of the final product. Control of intermediate products is equally essential. This is due to the fact that in each stage the product quality is ensured so as to produce the final product of desired quality [6].

2.4 Quality inspection and testing

Quality inspection: Industrial activities which ensure that manufactured products, individual components, and multicomponent systems are adequate for their intended purpose. Whereas *inspection* is the activity of examining the product or its components to determine if they meet the design standards, *testing* is a procedure in which the item is observed during operation in order to determine

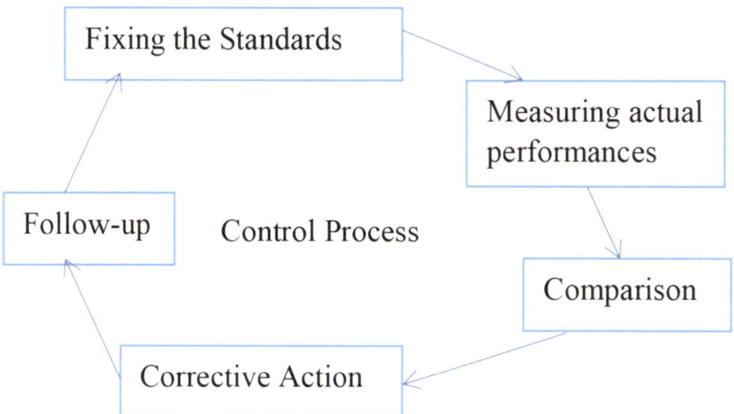

Figure 1.
Process control.

whether it functions properly for a reasonable period of time under given stress conditions. Inspection and testing are performed before, during, and after manufacturing to ensure that the quality level of the product is within acceptable design standards.

There are also various types of inspections. The following categories are used in leather goods and garment production:

- **Incoming materials inspections:** checking the quantity (finished leather, accessories, etc.), quality, rejection allowances, verification as per purchase order, lead time, etc.

- **First-article inspections:** QC inspects first-article samples prior to volume production. This verifies that product specifications are being met and avoids unnecessary re-engineering work later.

- **In-process inspections**: these on-site inspections evaluate samples of the products selected during the manufacturing process. This confirms the quality of the product and allows any necessary changes to be addressed early on reducing, hence, rework time and costs.

- **Pre-shipment inspections**: during a pre-shipment inspection, engineers verify that finished goods conform to set specifications.

- **Sample inspections**: samples are taken from inspection lots for end user evaluation, laboratory testing, or customer approval randomly, and processing QC can help for inspection. After this type of inspection, one can offer rapid service at a very affordable rate.

International standards are preferred to be used for testing leather products, especially in the garment industry. **Table 1** shows this standard.

No.	Items	Standard
1	Elastic tapes	IS 9686
2	Metal buckles	IS 96986:1980
3	Threads	IS 1376/1803
4	Leather garment sizing system	IS 10397
5	Metallic slide fastener	IS 3148:1983
6	Garment quality guide	IS 12675
7	Leather for garments	IS 12718
8	Fur leather	IS 3840/2961
9	Fusible lining	IS 12806
10	Zip fasteners	IS 8894/3184/4829

Source: Leather Industry Development Institute, Advanced Garment Production, Level IV Training Materials, June 11, 2016.

Table 1.
Standards related to leather garments industry and related items.

3. Results and analysis

3.1 Leather goods common quality parameters

In addition to the eight quality parameters of any product like durability, feature, performance, conformity, esthetics, serviceability, perceived quality, and reliability, there are also other leather goods-specific quality parameters.

3.2 Most commonly used types of testing

Leather testing: it includes wet rub fastness, dry rub fastness, tool test, stress strain test, and plaster test fastness.

Leather goods and garments testing: it encompasses handbags and small luggage, wherein the strength – say – of strap fastenings is an important consideration in the quality assessment of handbags and luggage. A large number of companies in Ethiopia are able to carry out all strength tests utilizing state-of-the-art equipment to assess the risk of strap failures, whether at fastenings (e.g. buckles) or where the strap is attached to the body of the item itself. The other one is the leather belt testing, from an assessment of the components of a belt for labeling purposes. Also, specialty companies can perform further tests in order to satisfy all clients' requirements such as the color fastness (wet and dry rub fastness test) to tarnishing of buckles and metal components, to ensure the products are fit for the purpose they are intended for.

A few examples of tests used in leather products manufacturing firms are as follows:

Smell test: the smell test is an important part of every inspection. To avoid illegal toxins, the most reliable way to check it is to perform chemical tests as per ASTM D1296 in an accredited Leather Industry Development Institute (LIDI) laboratory.

Function test: the objective is to check if the product works as designed or anticipated. In the case of the leather bag, an inspector will wear it and test the zippers' direction and strength.

Color fastness check on leather: excessive dye may be rubbed off during a color fastness check. On leather, this is a frequent problem. The test may be repeated 10 times with a dry cloth and 10 times with a wet cloth.

Abuse and fatigue tests: pulling on straps and zippers with stronger-than-usual force helps to understand the manufacturing quality of leather bags.

Seam strength test for leather bags: this test is similar to the abuse test but focuses on the seams. It uses a tension gauge to check seam strength.

Load test: the inspector loads the leather bag with weights (depending on the model between 2 and 20 kg for backpacks (bag type) most of the time. Then the bag is lifted at least 20 times and is hanged on a hook for 4 h. This is an internal company policy similar to that of color fastness check.

Zipper twisting test: this type of test is used to check both the strength of the zipper and the seams holding it in the open middle and closed position. The QC pulls the zipper sideways for 10 s in each direction. Low-quality zippers tend to open and bend beyond repair. Extensive laboratory equipment test products (e.g. opening and closing zippers 5000 times) could be used also. However, most of small and medium leather products manufacturing companies use the manual test.

Carton humidity check: This test is performed in order to assess the behavior of the product in rainy conditions, while avoiding the buildup of mold or fungus, aiming at maintaining a humidity level below 12%. In particular, during the rainy season, the inspector should check the humidity of the export cartons with a humidity tester. As such, it ought to be ensured that sufficient desiccant (calcium oxide absorb water) is placed in the right spots.

3.3 Factors that influence the quality of leather goods and garments

Factors that influence quality aspects make bags and garments good and/or cheap. The following aspects are commonly experienced in leather products manufacture:

3.3.1 Designs and materials

The design room is where quality starts in leather goods manufacturing companies. Bag design is a system, which is not only the combination of the technique, knowledge, and the art, but also the connection of design and craft from the choice of the theme to grasp the inspiration and the accomplishment of the finished product. Through the design effect of a product, a bag or a garment should become a bridge between designers, technicians, and consumers. In that sense, it would be common language among them. The designers should identify the materials like type of leathers, accessories, colors, and hardware that will be used in the production process. A well-designed bag or a garment should include all information about its design. As an example, a good leather goods design should have at least the following information:

- leather type (color, thickness, and feeling),
- origin (cow, sheep, goat, buffalo, etc.)
- reinforcement (EVA sheet, water proof, fusing, foam, etc.),
- lining (velvet, cotton fabric, nylon, and polyester),
- accessories (eyelet and rivet),
- zipper, in terms of size (3, 5, and 8 mm), finishing (silver, gold, and bronze), and type (metallic, plastic chain, and plastic molded),
- stitching (seam type and seam length per centimeter),
- thread size (for needle thread and bobbin thread), type (cotton, nylon, polyester, silk, and polyester spun cotton), each dimension, (volume, height, and base, handling length, and width),
- edge finishing (raw edge, folded, and edge color) and hard ware's (buckle, color, and adjustable size) [7].

3.3.2 Material selection

Material selection refers to the materials selected for the manufacture of – say – a bag including the hardware and the accessories, as well as the processes involved.

3.3.3 Pattern making and cutting

The pattern making, which is also referred to as a sample making process is an important aspect and is regarded as a bridge of transforming the graphic designs into the products. The maximum permissible error (acceptable level) of pattern is 1/32 inches (1 inch error of 32 inches length), as by reference to any bag, or in accordance with international standards, such as SATRASumm, which is an industry standard package concerning the efficient cutting of leather and synthetic materials. In pattern making, usually major parts (shape and size of the bag) are made first, and then relatively smaller parts are followed and so on.

3.3.3.1 Fixing product size standards

Unlike leather garment and foot wear products, leather bags have no fixed specifications such as height, depth, and width and may therefore be easy to categorize as small bags, medium bags, and large bags. One could remember Galileo Galilei's quote: "*Measure what can be measured, and make measurable what cannot be measured.*" From a quality management point of view, this means that "we cannot manage what we

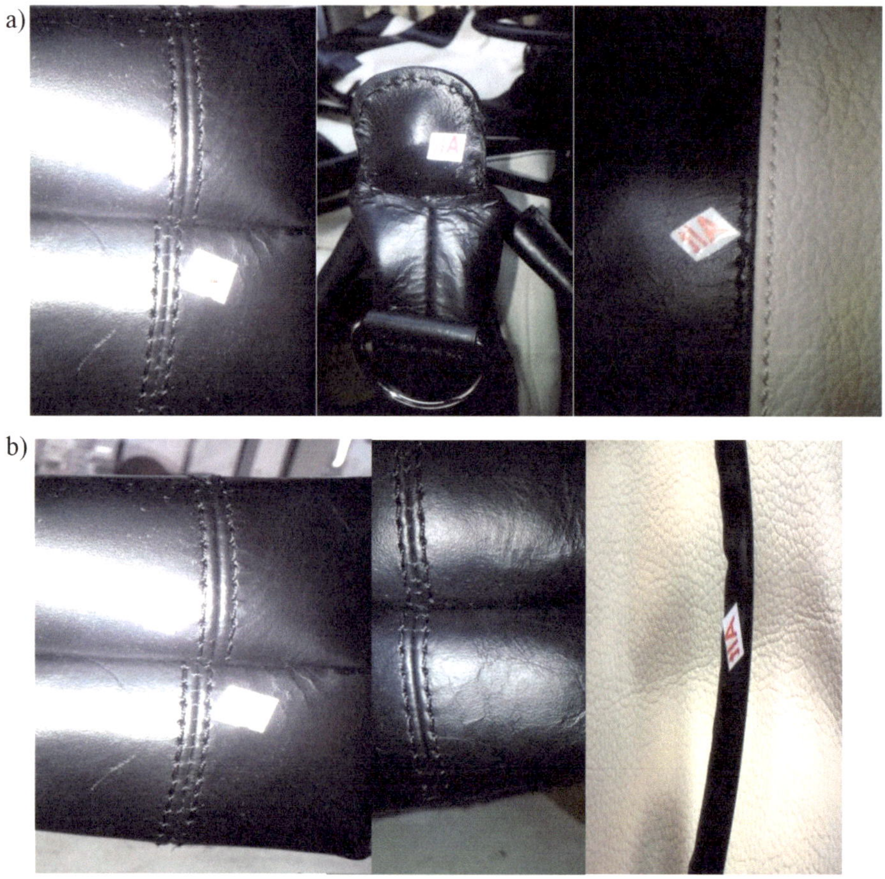

Figure 2.
Pattern alignment variations: (a) and (b).

cannot measure." In short, the clearer the specification, the better the possibility of creating and delivering quality products.

3.3.3.2 Construction

Construction is the matter of how everything or patterns have been put together. **Figure 2** shows pattern alignment variations.

3.3.3.3 Technology

The manufacturing process is a key factor to leather products' quality. A different technology represents a different style of leather products. The quality of bags is as good as the people that make it. That is why the best stitchery, leather workers, and quality control technician or experts are required. In order to have best-quality products and workers, it is important to put a lot of resources into training or hiring the most qualified staff and paying them well fostering their commitment and creative minds.

3.4 Defects in goods and garments manufacturing

3.4.1 Defects and their types

Defects are deviations/nonconformities of processes, products, or materials from the requirements/standards. Causes of defects may be man-made (assignable causes) or common/natural causes. Assignable causes can be removed, while common causes can only be reduced. For example, a poorly build knife maybe a cause for cutting defects/human fault, while loose leather is a cause for less durability of the garments.

3.4.2 Methods of identifying and isolating faulty pieces

- Defects in the cutting section can be identified by various bodies operating therein that are briefly presented as follows:

  A. Cutting supervisors: they are the cutting supervisors that issue leather from raw material store where defects like loose leather, under substance, wrong color/shape, poor nap on nubuck, poor color fastness, and poor knife can be visually identified, and the leather is thereafter sorted accordingly. Only leather bundles that meet specifications are issued and allocated among cutting operators by the supervisors.

  B. Cutting operators: they can identify during cutting minor defects like grains not matched pair wise, wrong direction of cutting, cuts/flaws in component, open defect, wrong size cut, and color variations to name but a few. These operators, in addition to cutting operations, have the responsibility to take care of component quality. As such, items ought to be cut in line with the parameters stated earlier and the data be posted to the operators.

- Defects in the stitching (sewing) section can be identified by various bodies operating therein.
    - Bench workers can identify defects like notch marks not matched, edge folding inaccurate, improper alignment, wrong components placement, too much hammering, and too much glue.

  C. Stitching operators: they can identify minor defects like uneven stitching length, skipped stitches, stitches not locked at the end, wrong needle/thread used, stitches too far or too close to the edge, stitches not as per the marking, broken stitches, top tension tight, and seam puckering.

Possible defects during the final inspection stage may be:

- trimming,
- thread burning,
- glue erasing,
- leaving uneven stitching length,
- pattern vs. assembly correspondence,
- measurement and alignment,
- grain structure checking,
- component checking,
- color and size matching,
- ironing dimension,
- seem puckering,
- proper feeding system,
- thread tension,
- leaving broken stitches and skipped stitches,
- stitches too far or too close to the edge,
- top tension tight, thickness, and not ±0.2 mm allowances [8].

3.5 Part five: Finishing in leather products manufacturing

Finishing is the final process given to a garment or goods in order to achieve good appearance, desirable feel and look and to impart some important, and durable and functional properties.

3.5.1 Classification of finishing

Finishing in leather products manufacturing can be classified according to the nature of the finish such as Kawabata's Evaluation System for Fabric-KES-FB and the degree of performance (ISO11644:2009). **Figure 3a** and **b** show these classifications.

3.5.2 Edge coloring

Sand edges: this is done by using emery paper to sand the edges and to arrange many belts/straps of the same size side by side on a flat table and sand simultaneously. This will ensure that all the leather layers are even and square and that any residual glues or finishes have been removed. **Figure 4** shows edge coloring in industry.

Applying color: this is done either by using *a* machine or manually. For manual operation, *the* use of dye box like Fiebing's dye will make it simpler. Keeping the dyed edge by facing up for air-drying before applying on the opposite edge is worthy. After the other side got dried, one can paint the opposite one and keep the same way one has done previously. It can also be applied during the second round if necessary. This method is used everywhere globally even though manual coloring is preferably practical in Ethiopian leather products manufacturing firms. **Figure 5** indicates the application of color with the aid of a machine.

Applying filler: the leather filler paste is a white compound that can be air- or heat-dried and requires re-coloring with a leather repair pigment after its application. The

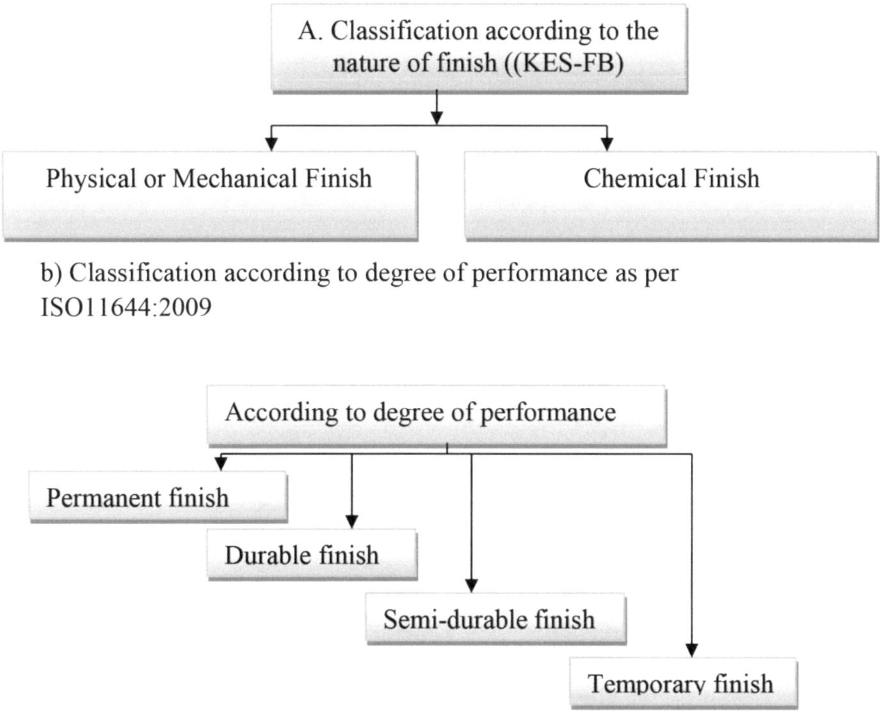

Figure 3.
Classifications based on (a) the nature of finish (KES-FB) and (b) the degree of performances as per ISO11644:2009.

leather filler remains flexible, durable, and natural to the existing leather surface. It is used to fill the edge surface and results in smoothness to the edge's surface (see **Figure 6**).

Wet and soap: the edge of the leather can be wetted using a sponge or piece of trimmed woolskin. The outcome will be a slick/polished rounded edge.

Burnishing: this is accomplished by briskly rubbing the canvas against the edge of the belt until the edge is smooth. A canvas wrapped around a motorized wood

Figure 4.
Edge coloring in practical.

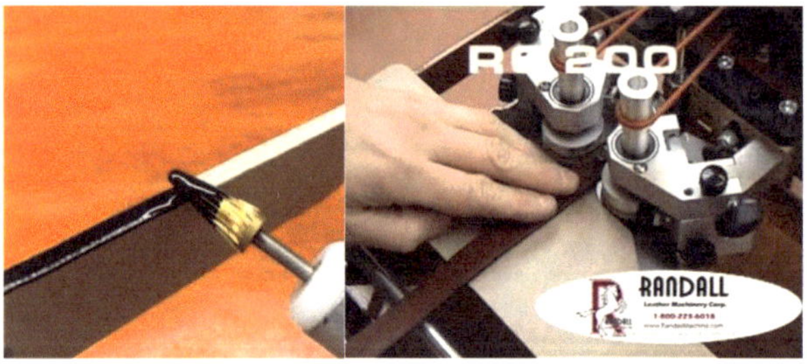

Figure 5.
Applying color with machine.

Figure 6.
Leather filler paste.

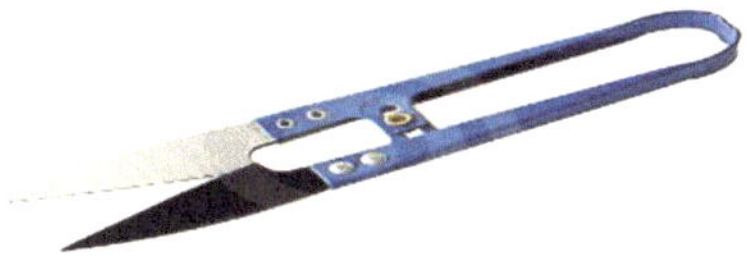

Figure 7.
Scissors for trimming.

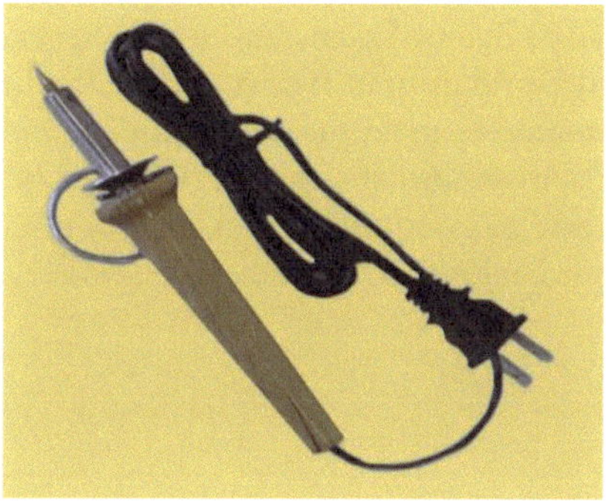

Figure 8.
Soldering iron used for thread burning.

burnishing wheel which speeds up the process could be used herein. However, care should be taken not to over-burnish, which will result in a rough edge.

Hand burnishing: this is done by means of a clean cloth that rubs the edges removing hence, any residual dye and determining thus, if the second coat is necessary.

Polish: at this point, paraffin is applied to the edge of the belt and burnished again and again. Denim works well here if burnishing is done by hand. Once one is satisfied with the finish, one can polish to a high luster with a dry cloth [9].

Finish: after the edges are polished, final finish is applied.

3.5.3 Trimming: Hand trimming and trimmer machine

It is preferred to use thread trimmer machines as it reduces trimming costs, increases production, uses unskilled help, eliminates scissor damage, keeps trimming area clean, and reduces cleanup cost. One can choose between different clipper blades and motor control for diverse material. Scissors can be used for trimming (**Figure 7**).

Thread burning: it can be manual with a candle or by means of soldering iron (**Figure 8**).

4. Discussion

The quality control concept is very useful especially for exports of branded products. Apart from the general knowledge and experience of the author in the sector, secondary sources were used from institutions such as the Ethiopian Leather Industry Development Institute (LIDI), Ethiopian Leather Industry Associations, and medium- and large-scale leather products manufacturers. The LIDI laboratory was accredited from SANAS (South African National Accredited System) so as to support the leather sector with various laboratory testing (i.e. physical, mechanical, and chemical) services in 2012. Furthermore, the LIDI laboratory was also accredited from the Ethiopian National Accreditation Office (ENAO) in the same year.

With this responsibility, LIDI has been serving Ethiopian leather manufacturing firms by laboratory testing, technical training, quality management system (QMS), and quality control and quality assurance tools implementations. Under the Twinning program, which was made between LIDI and the Federal Democratic Republic of Ethiopia, FDRE, Ministry of Industry on Ethiopian side, and CSIR – Central Leather Research Institute-Council of Scientific and Industrial Research, India, in association with Footwear Design and Development Institute (FDDI), India, LIDI's R&D laboratory state of the art was created to meet the requirements and demands of leather and leather products in order to meet and ensure international quality standards [10].

Most of leather goods and garment manufacturing companies in Ethiopia use smell test, function test, and color fastness check on leather accepting it as internal

S.No.	Types of test	Test method
1	Determination of thickness	ISO 2589:2002
2	Determination of apparent density	ISO 2420
3	Determination of tensile strength and percentage elongation	ISO 3376
4	Determination of tearing load (single and double)	SO3377---1/3377---2
5	Determination of distension and strength of grain ball burst	ISO 3378
6	Determination of flex resistance by flexometer method	ISO 5402
7	Determination of shrinkage temperature up to 1150°C	ISO 3380
8	Water absorption (Kubelka) after 2 and 24 h	SATRA TM/ISO SLP 19
9	Determination of water resistance test for light leather	ISO 5403
10	Determination of water resistance of heavy leather	ISO 5404
11	Determination of water vapor permeability	ISO 14268
12	Determination of cold crack resistance leather finish	SLP 34
13	Determination of sole/upper adhesion tester	Internal
14	Measurement of shoe flex (walk meter)	Internal
15	Determination of dry heat resistance of leather	Internal
16	Determination of adhesion of leather finish	SLF 11
17	Color fastness to artificial light (xenon)	SLF 401
18	Determination of fastness to water spotting	ISO 11642
19	Color fastness to perspiration	ISO 11641
20	Determination of fastness of leather finish to (to and from rubbing)	ISO 11640
21	Determination of fastness to ironing (fastness to heat)	IUP 470
22	Determination of static water absorption	ISO 2417

Source: Leather Industry Development Institute's Physical Laboratory, March 23, 2016.

Table 2.
Leather physical testing.

company policy, as it matches with some of global/international standards in this aspect (ASTM D1296, ISO 11640, ISO 11641, SLF 401, IUP 470).

Apart from common quality parameters, there are also other leather goods-specific quality parameters. Items made from real leather or imitations, such as PU, which are very popular, should be treated accordingly. However, leather goods and garment manufacturing companies prefer to use simpler (by observation and manual tests) methods, whereas other manufacturers in footwear subsectors could use more test methods as per international standards in order to check, for example, grain structure, thickness, apparent density, shrinkage, flex resistance, water resistance, and so on as per ISO 2589:2002, ISO 2420, ISO 5402, ISO 3380, ISO 5403. That is because, footwear products are highly vulnerable to damage, and hence, their suitability to use needs to be assured before reaching the end users. **Table 2** provides a summary of physical testing standards for leather that could be recommended by the author to be used so as to improve productivity and reduce defect rates, rework, and waste.

Regarding the effect of human factors in product quality, it is the author's view that most quality problems are caused primarily by a lack of interest or care on the part of the worker in the production department. However, it is usually not only the worker who is responsible for this but also the conditions necessary to carry out the work correctly often do not exist. For example, instructions may be inadequate, the incoming material may be defective, the machines may not be capable of producing goods of the required quality, and proper conditions for conducting inspection of the product are not given to the workers, and so on. The study done by joint consultancy of Ethio-Indian twinning project in collaboration with the Leather Industry Development Institute (LIDI) and the Footwear Design and Development Institute (FDDI) of India approves this fact [11]. **Figure 4** (in Section 4.1) shows that inadequate instructions, which accounts for about 28% caused the rest effects. Effective understanding of the worker to the instructions in every step of production will surely lead to more pleasant effects on the product quality. However, although workers may not have control over these factors, they may though lead to defective work. **Figure 9** shows Pareto analyses of one factory.

In Japan, it is generally believed that 40% of quality problems are caused by poor product design, 30% of quality problems are due to wrong or defective materials being purchased from suppliers, and the remaining 30% are due to errors made during the manufacturing process [12]. One could argue that any other quality problems

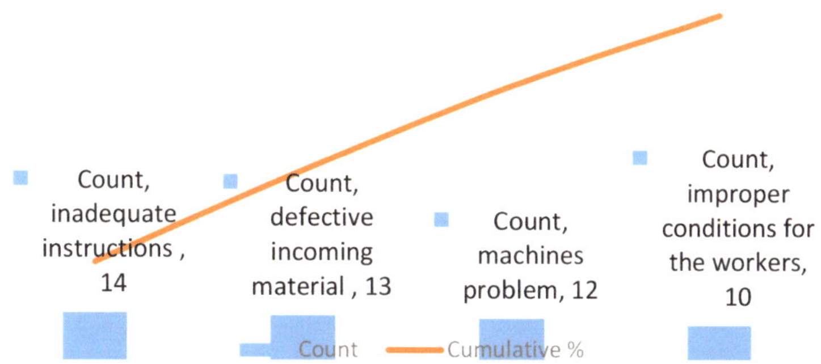

Figure 9.
Result of Pareto analysis for ELICO-universal leather products unit. Source: Twinning report [11].

in manufacturing are caused in equal proportion by managers (by not providing adequate training for workers) and by workers (by not paying adequate attention to machine settings).

Regarding defects observed in goods and in garments manufacturing listed in the following section are common examples of deficiencies in leather products manufactured in Ethiopia:

Sewing defects: open seams, wrong stitching techniques, non-matching threads and missing stitches, improper creasing of the garment, erroneous thread tension and raw edges are some of the sewing defects which can affect the garment quality adversely. Firms mitigate these types of defects by providing continuous on-the-job trainings for sewing operators.

Color defects: this category includes color variations between the sample and the final garment, wrong color combinations, and mismatching dyes' that should always be avoided. Leather issuers check this in store for every order with the help of leather sorter or in-process quality inspector.

Sizing defects: this refers to wrong gradation of sizes and difference in the measurement of various parts of garment-like sleeves of XL size for a body of L size garment that can deteriorate the garments beyond repair. Though tanneries use leather grading machine during production, Ethiopian leather goods and garment manufacturing firms usually identify and mitigate these defects by cross-checking cut components visually.

Other defects: this group entails broken or defective buttons, snaps, stitches, different shades within the same garment, dropped stitches, exposed notches and raw edges, fabric defects, holes, faulty zippers, loose or hanging sewing threads, misaligned buttons and holes, missing buttons, needle cuts or chews, pulled or loose yarn, stains, unfinished buttonhole, short zippers, inappropriate trimmings, etc. These defects, unless tackled at the very beginning, and/or quality assurance is undertaken in every step, could lead leather products manufacturing companies to be less competitive and affect in turn their existence. Due to globalization and acceptance of Ethiopian leather products to export markets, manufacturers of leather products are obliged to implement various quality improvement tools including QC/QA. Thus, respective process and final quality checking parameters have been posted in front of operators in each section along with visual defective and free cut components. This method encourages operators to think about quality issues in addition to their duties of – say – cutting, table work, sewing, and finishing.

Concerning defect control at the finishing section, various final quality control parameters are used that include trimming, thread burning, glue erasing, pattern vs. assembly correspondence, thread tension, leaving broken stitches and skipped stitches, and stitches too far or too close to the edge.

As for Ethiopian leather products manufacturers, the defect control parameters during the final inspection stage include aspects such as:

- leaving uneven stitching length,
- measurement and alignment,
- grain structure checking,
- component checking,
- color and size matching,

- ironing dimension,
- seem puckering,
- proper feeding system,
- top tension tight and thickness, etc., are inspected prior to this stage.

5. Conclusion and further work

The leather sector's contribution is very high with respect to export incomes and economic development, especially on creating job opportunities. For instance, according to the Central Statistical Agency (CSA) of Ethiopia, export of leather and leather products, which was US $23 million in 2013, reached US$133 million in 2018. Hence, leather goods and garment to be exported need high care during all manufacturing stages in order to increase competitiveness in the global market.

This book chapter discussed quality control concepts and quality standards for leather goods and garment. In doing so, it highlighted applicable procedures and documents enabling supervisors, quality controllers, and operators in those companies to get detailed knowledge about quality parameters and control mechanisms so that defect-free products reach end users. Furthermore, it will allow readers to familiarize themselves with quality concepts in this sector. It is the author's view that this research work may prompt readers to confront themselves with quality control aspects and to research more about these aspects in this specialized manufacturing area.

Moreover, leather goods and garment quality parameters and factors that influence the quality of leather goods and garment were included. In addition to the earlier-mentioned ones, commonly occurring defects, methods of identifying and isolating faulty pieces, and some finishing types in leather products production were discussed.

Studying the application of QC/QA on the whole leather sector (leather processing, footwear industry, glove making, and other related subsectors) will be the next tasks of the researchers and book writers. This may include subsector-specific inspection and control mechanisms starting from designing, cutting, table work (preparation), sewing, inspection and testing, packing, and shipping that need to be further analyzed.

Acknowledgements

The author would like to express his gratitude to his wife for her patience and assistance.

Further reading

To get more practical explanation of quality aspects, readers are advised to read of the work of David Garvin (1988) – *Eight Dimensions of Quality*.

Author details

Abduletif Hebo
Bio and Emerging Technology Institute, Emerging Technology Center, Reverse Engineering Directorate, Addis Ababa, Ethiopia

*Address all correspondence to: abduinda@gmail.com

References

[1] Worku G, UNCTAD. Is the Ethiopian leather industry on the right track? An empirical investigation. Ethiopian Journal of Economics. 2000, 2001;**X**(2):16

[2] Girum A, Florian S. High Hopes and Limited Successes: Experimenting with Industrial Polices in the Leather Industry in Ethiopia. Working Papers. Addis Ababa: Ethiopian Development Research Institute; 2014

[3] Annual Leather Sector Performance Report for LIDI. Addis Ababa, Ethiopia: Leather Industry Development Institute; 15 October 2019

[4] Juran JM, Blanton GA. Juran's Quality Handbook. 5th ed. New York: McGraw Hill; 1998. pp. 20-28

[5] Alemu T. Leather Goods Design, Manufacturing & Quality Book. Maude Avenue, Sunnyvale, USA: LAP LAMBERT Academic Publishing; 2018. pp. 46-58

[6] Application of Quality Standards, College Level-III Training Material. 2019. pp. 15-39. Available from: https://www.elidi.com [Accessed: 2021-06-12]

[7] The innovative garment accessories. Available from: https://www.fibre2fashion.com/industry-article [Accessed: 2021-05-10]

[8] Engineer's Manual No. E356-01. p. 23

[9] Mechanical finishing, polishing vs. buffing by Pat Wenino. Available from: https://www.pfonline.com [Accessed: 21 September 2021]

[10] Leather Industry Development Institute. Compendium of Outcome and Achievements of Twinning Program, 2011-2014. September 16th, 2014

[11] Twinning Project. Unpublished Company Report. ELICO Universal Leather Products Factory. 2014

[12] Product Quality, A Guide for Small and Medium-Sized Enterprises. Working Paper. Vienna: United Nations Industrial Development Organization; 2006

Chapter 5

Development of a Quality Gate Reference Model for FDM Processes

Marcel Randermann, Timo Hinrichs and Roland Jochem

Abstract

Additive manufacturing (AM) enables industries to accomplish mass customization by creating complex products in small batches. For this purpose, fused deposition modeling (FDM) is widely used in 3D printing where the material is applied layer-by-layer from a digital model to form a three-dimensional object. There still exist problems in FDM processes regarding the failure rate of printed parts. Failures vary from deformed geometry, clogged nozzles, and dimensional inaccuracies to small parts not being printed that may be attributed to various process steps (e.g., poor quality CAD models, converting issues, overheating, poor quality filament, etc.). The majority of these defects are preventable and are caused by imprudent try-and-error print processes and troubleshooting quality control. The aim of this chapter is to propose a quality gate reference process with defined requirement criteria to prevent the occurrence of defects. The framework shall be applied in quality control and in-situ process monitoring to enhance overall manufacturing quality.

Keywords: additive manufacturing, fused deposition modeling, reference process, quality gates, quality control

1. Introduction

Customer demands on products and services are constantly increasing in the global and local market and the competitive conditions of companies are in constant flux. As a result, companies are faced with changing market conditions in which they have to assert themselves against increased competitive pressure, greater product complexity, and an ever-increasing variety of products [1].

Industrial 3D printing or additive manufacturing (AM) is considered the key technology for mass customization. This allows the individual production of complex components and offers various possible solutions for increasingly complex requirements [2]. Additive manufacturing is an umbrella term for manufacturing processes in which components are built up element by element or layer by layer directly from computer-aided design (CAD) data without component-specific tools [3].

Fused deposition modeling (FDM) belongs to the AM technologies, which enable incorporation of cavities in a part's design and have little changeover cost compared to conventional manufacturing processes, potentially enabling individualized production and new possibilities for light-weight products [4].

The first step of state-of-the-art FDM processes involves a so-called slicing software that is used to generate machine-executable instructions (G-code), approximating a virtual product in the form of a CAD model. The desired product geometries are decomposed (sliced) into stacked layers of equal height along a specified axis, called the build orientation. Furthermore, for each of the layers, a closed two-dimensional path is planned, incorporating print head velocities set by the user. In the production step, the FDM machine follows the defined path while extruding heat-liquified raw material threads. Starting on the build plate, for each layer, the respective trace is followed by the machine, and so the whole part is fabricated. During the process, the machine must highly accurately control the correct material flow, the build plates state, and the correct positioning and velocity of the print head [5, 6].

While the concept seems straightforward to realize, practitioners long since report that reproducibility and reliability issues persist, demanding effective quality control measures [7–9]. Achievement of the aforementioned demand requires controlling for influences from the 5M-domains during pre-process and in-process stages: man, machine, milieu, material, and method affect the success of a print [10].

To achieve a successful production result, a set of numerous interrelated process parameters must be determined, some of which have been mentioned above. Finding appropriate parameters can pose a challenge to beginners, leading to failure in almost every second print [11]. Even expert knowledge does not necessarily lead to a good print result, but their experience helps them to avoid easily preventable mistakes. Many man-hours and unnecessarily wasted material could be saved by the prevention of simple mistakes. The reason lies in the fact that there is almost no recognized or approved reference process in which defined requirements serve as a quality control measure.

As of now, there are only few works that serve as a reference process for additive manufacturing. There is a lack of standards and norms that ensure high process and product quality. A lot of previous academic literature focuses on particular printing defects such as warping or oozing. Performed research indicates that comprehensive guidelines regarding failure prevention in the overall printing process ought to be developed. Additionally, there is a lack of in-depth documented requirements to achieve high quality in process and printed products.

This paper proposes a reference process model including 10 quality gates that serve as documented requirements to prevent defects and failure prints beforehand instead of costly troubleshooting. Section 2 describes related work and shows up the gaps upon which this present scientific work further elaborates. In Section 3, various failure types are introduced and considerations to prevent them. Section 4 contains the proposed reference process model including quality gates. Section 5 summarizes the results and discusses the advantages of the proposed approach and concludes with future research potentials.

2. Related work

An extensive list of FDM print issues and their causes have been published by Loh et al. [12]. Qualitative expert knowledge has been formulated in natural language and lacks precise arguments. Each failure is assigned to a category which is either "printer-associated," "deposition-associated," or "print quality," but no reason nor meaning for an assignment is given. Livesu et al. provide a detailed description of the main process starting with the CAD model and ending in the G-code, but they focus mainly on software issues [13]. Baş et al. describe print conditions that sufficiently lead to faults by the application of a fault tree analysis (FTA) [14]. Many faults are described and their dependencies are formally expressed. The German Norm DIN SPEC 17071 proposes a print flowchart, leaving open the actual events of quality checks [15]. Oropallo et al. name error control in a list of 10 challenges in 3D printing. A distinction is made between errors during printing and errors before printing, which is partially avoidable [16]. Bähr and Westkämper divide a print into three stages: pre-process, in-process, and post-process. The importance of cooling is emphasized and divided up into a sinter phase, crystallization phase, glass transition phase, and a shrinkage phase, which are bounded by corresponding temperature values. Additionally, a table is provided that relates process parameters to their manifestation in component properties. Martinez-Marquez et al. developed a detailed quality control procedure including 18 quality gates but tailored to the production of patient-specific medical implants [17]. The process assumes the use of a laser-based AM system and error control is only briefly described. Fu et al. provide literature research and an overview of sensor technologies for in-situ monitoring of FDM processes [18]. Their flowchart is limited to in-situ printer health and product quality monitoring. Oleff et al. do systematic literature research in order to find quality-related research gaps, giving examples for a few FDM-process errors [19]. Song and Telenko examine FDM-print failures in a university makerspace [20]. They categorize these into user errors, machine errors, and designer errors. Also, a poll has been carried out to determine failure rates dependent on the user's experience level. The results show wastage levels of about 34% of the total material and a print failure rate of 41.1%. Gibson et al. provide a rough overview starting with the CAD model, ending in the application. An in-process view, as well as defects, is not considered.

To the best of the authors' knowledge, no publication at present exists wherein a generic reference process is determined in which quality gates serve as requirements for quality control to prevent printing defects.

3. Defects in additive manufacturing

Defects that occur later in the process chain are harder to assess, as this presumes that no defect has occurred in a preceding process step.

In the following, examples of failures are explained which can either be pre-process, in-process, or post-process. Also, dependencies among failures are illustrated and research hypotheses for their assessment are formulated. Most of the enumerated defects have already been explained by Loh et al., whose work is extended in the following.

3.1 Pre-process defects

The following shows defects that may occur in the pre-process steps and are possibly preventable through quality control measures.

3.1.1 Tangled filament

If the end of the filament thread on a roll has been guided through under itself, a knot will eventually form on this roll, making proper unwinding impossible. This can happen after a user has unloaded filament from a printer. A proper loading process of filament should therefore be examined and verified.

3.1.2 Gaps

In all instances of this kind of defect, print segments are not properly connected and small gaps are recognizable by the naked eye. Loh et al. distinguish between three kinds "walls not touching," "gaps between Infill and outline" or "gaps between thin walls." Such gap appearances are introduced by the slicing software, affected by the extrusion line width. Thus, gap defects are avoidable if slicing errors are being determined.

3.1.3 Small features not printed

This defect highlights noticeable differences between the provided CAD model and the production instructions executed by the printer. Two distinctions between affected features can be made: A vertically standing wall whose width is smaller than the extrusion line width and a feature parallel to the build plate, whose height is smaller than the layer height. Material waste can be prevented if the slicing software informs the user about deviations between the CAD model and the generated G-code.

An example part that is susceptible to these kinds of defects is given in **Figure 1**, along with G-code paths generated by a slicing software using different parameter settings. The part consists of a block, a thin wall whose depth is 0.35 mm on its top, and a thin feature whose height is 0.1 mm in parallel to the build plate and is shown in (a). If a layer height smaller than the height of the thin feature of the CAD model and a line width smaller than the wall depth is chosen, then both the wall and the thin feature are not included in the generated G-code path, as demonstrated in (b). Conversely, if the layer height is smaller than the thinnest feature and the line width is smaller than the wall depth, then the sliced result matches the expectation of the

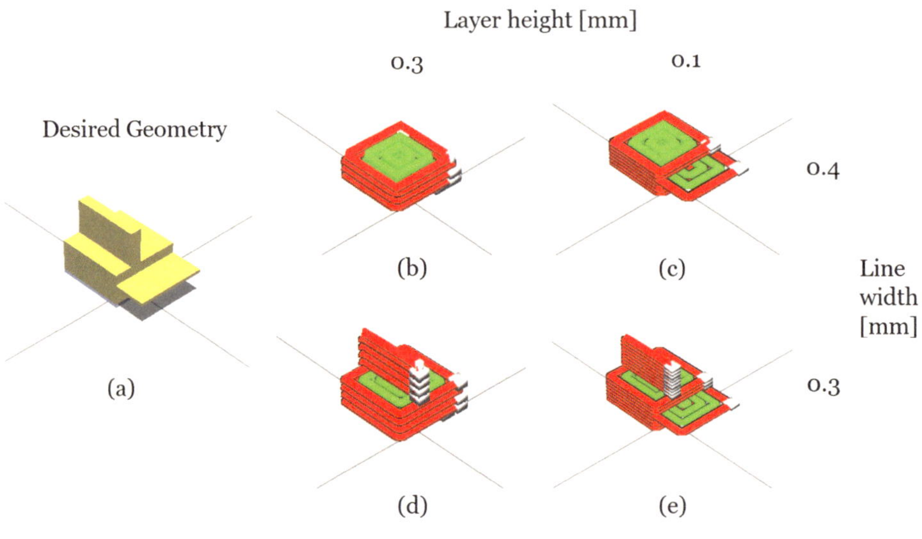

Figure 1.
Variations of parameter settings in slicing process of a complex cuboid object.

user, illustrated in (e). If either layer height or line width is not chosen appropriately, corresponding results will be sliced, as can be seen in (c) and (d), respectively.

3.2 In-process defects

3.2.1 Warping

The occurrence of warping (see **Figure 2**) manifests in an up-curl and a detachment of corners that touch the print bed at the bottom of the part in production due to tension arising from a non-uniform thermal gradient and shrinkage effects in the part. Deposition of additional layers on a corner that has already started to warp can negatively amplify the situation and may lead to an extension of the detached area. In the worst case, the part completely detaches from the bed; at this point, a print job should be canceled to avoid damage to the printer and further material waste. Scholars indeed argue that inclusion of brims or rafts or a reduction of infill percentage can help to avoid warping. As warping is irreversible, its occurrence should trigger a cancelation of the print process.

3.2.2 Detachment

The adhesion of the parts' first layer to the bed is essential for a successful result. If a part detaches from the bed, the print heads' movement will shift the part through its slight connection by the deposition strand. Thus, the material cannot be deposited at the correct location and the process should be stopped. The reason for detachment is an insufficient adhesion between the part and the build plate. Like warping, this defect is practically irreversible and the print process must be stopped.

3.2.3 Shifted layers

The path specified by the G-code must be executed accurately by the machine. Unsuccessful movement execution, for example, caused by missed-out steps of an axis motor, or a detached part creates positional deviations that lead to shifted layers if not compensated for.

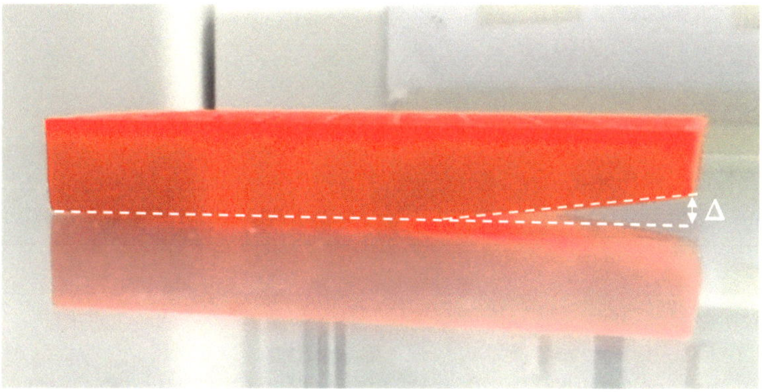

Figure 2.
Warping of a cuboid print model.

3.2.4 Clogging

Over the course of a print, all of the final parts' material must pass through the nozzle. Dust and undesired objects in the raw material can accumulate in the nozzle and lead to obstruction so that no material is deposable. Other causes could be burned residual material inside the nozzle as a result of an excessive extrusion temperature.

3.2.5 Nozzle cake/extruder blob/head flood

If material is continuously fed into the print head but cannot leave through the nozzle, then there is a risk of an occurrence of a head flood. The material sticks to material residuals that have previously been attached to the nozzle. Over time, more and more material accumulates around the nozzle. If the problem is not noticed at an early stage, the printer will likely be damaged severely. Removal of the accumulated material is time-consuming and a downtime eventuates.

3.2.6 Grinding

In FDM printers, a knurled ring drives the filament toward the hot end. If it loses grip on the filament, a small amount of material is removed from the filament's surface. If this happens for some time, then a groove forms in the filament thread, making transportation to the hot end difficult or impossible. Aggressive retraction settings can be the cause.

3.2.7 Overextrusion/underextrusion/missing extrusion

There are multiple causes for a missing extrusion: clogging, nozzle cake, or grinding. A camera, mounted at the nozzle's height, is used by many authors to monitor if the specified extrusion amount is matched.

3.2.8 Overheating

A central concept of extrusion-based 3D printing is the use of heat to extrude the raw filament. Overheating happens when a new layer is deposited on a lower layer, whose temperature has not cooled down to a certain level. Uneven printed parts may occur as a result (see **Figure 3**).

3.2.9 Curling

Another temperature-related defect caused by overheating is curling, which becomes apparent on topmost corners. This is similar to warping, but on the top instead of the bottom.

3.2.10 Pillowing

The exhibition of blisters or undesired holes in the topmost layers is termed pillowing and is caused by overheating. It can be detected the earliest after the last layer has finished and cooled down.

Figure 3.
Effects of overheating on print.

3.2.11 Stringing and oozing

The hot-end provides a continuous stream of liquid material whose flow is influenced by the feeder. As the hot-end consists of metal parts, rapid cooling down below the glass transition temperature of a given material to solidify inside the nozzle is infeasible. Reasons for stringing can be too high extrusion temperature or unapplied retraction settings. Once strings have appeared, there is no way to remove them during the print process. Stringing can be corrected by a post-process heat treatment.

3.3 Post-process defects and quality measures

After a successful production, the quality-related requirements may be assessed.

3.3.1 Blobs and zits

The occurrence of small bulges on the side of a part is termed blobs and zits. The interplay between start/stop position and retraction settings causes this defect. **Figure 4** shows a corresponding example, which demonstrates the development of blobs and zits that appear at the start/stop positions (right), which are marked by white dots on the left figure.

3.3.2 Porosity/voids

Due to the stacked deposition of round material beads, part-internal voids are a natural consequence. Such internal properties may be examined either by a dissection of a parts' region-of-interest or by nondestructive metrology like microcomputer tomography.

3.3.3 Vibrations and ringing

High print speeds induce vibrations that propagate through the printer's frame and cause small deviations in the head position. Hence, patterns according to these deviations appear on the parts' surface.

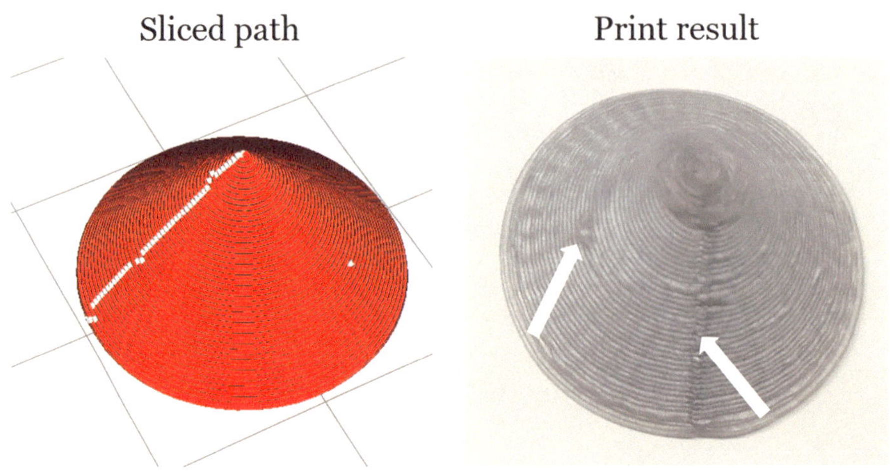

Figure 4.
Blobs and zits on a conical print object.

3.3.4 Dimensional accuracy

Overall shape deviations can be assessed by a separate measurement instrument.

3.3.5 Surface quality

The surface quality is the first aspect of an object that a customer perceives during an inspection of the object. This quality criterion can be checked independently of the actual model.

3.3.6 Mechanical properties

After quality checks of dimensional accuracy and surface quality, mechanical quality checks may be performed. Destructive assessments include tensile, shear, and compressive strengths, while part density can be measured using a scale.

4. Reference framework and quality gate process

The lack of norms and standards often leads to manufacturing processes that are defined from scratch for each individual production run, which provokes plenty of try-and-error operations. Indicative of this are numerous troubleshooting guides that help individuals cope up with problems that occur during the printing process as well as frequently discussed issues in community-based online forums (see for example [21–24]). On the other hand, there are very few references on how to plan quality and prevent easily avoidable defects beforehand. This leads to the overall conclusion that a lot of quality issues can be prevented if a reference process with criteria-based quality gates guides through the manufacturing process to ensure high process and product quality.

A reference process supports process requirements so that the process quality and resultant product quality remain consistent and reproducible at all times. This paper proposes a generic reference process model for additive manufacturing that describes the common sequence of activities for fused deposition modeling. Furthermore, this work suggests a model that contains pre-process, in-process, and post-process steps and starts with the CAD design and ends in machine and product post-processing (see **Figure 5**). The reference process model is based on the standard DIN SPEC 17071:2019–2112 [15] that represents a process chain for additive manufacturing that can be seen in **Figure 6**.

This blueprint of an additive manufacturing process chain is further specified in the reference process model in **Figure 5** and quality gates are added. A quality gate specifies criteria in process steps as well as quality-relevant characteristics and factors that have to be met in order to continue the process flow. It enables to perform corrective and/or preventive action to ensure high quality [25].

The reference process starts with the CAD of the product that is going to be manufactured. After that, the pre-processing steps will ensure the material selection, preparation, loading, and build chamber preparation. Moreover, the build orientation and strategy, as well as the generation of support structures, will be determined. In the manufacturing process itself, the production of the first layer of the build is a crucial part and is a decisive factor for the continuous process. After the build is finished in printing, a cooldown process will harden the material. In post-processing, the build product has to be removed from the build platform and both the machine and the product itself need post-processing. The machine is cleaned-up and restored to the initial state in order to be prepared for following production runs. The support structures are removed from the printed parts and a surface finish is performed where required.

There are nine quality gates in the proposed reference process model that serve in the course of the manufacturing process as points at which a decision is made on the progression to the next process step on the basis of quality criteria clearly defined in advance. Each criterion may be checked to prevent quality issues in the succeeding process steps. **Table 1** gives an overview of all nine quality gates and the respective criteria.

In the following, an example will show how the quality gates may prevent printing issues and may ensure the overall process quality. Therefore, a 3D-printed door hinge was manufactured according to the reference process model, and after each process step the quality gate criteria are reviewed and verified.

The first quality gate is actually positioned before the pre-processing of the additive manufacturing process and verifies the CAD design of the print part (QG 0). First of all, the manufacturability in regard to printer settings and the adherence to design rules can eliminate severe quality issues that may occur during printing' as an example thereto, if the door hinge cannot be assembled after printing because of poorly placed through-hole positions.

In QG 1, storage and material validation should be performed during the material preparation to prevent material-related quality issues. Filament could be damaged because of humidity or temperature-related variations in the storage area and may provoke damage during the printing process. In addition to that, there should be sufficient filament supply for the print that has to be printed as well as a coherent diameter of filament.

After the material was loaded to the feeder of the printer, QG 2 ensures that the orientation of the filament feed is adequate and the nozzle of the printer is unclogged. Moreover, the filament tubes should be empty and the overall filament feed rate is sufficient.

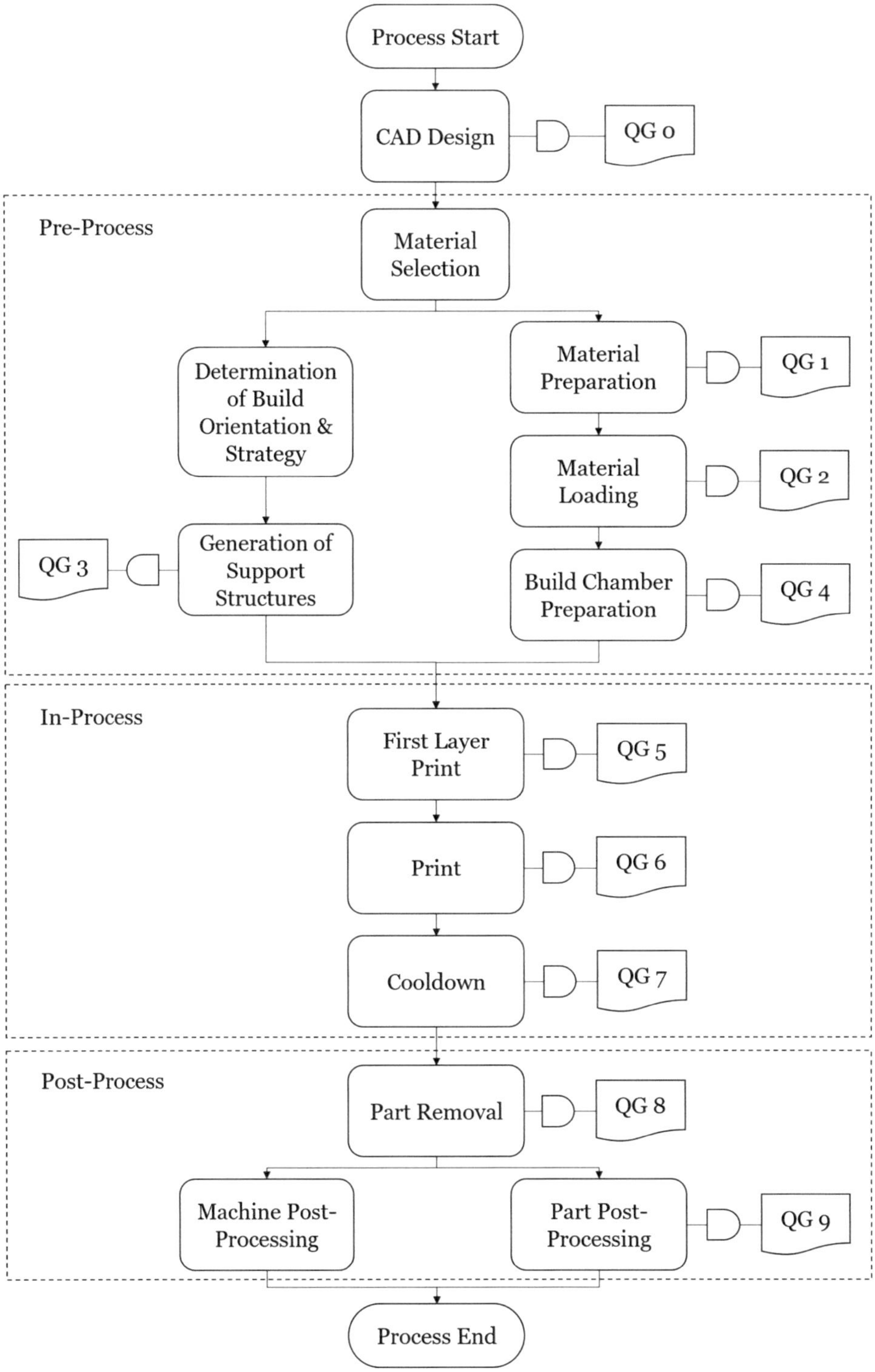

Figure 5.
Reference process for additive manufacturing.

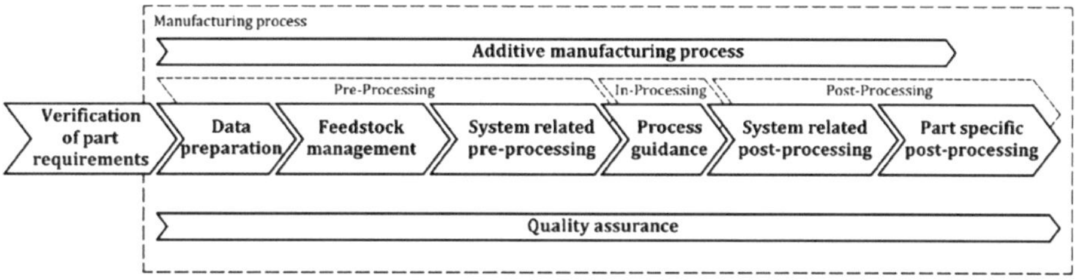

Figure 6.
DIN SPEC 17071 process standard for additive manufacturing [15].

Quality gate	Description	Criteria
QG 0	Verification of requirements and design approval	• Verification of manufacturability • Adherence to design rules • Avoidance of mass accumulation • Verification of process requirements • Verification of product requirements • Customer design approval • Wall thinness according to nozzle size • Solid bottom and support structures
QG 1	Storage and material validation	• Humidity of storage area • Temperature of storage area • No filament tangling • Age of filament according to manufacturers' instruction • Sufficient filament supply according to design • No visual material damages • Coherent diameter of filament
QG 2	Loading quality validation	• Sufficient filament feed rate • Adequate filament feed orientation • Unclogged nozzle • Empty filament tubes
QG 3	Software and file validation	• Error-free STL-files • Correct infill density • No infill overlaps • Error-free slicing process • Updated slicing software • Updated printer firmware

Quality gate	Description	Criteria
QG 4	Build chamber validation	• Free belt movement • Free extrudement wheel • Clean and dry build platform • Adequate build platform temperature • Leveled and flat build platform • Aligned rod • Fixated printer position • Adequate nozzle height from build platform • Adequate retracting settings • Adequate cooling settings • Primed nozzle
QG 5	First layer validation	• Verification of extrusion process • Adherence to build platform • Verification of geometric and dimensional proportions
QG 6	Process monitoring	• Geometric stability of build and filament • Coherent filament flow • Adherence to build platform • Adequate extrusion process • Adequate extrusion temperature • Adequate build platform temperature
QG 7	Cooldown validation	• Geometric stability (no warping or curling) • Verification of layer adhesion • Adequate part temperature • Adequate in-part temperature gradient (to avoid material stress)
QG 8	Visual inspection	• Verification of geometric stability • Verification of nondestructive detachability • Verification of transportability
QG 9	End of line quality validation	• Elimination of porosity/voids • Verification of surface quality • Verification of geometric form • Verification of tolerances • Verification of mechanical, chemical, thermal properties • Verification of density/mass

Table 1.
Quality gates for the reference process model.

QG 3 states that on the other hand, the slicing files, as well as the whole slicing process, should be error-free. Printer firmware and slicing software should load the latest update to prevent failure. The correct infill density and no infill overlap should be checked.

A lot of process failures can be associated with a build chamber that has not been calibrated for error-free printing. QG 4 examines if there is a free belt movement and a free radial movement of the extrudement wheel. Moreover, a leveled and clean build platform that has an adequate temperature can ensure a consistent printing process. The printer should have a fixed position because the printing process may cause vibrations and an unintended re-orientation of the whole system that may, in turn, interrupt the filament feed.

The first layer of printing is a crucial step for the whole printing process. Therefore, QG 5 should verify the extrusion process and the adherence to the build platform. Moreover, a visual inspection of correct geometric and dimensional proportions should be performed.

After the first layer print, the continuous layer-by-layer printing should be in-situ monitored (QG 6). The filament flow and the extrusion process should be closely monitored, as well as the geometric stability of the print. Temperature sensors may observe the extrusion temperature and build platform temperature.

When the printing process is completed, the cooldown process is also a quality-relevant aspect. To prevent warping or curling of the print due to material stress, the temperature should be lowered slowly (QG 7). In addition to that, layer adhesion should be verified.

A visual inspection can be performed as soon as the print is removed in a nondestructive manner from the build platform (QG 8). The geometric stability and the transportability should equally be verified.

Lastly, an end of line quality check should be performed after the post-processing (i.e., surface finish, removal of support structures). Verification of the surface quality as well as the geometric form including all relevant tolerances should also be performed. Finally, the mechanical, chemical, and thermal properties should be checked.

5. Conclusion and further work

The proposed reference process model including the criteria-based quality gates to prevent printing issues serves as a guideline to achieve high process and product quality. Opposed to common troubleshooting that is carried out during the occurrence of printing issues, the presented model herein allows executing corrective or preventive action. The lack of norms and standards in additive manufacturing as well as rudimentary reference processes makes it difficult to meet process and product requirements *per se* and achieve a planned quality level.

This research work has introduced a reference process including pre- and post-process activities with the aim to standardize the printing process of FDM 3D printing. These process steps are sub-divided by quality gates that ensure the fulfillment of requirements to ascertain the prevention of quality issues. There are nine quality gates that have quality descriptions in form of documented requirements that have to be met.

In view of the above, it may be stated that the proposed process model requires some effort for the verification steps in terms of an operational measuring system. Regarding the in-situ monitoring of the printing process, temperature sensors for the extruder system as well as the build platform and a feed rate sensor need to be installed. Additionally, the calibration of the printer settings and the validation of the build chamber is a time-intensive procedure and therefore extends the printing process not inconsiderably. The availability of all required sensors is a valid difficulty.

There are several limitations to the introduced reference process model. First of all, it only addresses manufacturing processes that are based on thermoplastic materials. Powder bed fusion like selective laser melting is not considered in this reference process or in the quality gates. Therefore, this reference process model has to be adjusted accordingly in order to allow for these alternative additive manufacturing processes.

Further research activities have to be performed to achieve a more concise insight into how to prevent quality issues during additive manufacturing processes. First of all, this model needs further verification and validation in order to define the degree at which it can prevent relevant quality issues. Qualitative and quantitative studies may focus on what the overall benefits of this quality gate process are in terms of not aborted production runs or customer-relevant requirements. Moreover, the list of documented requirements in the quality gates is not exhaustive and quality criteria may be composed through further research to generate a complete reference model.

Acknowledgements

The Industrial Collective Research (IGF) project (No. 21675 N) of the Federation of Quality Research and Science (FQS), August-Schanz-Straße 21A, 60433 Frankfurt am Main, Germany, was funded by the Federal Ministry for Economic Affairs and Climate Action through the German Federation of Industrial Research Associations (AiF) under the IGF program on the basis of a decision by the German Bundestag.

Author details

Marcel Randermann, Timo Hinrichs* and Roland Jochem
Institute for Machine Tools and Factory Management, Technische Universität Berlin, Berlin, Germany

*Address all correspondence to: timo.hinrichs@tu-berlin.de

References

[1] Colosimo BM, Huang Q, Dasgupta T, Tsung F. Opportunities and challenges of quality engineering for additive manufacturing. Journal of Quality Technology. 2018;**50**:233-252. DOI: 10.1080/00224065.2018.1487726

[2] Ngo TD, Kashani A, Imbalzano G, Nguyen KT, Hui D. Additive manufacturing (3D printing): A review of materials, methods, applications and challenges. Composites Part B: Engineering. 2018;**143**:172-196. DOI: 10.1016/j.compositesb.2018.02.012

[3] Gebhardt A, Hötter J-S. Additive Manufacturing: 3D Printing for Prototyping and Manufacturing. Munich, Hanser Publishers, Cincinnati: Hanser Publications; 2016

[4] Campbell I, Bourell D, Gibson I. Additive manufacturing: Rapid prototyping comes of age. RPJ. 2012;**18**:255-258. DOI: 10.1108/13552541211231563

[5] Gibson I, Rosen D, Stucker B, Khorasani M. Additive Manufacturing Technologies. Cham: Springer International Publishing; 2021

[6] Cunico M, de Carvalho J. Design of an FDM positioning system and application of an error-cost multiobjective optimization approach. Rapid Prototyping Journal. 2013;**19**:344-352. DOI: 10.1108/rpj-11-2011-0117

[7] Moreau C. The State of 3D Printing. Villejuif, France: Sculpteo; 2021

[8] Moilanen J, Vadén T. Results from 3D Printing Survey 2013. Canberra, Australia: Journal of Peer Production; 2013

[9] Knoop F, Lieneke T, Schöppner V. Reproduzierbarkeit der Maßhaltigkeit im fused deposition modeling. In: Kynast M, Eichmann M, Witt G, editors. Rapid. Tech – International Trade Show & Conference for Additive Manufacturing. München: Carl Hanser Verlag GmbH & Co. KG; 2017. pp. 51-66. DOI: 10.3139/9783446454606.004

[10] Bähr F, Westkämper E. Correlations between influencing parameters and quality properties of components produced by fused deposition modeling. Procedia CIRP. 2018;**72**:1214-1219. DOI: 10.1016/j.procir.2018.03.048

[11] Song R, Telenko C. Causes of desktop FDM fabrication failures in an open studio environment. Procedia CIRP. 2019;**80**:494-499. DOI: 10.1016/j.procir.2018.12.007

[12] Loh G, Pei E, Gonzalez-Gutierrez J, Monzón M. An overview of material extrusion troubleshooting. Applied Sciences. 2020;**10**:4776. DOI: 10.3390/app10144776

[13] Livesu M, Ellero S, Martínez J, Lefebvre S, Attene M. From 3D models to 3D prints: An overview of the processing pipeline. Computer Graphics Forum. 2017;**36**:537-564. DOI: 10.1111/cgf.13147

[14] Baş H, Elevli S, Yapıcı F. Fault tree analysis for fused filament fabrication type three-dimensional printers. Journal of Failure Analysis and Prevention. 2019;**19**:1389-1400. DOI: 10.1007/s11668-019-00735-6

[15] DIN SPEC 17071:2019-12. Additive Manufacturing: Requirements for quality-assured processes at additive manufacturing centres. Deutsches Institut für Normung. 2019

[16] Oropallo W, Piegl LA. Ten challenges in 3D printing. Engineering with

Computers. 2016;**32**:135-148. DOI: 10.1007/s00366-015-0407-0

[17] Martinez-Marquez D, Jokymaityte M, Mirnajafizadeh A, Carty CP, Lloyd D, Stewart RA. Development of 18 quality control gates for additive manufacturing of error free patient-specific implants. Materials. 2019;**12**(19):3110

[18] Fu Y, Downey A, Yuan L, Pratt A, Balogun Y. In situ monitoring for fused filament fabrication process: A review. Additive Manufacturing. 2021;**38**:101749. DOI: 10.1016/j.addma.2020.101749

[19] Oleff A, Küster B, Stonis M, Overmeyer L. Process monitoring for material extrusion additive manufacturing: A state-of-the-art review. Progress in Additive Manufacturing. 2021;**6**:705-730. DOI: 10.1007/s40964-021-00192-4

[20] Song R, Telenko C. Material waste of commercial FDM printers under realistic conditions. In: University of Texas at Austin, editor. 2016 International Solid Freeform Fabrication Symposium; 2016

[21] 3D Printing Forum - 3D Print Board. 2022. Available from: https://3dprintboard.com/ [Accessed: February 10, 2022]

[22] 3DPrintingForum.org. General Forums—3DPrintingForum.org. 2022. Available from: https://www.3dprintingforum.org/forum/general [Accessed: February 10 2022]

[23] gCreate 3D Printing Forum. Community Troubleshooting Forum. 2022. Available from: http://forum.gcreate.com/index.php?forums/community-troubleshooting-forum.12/ [Accessed: February 10, 2022]

[24] Ultimaker Community of 3D Printing Experts. Ultimaker Community of 3D Printing Experts. 2022. Available from: https://community.ultimaker.com/ [Accessed: February 10, 2022]

[25] Prefi T. Qualitätsmanagement in der Produktentwicklung. In: Pfeifer T, Schmitt R, editors. Masing Handbuch Qualitätsmanagement. 1st ed. s.l.: Carl Hanser Fachbuchverlag; 2014. pp. 402-440

Chapter 6

Quality Management Costs in Logistics

Marieta Stefanova

Abstract

The minimization and elimination of deviations from quality that could cause a failure in the logistics system should be identified at an early stage in order to reduce the costs for recovering the system to its normal operation. The objective of this study is to analyze the contribution of prevention costs related to quality management to the total costs by focusing on the need to undertake priority preventive actions to ensure logistics services that meet the customer's quality requirements. The methodology of the study includes the integrated application of conventional scientific methods for comparative analysis and Taguchi's design for accounting regarding the primary costs for quality management with the predominant use of qualitative analysis. By applying these methods, the following groups of costs have been analyzed: prevention and avoidance of nonconforming quality; quality evaluation and control; and covering the costs for nonconforming quality of the logistics services. The contribution of the three groups of costs has been studied. Based on the analyses, this paper comes to the conclusion that the management of those costs by groups of factors for incurring them has the potential to contribute to the improvement of the quality of logistics.

Keywords: costs, logistics, nonconformities, quality

1. Introduction

The costs for maintaining the quality of the logistics services at the level of the customer's expectation are associated with the achievement of high efficiency of the processes [1, 2], better quality of the incoming material flows [3], and performance of the equipment and inventories without failure [4, 5]. The achievement of the "Just-in-time", JIT principle for all products delivered requires targeted managerial efforts for maintaining the continuity of the logistics processes [6–10], ensuring efficient human resource management [11–13], and optimal utilization of the warehousing potential [14–17]. The identification, minimization, and elimination of deviations from set quality levels and the causes for failure of the system should be identified at the earliest possible stage in order to reduce the expenses for recovering the system to its normal operation. The level of quality has a positive impact on the implementation of the selected logistics strategy [18–21], whereas the low level of quality is an indicator of the poor efficiency of the supply chain [22]. The main objective of the quality management processes is to ensure the effective performance of the logistics services. It has been concluded that when the quality management processes are operated as a system, they have a much more positive impact on the performance of the individual components of the system rather than the contrary. It has also been postulated that certain key areas of the logistics operations have a decisive impact on the efficiency of the logistics system, such as, for example, transportation and storage operations.

To analyze the costs for maintaining required quality levels, the components of those costs (**Figure 1**), need to be clarified:

Improvement of the logistics system efficiency is a key factor for ensuring products that meet demand and the flawless management of the organization [23–27]. Logistics operators that organize the supply chain by setting targets and results based on a limited budget manage to achieve growth in their revenues and assets, and, at the same time, reduce their operational costs [28, 29]. Due to the constantly emerging risks in the operations in the real economy where the logistics operations take place, it is often necessary to respond to the specific situation for overcoming the bottlenecks without considering the strategic guidelines for business development [30]. This is why the systems that are very flexible and capable of changing together with the market and the changes in the external environment are the most rapidly developing ones. In essence, the purpose of logistics operations can be defined as the effective and timely movement of goods to the places where the customer needs them at a reasonable price [31]. However, there are often restrictive conditions for the fulfillment of those purposes, and the appropriate equipment for loading and moving the transport vehicles is not always available when they are needed. In addition, the capacity of each logistics warehouse is strictly confined and fixed. There is competition in the sector of logistics, too. Therefore, the business needs to focus on its main competence and outsource to external contractors those services, the operation of which causes unjustified losses of resources, in order to achieve effective management of the costs for quality improvement. Actually, the main

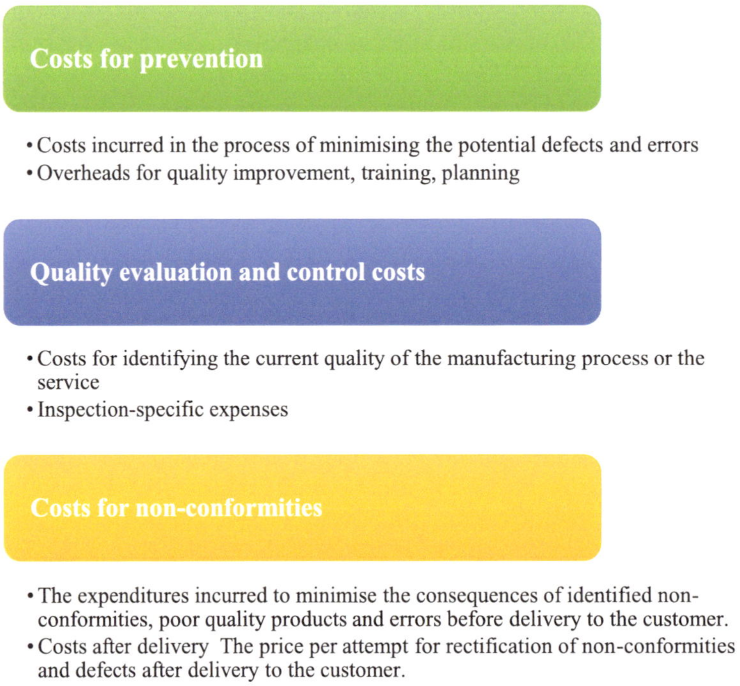

Figure 1.
Quality assurance-specific costs.

purpose of cost optimization is to practically [32, 33] satisfy the customer's requirements by reducing the time for making the delivery. The main impediment to achieving this purpose is caused by the limited possibility to transfer and receive information about the actual demand in real-time.

The main problem about minimizing the costs for achieving the logistics purposes is related to the understanding about the management of the system itself as one that distributes the cargo ("push") and one that requires distribution of the cargo where necessary and based on a customer's order ("pull") [34]. The first type of system is more appropriate in the case where no exact information about the need of goods is available. However, in this case, if demand is significantly higher than supply, distribution of scarce goods and priority servicing of selected customers is needed. The use of mathematical methods for planning routes, occupation of the warehousing facilities, and temporary hiring of workers can help to reduce costs. When accounting for the total operational overheads, the expenses for handling, storage, and transportation of the goods should be accounted for based on the main cost items. This can be done by identifying all the resources (including human resources), the packaging and repackaging operations performed, the processes, and the methods used for evaluation and control in order to ensure the overall performance of the process. In other words, the total costs for logistics are the sum of all costs incurred for the management and implementation of all processes and operations related to the logistics operations. Generally, the total costs can be divided into the following three groups:

- costs associated with the operations,
- costs related to with the management of the logistics system and,
- costs associated with the application of possible logistics risks.

There is an interesting approach in the control of quality management costs, which was developed by Taguchi [35]. This method focuses on the causes for deviations from the quality and on establishing clearer criteria for defining the critical boundary that distinguishes between conforming and nonconforming services [36]. Taguchi's contribution to quality management is related to the following principle that any variations and deviations in the function of quality are the results of random and nonrandom factors and losses are observed when the variation results in conditions where the product or service is on the exact boundary of the target conformity value [37]. This is the quadratic loss function since it is assumed that when the product or service is at its target value, the loss will be zero. According to Diallo, Khan, and Vail, the relationship between quality improvement by decreasing the variations and the costs can be analyzed by using Taguchi's function. Many researchers have also applied Taguchi's method in the field of logistics services [38–41].

The contribution to the reduction of the total operational costs for prevention of nonconforming logistics services as compared to the increase in the costs for their management and their relationship to the costs for monitoring and control in logistics services has not been studied. A number of logistics organizations have not focused on this analysis and, as a result, perform restructuring or investments which do not yield the expected positive outcome. It is the author's view that logistics service providers should draw their attention to investment in quality management related to the prevention of nonconformities. At the same time, logistics service providers should exercise more effort on the potential opportunities for the development of the actual logistics services and the service processes. This study offers a practical solution for management and analysis for measuring these qualitative changes.

The objective of the study is to analyze the contribution to total costs of the expenses for prevention related to quality management in logistics by focusing on the need to undertake priority preventive actions to ensure the provision of logistics services that meet the customer's quality requirements. By using the applied methods, this study analyses the contribution to quality improvement of the costs attributed to the prevention and avoidance of nonconforming quality, for quality evaluation and control, and for covering the expenditures for nonconforming quality of the logistics services. The relationship between these groups of costs in quality management has been identified by means of structural modeling, which helps to establish the contribution of the costs to the achievement of sustainable quality of the logistics services.

2. Research methods

The primary method for data analysis that has been used is Taguchi's method, which defines quality from the perspective of cost minimization and the subsequent loss to society. Based on his definition about quality management, continuous, consistent, and targeted actions are required to achieve minimum variability of the logistics services offered. According to Taguchi, the efforts should focus on the following two aspects: defining the combination of factors that have the lowest impact on any deviation from quality, and adjusting those factors that are the cause for the deviation from the set target of the logistics services. Based on the results obtained from Taguchi's loss function, the contribution of the different factors that could have an effect on the deviation from the customer's expectations for high-quality logistics services can be quantified. This can be used for initiating improvements that could have a positive impact in terms of satisfying those expectations.

2.1 Stages of Taguchi's method application for this study

Taguchi's method was applied in two stages:

1. A model generation stage, which allows the selection of those controllable levels of the factors that have the greatest contribution to the achievement of the logistics services quality level expected by the customers (studied dependence) and the respective significance levels.

2. Performing the actual analysis (Taguchi's design) to identify the parameters of the analyzed factors that minimize the variation in the deviations. The calculation of the tolerances that contribute to the reduction of deviations from the expected quality level of the logistics services is performed using the software Microsoft Excel XLSTAT 2021® [42].

2.2 Method for collection of data for analysis

The proposed data to be evaluated have been taken from the annual financial statements of an operating logistics company in the food sector and have been subsequently divided into three main groups: for prevention and avoidance of nonconforming quality; for quality evaluation and control; for covering the costs for nonconforming quality

Short name	Nbr. Of categories	Period of time before implementation of the changes in the cost structure	Period of time after implementation of the changes in the cost structure
Prevention costs	2	450 (in thousand euro) (450 k€)	500 (in thousand euro)
Evaluation and control costs	2	200 (in thousand euro)	250 (in thousand euro)
Cost of nonconformities	2	100 (in thousand euro)	50 (in thousand euro)

Table 1.
Variable information.

of the logistics services. The information collected about the last two-year period has been summarized in tables in order to visually illustrate the potential impact and the actual improvement of the economic result. After the final data from the studied two-year period were collected (before and after the introduction of the changes in the cost structure), those data were summarized and presented in **Table 1**.

By observing **Table 1**, it can be seen that the costs for improvements after the introduction of the changes are two times greater than the costs for prevention and control, whereas the costs for covering losses as a result of nonconforming logistics services have decreased by half as compared to the period before the implementation of the changes. The change in the cost structure based on the pre-defined three groups has allowed for the practical application of Taguchi's principle that the nonconforming logistics service cannot be improved through the process of control or covering the losses from the nonconformity after the service has been provided. The application does not have the potential to create a conforming service, but just to identify the conforming and nonconforming services. Based on the data obtained, the experimental design was built and a questionnaire was generated.

2.3 Evaluation collection method and discussion method

The data collection for the study was performed via telephone and online meetings in focus groups by taking into account all the restrictions imposed in relation to the pandemic. All participants in the study are currently managers in organizations where the main scope of business is the provision of logistics services in the field of trade and delivery of food products to wholesalers.

The participants in the study were selected based on their management experience and, in particular, their experience in the field of logistics services quality management costs. The required criterion for participation was at least 10 years of experience. Initial informative telephone conversations about the study and its methods, including the observation of all requirements of the relevant legislation related to personal data protection, were performed with potential participants in the study. Only 5 out of a total number of 20 potential participants did not agree to participate. The participants who confirmed participation received a questionnaire. The main purpose of this questionnaire was to study the potential attitudes and evaluations of the participants regarding the need of change in the structure of quality management costs. The study was performed in two consecutive panels in online meetings with a discussion in focus groups held in-between. The evaluation of the participants' opinion was performed based on a 100-point scale ranging from 1 to 100. The questionnaire of the study is presented in **Table 2**.

By using the 100-point scale (where 1 is the lowest value and 100 is the response with the highest value), please, evaluate which, in your opinion, would be the most suitable cost structure for logistics services quality management represented in 8 different categories.				
Observations	**Prevention costs**	**Evaluation and control costs**	**Cost of nonconformities**	**Respondents' answers in the two panels**
Obs1	450	200	100	
Obs2	450	200	50	
Obs3	450	250	100	
Obs4	450	250	50	
Obs5	500	200	100	
Obs6	500	200	50	
Obs7	500	250	100	
Obs8	500	250	50	

Table 2.
Questionnaire of the study.

After the study, the participants' responses were averaged and summarized for further analyses using Taguchi's method.

The method used allows on one hand a comparison to be made, while on the other to quantify the difference between the target function (optimum ratio between the quality management costs) and its actual manifestation. The objective was to find a solution for minimizing deviations from the target function for logistics services in the food sector.

3. Results

In the course of the study, Taguchi's principles and methods for quality management were used to identify the optimum ratio between the quality management costs. The first principle that was applied is related to the statement that quality should be designed in the logistics service before offering that service on the market and, respectively, a strategy should be undertaken to increase the prevention costs (designing conforming quality) at the expense of the other costs.

Based on the experimental design, further calculations were made to find the contribution of the increased or decreased share of certain overheads to the achievement of a conforming service impacted to the lowest possible extent by the other factors. The data obtained from the two focus group sessions held were averaged and entered in **Table 3**.

Based on the results from **Table 3**, the experts have given a significantly lower number of points to the ratio of costs in the cases where there is an increase in the costs for operations associated with the rectification of problems, rather than preventive actions. It was concluded that this was the right approach; however, despite this, it is the author's view that logistics organizations practically continue using their entire potential not for the development of the types of services offered on the market, but for the rectification of problems that have occurred in the course of providing those services. The reasons for that could be related to the fact that often when designing the actual services, the processes are dragged over time, which in turn, may

Observations	*Prevention costs*	*Evaluation and control costs*	*Cost of nonconformities*	*Response 1*	*Response 2*
Obs1	450	200	100	75.000	76.000
Obs2	450	200	50	80.000	82.000
Obs3	450	250	100	78.000	80.000
Obs4	450	250	50	84.000	85.000
Obs5	500	200	100	85.000	84.000
Obs6	500	200	50	98.000	97.000
Obs7	500	250	100	88.000	96.000
Obs8	500	250	50	99.000	99.000

Table 3.
Experimental design (response 1 and 2).

lead to a delay. Therefore, this process often needs to be compensated by reducing the time limits under signed contracts and by adjusting all the details and parameters related to the negotiation of the logistics service. As a result, certain logistics operations are skipped, which are subsequently performed without actually specifying their parameters. This, on the other hand, creates more favorable conditions for customer claims and undertaking actions to increase the control in order to avoid such nonconformity in the future. The higher level of control leads to an increase in costs and does not guarantee that the services will be conforming if the conditions that lead to the presence of claims remain unchanged.

The analysis of controllable factors that create conditions for deviations in the logistics services has been studied with respect to the contribution of costs for the different operations to the total operational costs. These costs include both the costs for planning the logistics services and the costs related to the control of those services and compensations to customers related to claims and returns, replacement, or repeated implementation of the logistics operations. It is the author's view that claims could be minimized by designing logistics services that are needed by the customer rather than services that the organization is capable to provide. A number of studies have come to the conclusion that the prevention of claims is more efficient than covering the costs once a claim has been filed and, respectively, could result in greater customer satisfaction [43–47].

Based on the data collected, Taguchi's model has been created, where the ratio LS means (Signal-to-Noise ratios) has been calculated. It defines the ratio between the mean value of the share of each cost from the total costs and the standard deviation. The variability of the analyzed indicators considered significant by the experts for the provision of a conforming service, defined by their standard deviation from the average value, is presented in **Figure 2**.

The results presented in **Figure 2** show that prevention costs have been evaluated as the most significant factor with positive impact, followed by the positive impact of the costs for control. The influence of the increase in the costs for nonconforming logistics services has been assessed as negative. The multiple criteria used by the logistics operators for calculation of the services are related to the satisfaction with their expected quality and are hard to quantify. The studies performed so far show that investing in the design of services has a significantly more positive impact on the expected quality than investment in a higher level of control on the performance of those services (the prevention costs and the costs for control are equally increased by 50 units).

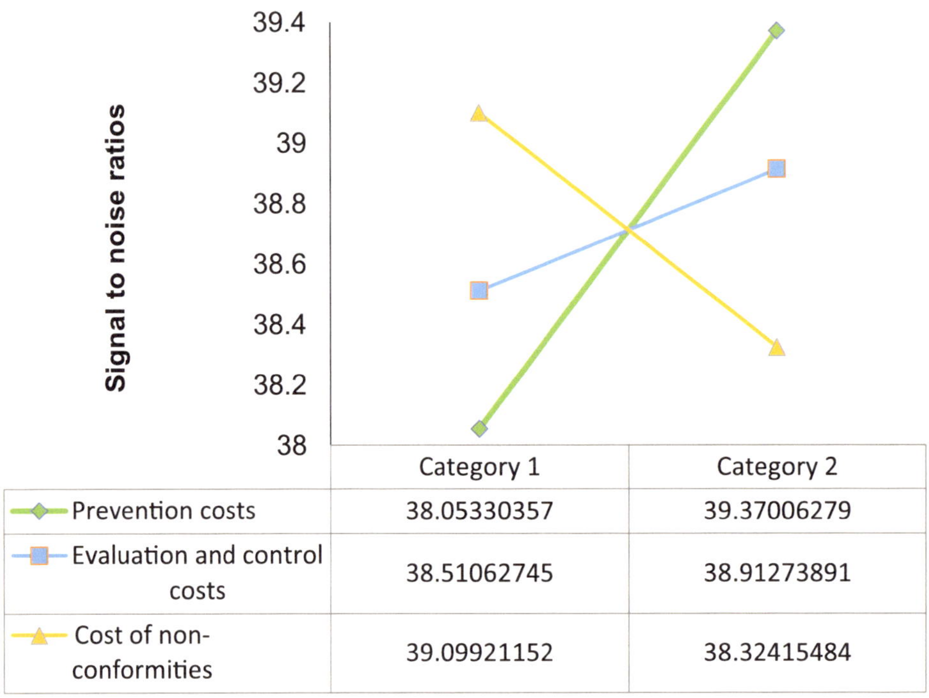

Figure 2.
Signal-to-noise ratios.

Taguchi's approach thus requires seeking an appropriate solution for reducing the variations applied to the expected quality of the logistics operations and provides an opportunity to find results for lower deviation from the target function. The model these decisions can be based on, so that the variations in the logistics services are lower than expected, is presented in **Table 4**.

The studied factors are statistically significant (at 0.05), which allows an optimum ratio to be set between the studied costs so that the deviation from the target function

Source	*Value*	*Standard error*	*t*	*Pr > |t|*	*Lower bound (95%)*	*Upper bound (95%)*
Intercept	39.184	0.136	287.649	**<0.0001**	38.805	39.562
Prevention costs-450	−1.317	0.136	−9.666	**0.001**	−1.695	−0.939
Prevention costs-500	0.000	0.000				
Evaluation and control costs-200	−0.402	0.136	−2.952	**0.042**	−0.780	−0.024
Evaluation and control costs-250	0.000	0.000				
Cost of nonconformities-50	0.775	0.136	5.690	**0.005**	0.397	1.153
Cost of nonconformities-100	0.000	0.000				

Table 4.
Model parameters (standard deviations).

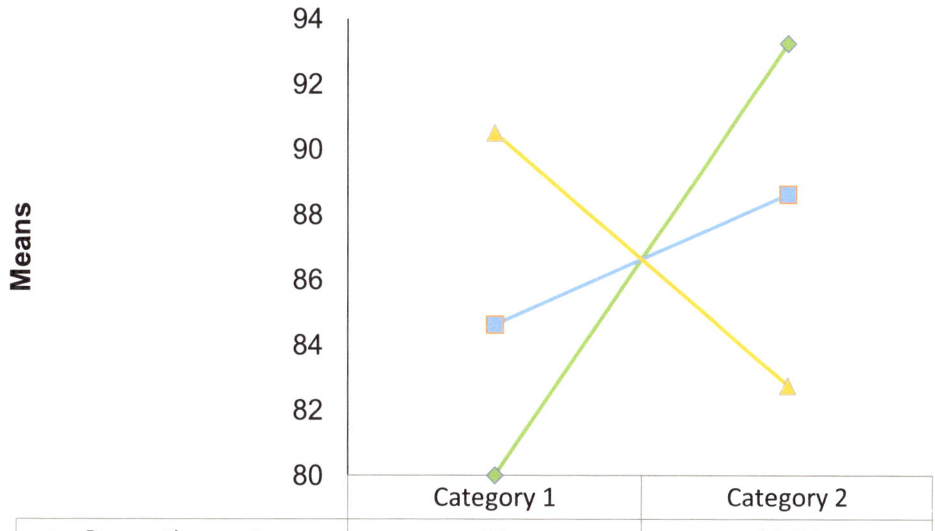

	Category 1	Category 2
Prevention costs	80	93.25
Evaluation and control costs	84.625	88.625
Cost of non-conformities	90.5	82.75

Figure 3.
LS means (means prevention costs, control costs, cost of nonconformities).

is as low as possible. **Table 4** equally shows the statistical significance of each type of costs and their contribution to the achievement of an optimum combination of those costs. According to the feedback provided by the questionnaires, the most important factor in quality management is cost prevention because the absolute value of this factor is the highest. It can be stated that the prevention costs and the costs for control on the processes in the specific case that was studied were increased by the same number of units; however, the prevention costs demonstrated a much higher effect on the target function.

These results are confirmed by the main effects graphs in **Figure 3**.

Logistics organizations should invest in operations for the prevention of nonconformities in order to decrease the variability in the target function, even if the causes of the variations are not eliminated.

It has been practically demonstrated that the costs for eliminating the variation in the target function are very high. A more feasible and practical solution is to just change the structure of the costs or to control the factors that are more significant and have a greater impact on the target function. This can be achieved without increasing the total expenses or with a minimum increase resulting just from the redistribution of costs in the proper direction. Furthermore, the overheads that do not have a positive impact on the variations in the target function could be decreased and thus invested properly, in areas where their impact could be more favorable. This is what the application of Taguchi's method on the structure of quality management costs in logistics allows the user to do – to calculate the contribution of the three cost groups in order to achieve an optimum effect in the target function.

4. Conclusion

During the study, the change in the structure of quality management costs was analyzed based on the significant factors for achieving customer satisfaction defined by the experts. Data from the expenses incurred by the logistics company operating in the food sector were analyzed, which were divided into three groups of costs and described in the methodology. The road to improvement was found to be associated with the following:

- cost reduction for nonconforming services after delivery,
- keeping a relatively stable level of the expenditures for control and,
- to distribute the highest share of expenses for prevention by investing in the improvement of the processes for designing conforming logistics services.

Based on the analyses, it was concluded that even a change in the cost structure could contribute to a higher level of customer satisfaction with the studied logistics services in the food sector. Reducing the variation around the target function could not only contribute to higher customer satisfaction; moreover, it could reduce the overheads for nonconformities after delivery which are caused in particular by such a variation. Indeed, higher customer satisfaction could be achieved, if there is less variation with respect to the service wanted by the customer and delivered on time.

Acknowledgements

First, I would like to thank my children, without whose support I would not have been able to realize this idea.

I also thank all the logistics companies and logistics managers who participated in the survey and shared their experiences.

Conflict of interest

The author declares no conflict of interest.

Author details

Marieta Stefanova
Department "Management and Logistics", Nikola Vaptsarov Naval Academy, Varna, Bulgaria

*Address all correspondence to: m.stefanova@nvna.eu

References

[1] Pauliková A, Lestyánszka Škůrková K, Kopilčáková L, Zhelyazkova-Stoyanova A, Kirechev D. Innovative approaches to model visualization for integrated management systems. Sustainability. 2021;**13**:8812. DOI: 10.3390/su13168812

[2] Dimitrakieva S, Kostadinov O, Atanasova K. Multilevel demand for sea transportation; correlation between Baltic dry index (BDI) and coaster shipping prices for sea routes between Baltic seaports and Mediterranean seaports. Pedagogika-Pedagogy. 2021;**93**:141-148. DOI: 10.53656/ped21-7s.12corr

[3] De Marco A, Mangano G. Relationship between logistic service and maintenance costs of warehouses. Facilities. 2011;**29**:411-421. DOI: 10.1108/02632771111146323

[4] Waeyenbergh G, Pintelon L. A framework for maintenance concept development. International Journal of Production Economics. 2002;**77**:299-313. DOI: 10.1016/S0925-5273(01)00156-6

[5] Stank T, Keller B, Closs J. Performance benefits of supply chain logistical integration. Transportation Journal. 2001;**41**:32-46

[6] Yang J, Xie H, Yu G, Liu M. Achieving a just–in–time supply chain: The role of supply chain intelligence. International Journal of Production Economics. 2021;**231**:107878. DOI: 10.1016/j.ijpe.2020.107878

[7] Taguchi G, Rafanelli AJ. Taguchi on robust technology development: Bringing quality engineering upstream. Journal of Electronic Packaging. 1994;**116**:161. DOI: 10.1115/1.2905506

[8] Hussein M, Zayed T. Critical factors for successful implementation of just-in-time concept in modular integrated construction: A systematic review and meta-analysis. Journal of Cleaner Production. 2021;**284**:124716. DOI: 10.1016/j.jclepro.2020.124716

[9] Si T, Li HX, Lei Z, Liu H, Han S. A dynamic just-in-time component delivery framework for off-site construction. Advances in Civil Engineering. 2021;**2021**:e9953732. DOI: 10.1155/2021/9953732

[10] Asikhia A, Osinowo O, Kassim S. Integrating just in time theory, resource based view theory, and rational choice theory in enhancing managements' efficiency. South Asian Research Journal of Business and Management. 2021;**3**:14-22. DOI: 10.36346/sarjbm.2021.v03i01.002

[11] Dimitrakieva S, Koritarov AK. The human element as a factor for sustainable development of the offshore oil industry. Scientific Works - Naval University N Y Vaptsarov. 2018;**32**:65-68

[12] Kanev D, Toncheva S, Terziev V, Narleva K. Specific Aspects of Motivation of Seafarers. Rochester, NY: Social Science Research Network; 2017. DOI: 10.2139/ssrn.3144743. Available from: https://web.archive.org/web/20220111123810/https://papers.ssrn.com/sol3/papers.cfm?abstract_id=3144743

[13] Mednikarov B, Tsonev Y, Lazarov A, Lazarov A. Analysis of cybersecurity issues in the maritime industry. Information & Security: An International Journal. 2020;**47**:27-43. DOI: 20200924053840237

[14] Trichai P et al. Developing warehouse employee performance by applying the principles of 7R in logistics. Turkish Journal of Computer and Mathematics Education (TURCOMAT). 2021;**12**:4121-4126

[15] Nantee N, Sureeyatanapas P. The impact of logistics 4.0 on corporate sustainability: A performance assessment of automated warehouse operations. Benchmarking. 2021;**28**:2865-2895. DOI: 10.1108/BIJ-11-2020-0583

[16] Maddikunta PKR, Pham Q-V, Prabadevi B, Deepa N, Dev K, Gadekallu TR, et al. Industry 5.0: A survey on enabling technologies and potential applications. Journal of Industrial Information Integration. 2021;**26**:100257. DOI: 10.1016/j.jii.2021.100257

[17] Stoyanov V. Psihichen Stres v Organizatsiyata. 1st ed. Varna, Bulgaria: VSU Chernorizets Hrabar; 2011

[18] Diallo A, Khan ZU, Vail CF. Cost of quality in the new manufacturing environment. Management Accounting (USA). 1995;**77**:20-26

[19] Dimitrakiev D, Molodchik A. Digital platforms as factor transforming management models in businesses and industries. Journal of Physics: Conference Series. 2018;**1015**:042040. DOI: 10.1088/1742-6596/1015/4/042040

[20] Zafirova T. Impact of changes in the environment on the strategic aspects of crisis management. Izvestia Journal of the Union of Scientists - Varna Economic Sciences Series. 2021;**10**:184-192

[21] Zafirova T. Integration of the processes of strategic and crisis management in the conditions of crisis. Известия на Съюза на учените - Варна Серия Икономически науки. 2021;**10**:158-168

[22] Stoykova T, Zlateva D, Pashova S. Commodity science in modern market conditions. Economic Science, Education and the Real Economy: Development and Interactions in the Digital Age. 2020:427-439

[23] Christopher M. Logistics and supply chain management: Strategies for reducing cost and improving service. International Journal of Logistics Research and Applications. 1999;**2**:103-104. DOI: 10.1080/13675569908901575

[24] Cooper JC. Logistics strategies for global businesses. International Journal of Physical Distribution & Logistics Management. 1993;**23**:12-23. DOI: 10.1108/09600039310041473

[25] Mckinnon A. Integrated logistics strategies. In: Brewer AM, Button KJ, Hensher DA, editors. Handbook of Logistics and Supply-Chain Management. Vol. 2. Bingley: Emerald Group Publishing Limited; 2017. pp. 157-170

[26] Rao KR, Stenger AJ, Wu H-J. Training future logistics managers: Logistics strategies within the corporate planning framework. Journal of Business Logistics. 1994;**15**(2):249

[27] Kanev D. Commitment as a constraint to the pursuit of self-interest. Economics. 2017;**1**:3-20

[28] McGinnis MA, Kohn JW. A factor analytic study of logistics strategy. Journal of Business Logistics. 1990;**11**(2):41

[29] Waters D, Rinsler S. Global Logistics: New Directions in Supply Chain Management. London: Kogan Page Publishers; 2014

[30] Zafirova T. Strategic decisions in the crisis stages of the organization. Izvestia Journal of the Union of Scientists - Varna Economic Sciences Series. 2020;**9**:100-108. DOI: 10.36997/IJUSV-ESS/2020.9.1.100

[31] Tan K, Kannan VR, Handfield RB, Ghosh S. Supply chain management: An empirical study of its impact

on performance. International Journal of Operations & Production Management. 1999;**19**:1034-1052. DOI: 10.1108/01443579910287064

[32] Drury CM. Management and Cost Accounting. New York, NY: Springer; 2013

[33] Banerjee B. Cost Accounting: Theory and Practice. Delhi: PHI Learning Pvt. Ltd.; 2021

[34] NG T, Chung W. The roles of distributor in the supply chain–push-pull boundary. International Journal of Business and Management 2009;3:28-39. DOI: 10.5539/ijbm.v3n7p28.

[35] Taguchi G. Quality engineering in Japan. Communications in Statistics - Theory and Methods. 1985;**14**:2785-2801. DOI: 10.1080/03610928508829076

[36] Taguchi G. Robust technology development. Mechanical Engineering-CIME. 1993;**115**:60-63

[37] Taguchi G, Chowdhury S, Wu Y, Taguchi S, Yano H. Taguchi's Quality Engineering Handbook. Hoboken, N.J.: Livonia, Mich: John Wiley & Sons, ASI Consulting Group; 2005

[38] Shang JS, Li S, Tadikamalla P. Operational design of a supply chain system using the Taguchi method, response surface methodology, simulation, and optimization. International Journal of Production Research. 2004;**42**:3823-3849. DOI: 10.1080/00207540410001704050

[39] Shukla SK, Tiwari MK, Wan H-D, Shankar R. Optimization of the supply chain network: Simulation, Taguchi, and Psychoclonal algorithm embedded approach. Computers & Industrial Engineering. 2010;**58**:29-39. DOI: 10.1016/j.cie.2009.07.016

[40] Roy RK. Design of Experiments Using the Taguchi Approach: 16 Steps to Product and Process Improvement. New York: Wiley; 2001

[41] Kumar RS, Choudhary A, Babu SAKI, Kumar SK, Goswami A, Tiwari MK. Designing multi-period supply chain network considering risk and emission: A multi-objective approach. Annals of Operations Research. 2017;**250**:427-461. DOI: 10.1007/s10479-015-2086-z

[42] Addinsoft. XLSTAT 2021.3.1 Data Analysis and Statistics Software for Microsoft Excel. Paris, France: Addinsoft; 2021

[43] Christopher M. Logistics and Supply Chain Management. 1st ed. London: Financial Times/Irwin Professional Pub; 1992

[44] Kanev D. The Paradox of Information. Economics and Management. Vol. 14. BLAGOEVGRAD: Faculty of Economics, SOUTH-WEST UNIVERSITY "NEOFIT RILSKI"; 2018. pp. 59-73

[45] Sutrisno A, Andajani E, Widjaja F. The effects of service quality on customer satisfaction and loyalty in a logistics company. KnE Social Sciences. 2019;**2019**:85-92

[46] Thai V. Logistics service quality: Conceptual model and empirical evidence. International Journal of Logistics Research and Applications. 2013;**16**:114-131. DOI: 10.1080/13675567.2013.804907

[47] Wee HM, Blos MF, Yang W-H. Risk management in logistics. In: Lu J, Jain LC, Zhang G, editors. Handbook on Decision Making. Vol. 33. Berlin, Heidelberg: Springer Berlin Heidelberg; 2012. pp. 285-305. DOI: 10.1007/978-3-642-25755-1_15. Available from: https://web.archive.org/web/20180618045623/https://link.springer.com/chapter/10.1007%2F978-3-642-25755-1_15

Chapter 7

Managing Foodservice Quality in the Foodservice Industry

Lindiwe Julia Ncube

Abstract

Quality has become a value that enables businesses to survive and continue existing. Henceforth, food industries need to entrench quality into their business performance. Foodservice quality is characterized as a service that bears on its ability to satisfy stated or implied needs and service free of defects. Foodservice businesses are an integral part of social life, both biologically and socially, biologically as satisfying the nutrition requirements of the society and socially in terms of addressing socialization and esthetics-pleasure values. Therefore, by adopting quality approaches, food industry businesses may encourage customers' preferences for those businesses that diligently offer these services. Managing food service quality is a complex and challenging task requiring commitment, discipline, and emergent effort from everyone involved in food production processes. The task also requires the necessary management and administration techniques to continuously improve all processes (including quality control from raw material to finished product). Food industries need to be organizationally structured, establish policies and quality programs, measure customer satisfaction, use more quality tools and methodologies, embrace knowledge, apply techniques, and food safety programs to manage food quality. This chapter aims to describe the ISO 22000 system—widely used for quality management in the food industry.

Keywords: foodservice, food industry, quality management, customer satisfaction, food production

1. Introduction

Unsafe food is a risk for all, and consumers can become seriously ill; hence the food industry may face serious legal consequences. These constant problems call for additional strategies for decreasing and eradicating the risks. As food safety is a joint responsibility for all participating parties, communication and raising awareness of potential hazards through the entire food chain is crucial [1]. Recent research suggests that most microbial food contamination in the food market happens in developing countries than in developed countries. Indeed, most chemical food contamination and food adulteration occur in developing countries. Additionally, the misuse of food additives was a common problem in developing countries; and in developed and developing countries, mislabeling was a problem. Furthermore, the selling of outdated foods occurred in developing countries than in developed [2].

Food products are produced from farms or from food originating from farms. For example, food items such as bread, milk, meat, fruit, vegetables, and sugar originate from agriculture. Farmers grow, harvest, store, and transport food and food products raw materials to markets or processing plants, and transform them into various food items and products [3]. However, since the original standard was published over a decade ago, there have been substantial changes in how food is grown, transported, manufactured, and consumed. A study by [4] on food safety management systems (FSMS) performance in African food processing companies reported high microbiological and chemical contamination levels in most African food products, which exceeded the acceptable (legal) limits. In developed countries, innumerable deficiencies that affect the performance of FSMS in Africa were found at government, sectoral, retail, and organizational levels. For example, most companies (except for the exporting and large companies) hardly implemented HACCP and ISO 22000:2005.

The authors further recommended the use of measures such as the construction of risk-based legislative frameworks, strengthening of food safety authorities, and using ISO 22000:2005 for food safety management in the food industry. Indeed, consumers' food safety training was projected to be implemented by the government. The food sector had to develop sector-specific guidelines and third-party certification, while the food retailers had to develop stringent certification standards and impose product specifications. Food companies had to improve hygiene, apply strict raw material control mechanisms and production process efficacy, enhance monitoring systems assurance activities, and develop supportive administrative structures. Globally, it has been an accepted norm that food safety management systems be based on Hazard Analysis Critical Control Point (HACCP) principles, which is an internationally accepted FSMS. However, the implementation of HACCP in South Africa has been driven by the requirements of international trade—where foods are exported to developed countries such as Europe and the United States of America. A regulation requiring HACCP implementation was publicized in South Africa in the year 2003. However, the foodservice industries are not compelled to comply. According to [5], there is currently no force that pressurize the foodservice industry to implement formal food safety management systems in South Africa. Hence, the growing need for international traveling and hosting of international sports events dissected this industry.

Urbanization, consumer changes in eating habits, and travel have increased the number of people buying and eating food prepared in public places. As a result, globalization has triggered growing consumer demand for a wider variety of foods, resulting in an increasingly multifaceted and longer global food chain [6]. Food safety is becoming more critical with the demand for food to meet the rapidly growing world population. The food-to-table progression put a significant focus on food contamination prevention and maintaining good food quality standards. Each food supply chain is highly regulated by government agencies such as the FDA and the newly implemented Food Safety Modernization Act. The population growth results in increased agriculture and animal production to meet the increasing demand for food, subsequently creating opportunities and challenges for food safety. Food safety ensures that products delivered to consumers do not negatively impact their health [1]. Hence, failing to comply with the food safety regulations may result in foodborne diseases. Climate change is also influencing food safety. These challenges put greater responsibility on food producers and handlers to ensure food safety.

2. The food supply chain and food contamination

Globally, about 600 million people fall ill after eating contaminated food, and 420,000 die every year, resulting in the loss of 33 million healthy life years (DALYs). Safe food supplies support national economies, trade, and tourism, contribute to food and nutrition security, and underpin sustainable development [7]. The Centre for Disease Control, CDC, estimates that roughly 48 million people get sick, 128,000 are hospitalized, and 3000 die from foodborne diseases each year [8]. Due to the speed and product distribution range, local incidents can quickly develop into international emergencies. Severe foodborne illness outbreaks have occurred in every continent in the past decade, often augmented by globalized trade. For example, the contamination of ready-to-eat meat with listeria monocytogenes in South Africa in 2017/18, brought about 1060 cases of listeriosis and 216 deaths. At the same time, contaminated products were exported to 15 other countries in Africa, requiring an international response to implement risk management measures [9].

Food can become contaminated at any stage of the food chain including the production and distribution stages, and the primary responsibility to prevent food contamination lies with the food producers. The authors [10] reported that foods that are not prepared properly, or foods mishandled at home, in foodservice establishments, or at markets contribute to the majority of foodborne disease incidents. In addition to this, most consumers and food handlers may lack knowledge and understanding of the practice of basic hygiene measures when buying, selling, and preparing food items hence, their health and that of the wider community may be at risk of foodborne illnesses. Tracking food through all the supply chain stages has become more complex and difficult as consumers are distant from the farm. Therefore, farmers must ensure food safety when growing and processing food, and during food preservation and transportation. Due to the fact that food items travel long distances, food products are exposed to a greater possibility of contamination or spoilage [11]. **Figure 1** below shows the flow of food and services that begin at the input and farm production sector and extend along the food supply chain until they reach the consumer.

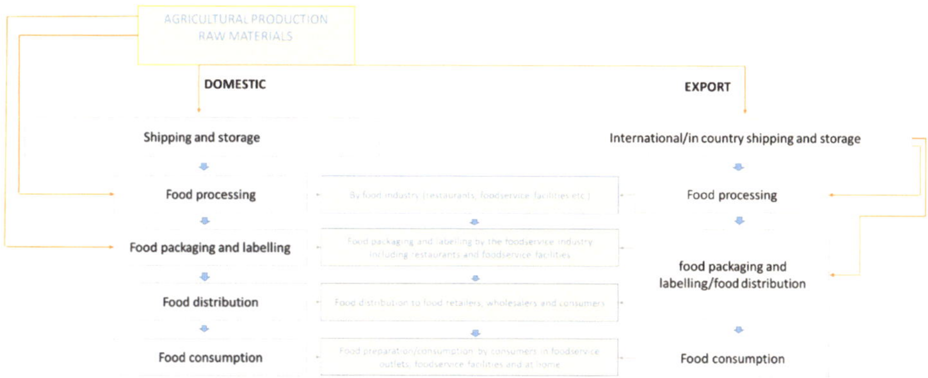

Figure 1.
The food supply chain.

2.1 The ISO 22000:2018 food safety management system

The ISO Food Safety Management System is malleable and can be utilized by all organizations involved in the food chain. Resulting from the usage, the food industry organizations may share common food safety language, thus reducing the risk of critical errors and maximizing the use of resources. Enterprises such as growers, transporters, packagers, processors, retailers, public and private catering companies, public and private food production and services units, bottlers, and restaurants can implement this standard [12].

The ISO 22000 was first published in 2005 to overcome the food crisis and facilitate harmonizing international food safety rules and regulations. The food industry has received the standard well, however the new food safety risks impelled for an updated version. Therefore, the latest edition was published on 19th June 2018 and upholds a concrete association with the Codex Alimentarius standards. The standard also addresses evolving food safety challenges and supports the organizational strategic direction with its Food Safety Management objectives.

The ISO 22000: 2018 is an internationally recognized food safety management system that can be used in any organization in the food chain. The latest ISO 22000:2018 is the newest food safety management standard bringing a common framework to all management systems. The ISO 22000:2018's framework can assist in aligning the different food safety management system standards, helps to keep uniformity, offer corresponding sub-clauses against the top-level structure and apply communal language through all standards. Also, the new standard in place, makes it easier for organizations to incorporate their food safety management system into the fundamental business processes and attract more participation from senior management [13]. The ISO 22000:2018 is more focused on top management to demonstrate leadership and commitment to the FSMS and food safety policies. Additionally, top management needs to ensure the consignment, communication, and understanding of all the responsibilities, across the organization. Also, top management has a responsibility to ensure that the adequate food safety importance is communicated and understood by all parties and that the FSMS achieves its intended outcomes.

2.1.1 Key changes of the ISO 22000 standard

The following critical changes of the IS0 22000 standard were identified by [14].

1. Organizational context: clause four (4)

This clause intends to provide a high-level, strategic understanding of the essential issues that can positively or negatively affect organizational food safety management. It countenances the organization to identify and understand factors and parties that affect the intended outcome(s) of the FSMS. It addresses the concept of preventive action, where organizations need to determine external and internal issues, problems, and risks relevant to their purpose. The issues should also include conditions that affect the organization, such as those highlighted in the general guidance on "issues" in Clause 5.3 of ISO 31000:2009.

The organization needs to identify the interested parties relevant to the FSMS. For example, these groups could include customers, consumers, suppliers, and non-government organizations. Determining their relevant needs and expectations is currently part of establishing the context for a FSMS. After the context has been established, the FSMS scope must be determined with various additional factors.

Finally, Clause 4 requirement is to establish, implement, maintain and continually improve the FSMS. This clause requires adopting a process approach. Although each organization may be different, documented information such as process diagrams or written procedures can be used to support this.

2. Clause eight (8): operation

With the exception of the HACCP Step one (Assemble HACCP team), Step 12 (Establishing documentation and record-keeping), which are addressed outside clause 8 (in clauses 5.3 and 7.5, respectively), Clause eight of ISO 22000: 2018 is more focused on the HACCP principles and steps. Internal audit—a verification procedure (Step 11), is covered in clause 9.2.

Clause 8.1—Operational planning and control

In clause 8.1, an organization's responsibility regarding processes required to meet requirements (plan, implement, control, maintain, and update) is highlighted in more detail. Examples of establishing criteria, implementing processes control, and demonstrating that processes have been carried out as planned are also provided, and a necessity to implement risk and opportunities assessment actions is introduced.

Clause 8.2—Prerequisite programs

For the effective implementation of any food safety system, prerequisite programs are crucial. The following differences were observed between the ISO 22000: 2018 and the ISO 2005 versions: (1) The word establish was replaced with the word "*update*" in the statement "establish, implement, maintain and update PRP(s)"; (2) Since the ISO/TS 22002 series prerequisites are not compulsory, the standard includes the terms "shall" which indicates a mandatory requirement and "should" which indicates a recommendation. This change denotes that the only prerequisites organizations must implement (mandatory) are presented in the standard (clause 8.2.4). (3). Hence, the prerequisites list in the ISO 22000:2018 is similar to those specified in the ISO 2005 version. The significant differences are the additional terms such as product information/consumer awareness and supplier approval (although it is apparent that most organizations may have some related procedure in place to address the purchased materials management presented in the ISO 2005 version).

Clause 8.3—Traceability system

The ISO 22000:2018 standard presents a list of topics to be considered when organizations establish a traceability system. For example, the reworking of materials/ products and the connection between received materials, ingredients, and intermediate products to the end products were not mentioned in the IS0 2005 version. The mandatory verification and testing of the traceability system's effectiveness exist in the ISO 22000:2018 version. Although this was not detailed in the ISO 2005 version, the guide for its application (ISO 22004:2014) encompassed making tests.

However, reference to quantities reconciliation (end products vs. ingredients) is a new requirement presented in the ISO 22000: 2018.

Clause 8.4—Emergency preparedness and response

Compared with clause 5.7 of ISO 22000:2005, the term "accident" was substituted with "incident." Clause 8.4 of ISO 22000: 2018 is more extensive than the one in the 2005 version. In the standard, it is currently compulsory to lessen the food safety emergencies, review, and update documentation. Additional examples of emergencies such as some new examples of emergencies were workplace accidents, public health emergencies, and interruptions of services like water, electricity, and refrigeration supply were added.

Clause 8.5.1, covers the preliminary steps to enable hazard analysis. In this clause, identifying raw materials, ingredients, product contact materials, end products (and their intended use), preparing flow diagrams, and describing processes are considered as the first step of hazard control.

Also, the information to be collected to conduct hazard analysis is more detailed in the IS022000:2018. It is well explained that, at a minimum, the information collected by the food safety team should include statutory, regulatory, and customer requirements, food safety hazards, and products processes and equipment. A new point requiring organizations to declare the source (e.g., animal, mineral, or vegetable) of their raw materials, ingredients, and product contact materials were also added. Therefore, the new word "place of origin (provenance)" replaced the wording "origin," which demands organizations to identify their product origin.

The IS0 22,000:2018 also specifies that the organization must include the introduction of processing aids, packaging materials, and utilities in the flow diagram. When describing hazards analysis, The IS0 22,000:2018 stipulates that the food safety team address the following expanded issues:

Clause 8.5.2—Hazard analysis

The implicit understanding is that the food safety team must conduct a hazard analysis based on the preliminary information covered in the ISO 22000:2005. However, it is explicitly stated at the beginning of the ISO 22000:2018 to highlight its importance.

Changes were also observed in the type of information used to identify food safety hazards. The ISO 2018 food safety standard requires organizations to use internal information such as epidemiological, scientific, and historical data, statutory, regulatory, and customer requirements to identify food safety hazards. Therefore, instead of only focusing on the steps preceding and following the specified operation, organizations must consider all steps in the flow diagram, including the people involved.

In conducting a hazard assessment, organizations must determine the likelihood of occurrence prior to applying control measures and evaluate the severity concerning the intended use.

After identifying the hazards, determining acceptable levels, and hazard assessment, the step that follows is to select and categorize control measures. The following aspects to consider when selecting and categorizing control measures are available in the ISO. However, most of them are similar to the ISO 2005 version. Notably, three issues are more critical: (1) assessing the practicability of creating assessable critical limits and measurable/observable action criteria. This is similar to what was also previously stated in the ISO 22004:2014: (2) To assess the viability of applying well-timed improvements in case of failure (3) and using the external requirements to select control measures.

Clause 8.5.4—Hazard control plan (HACCP/OPRP plan)

In the ISO 22000:2018 standard, information that was previously separated into two clauses: *Establishing the operational prerequisite programs* and *Establishing the HACCP plan is combined*. This assists in recognizing that the Hazard Control Plan must include a critical limit(s) at CCP and action criteria for Operational Prerequisite Programs (OPRP).

This standard presents the need to document the monitoring methods used in monitoring systems. The standard also augments the probability of utilizing *comparable methods* for calibration *to verify reliable measurements or observations for* OPRPs.

Clause 8.6—Updating information specifying the PRPs and the hazard control plan

The clause in the ISO 22000:2018 standard remains similar to clause 7.7 in the 2005 version. Above and beyond using a hazard control plan to substitute what was previously considered operational PRP(s) and HACCP plan, it announces that after establishing the hazard control plan, organizations need to update raw materials, ingredients, and product-contact materials characteristics.

Clause 8.7—Control of monitoring and measuring

The ISO 22000:2018 standard's clause 8.7 was adjusted to make it more explicit. it has been declared that for monitoring and measuring PRP's hazard control plan, organizations must provide evidence that specified monitoring and measuring methods and equipment are adequate to ensure the monitoring and measuring procedures. The clause is more demanding for monitoring and measuring software as it requires organizations to validate its adequacy prior to use and when it is changed/updated.

Clause 8.8—PRPs and the hazard control plan verification

There are three differences identified in this clause: (1) The list of the constituents of the verification activities in ISO 22000:2018 corresponds to clause 7.8 of the ISO 2005 except that implementation, and the PRP's effectiveness (s) must be confirmed. Hence, the rewording of operational OPRP(s) and HACCP plan to hazard control plan is also to be noted. (2) It is mandatory in the ISO 22000:2018 standard that organizations must warrant the objectivity of the person who does the verification activities (3) Every time nonconformity is found in testing final manufactured goods or natural process samples, the ISO 2018 version postulates the necessity to take corrective actions.

Clause 8.9—Product control and process nonconformities

It is well explained in the clause that organizations must ensure that process nonconformities are addressed. Clause 8.9.2.4. of this standard clearly explains the information reserved to describe corrections made.

In the ISO 2005, the clause indicated that only designated persons (with competence and authority) might evaluate nonconformities and initiate corrections and corrective actions, which was dispersed throughout the clause. However, in the ISO 22000:2018 version, the clause is placed at the beginning of the title. Organizations must also review nonconformities identified by consumers or in regulatory inspection reports. In contrast, only customer complaints were given as examples in the 2005 version.

Clause 8.9.4.3—dispositions of non-conforming products, it is required that any product that fails to remain within critical limits at CCPs not be released.

Clause 9—Performance evaluation

Clause 9 covers the evaluation of how the system performs. The clause covers the monitoring, measurement, and analysis; including valuation—a new item, which forces organizations to indicate what and when monitoring and measurement should take place, and how, when, and by whom will the results be analyzed and evaluated. Also, when conducting internal audits, and after introducing the audits program (which must be used to verify the FSMS against the food safety objectives and policy), the clause expects organizations to recognize the importance of integrating the changes in the FSMS and the results of monitoring and measurement. The clause also highlights the importance of reporting the audit results to the food safety team and pertinent management. Items such as nonconformities and corrective actions, the performance of external providers, adequacy of resources, and opportunities for continual improvement were added to the management review section of clause nine. The internal and external issues are covered as inputs for addressing any applicable change essential for the FSMS mainly, changes including decisions and actions related to output continual improvement opportunities.

Clause 10—Continuous improvement

The is a new sub-clause that was added to clause ten of ISO 22000:2018 which gives clear guidance on addressing nonconformity and corrective action. The sub-clause is similar to the one in ISO 2005 standard however, the need for an organization to continuously improve the effectiveness of the FSMS and its suitability and adequacy was added in the 2018 version. No relevant changes were found in the last clause of the system (*updating the FSMS*).

The standard considers these changes essential to help organizations reduce food safety hazards and beneficial in alignment with the organization's strategic direction for the food safety management system.

For effective implementation, ISO 22000:2018 is developed on a high-level structure and enables an organization to use a process approach (PDCA) cycle along with risk-based thinking. This high-level structure is beneficial in the integration of other management standards. This standard enables an organization to control food safety hazards along the food chain. This "Norm" also applies to all types and sizes of organizations in the food industry.

2.2 Process approach and risk-based thinking

In addition to making ISO 22000 and the resulting FSMS easier to integrate with other ISO management systems, the IS0 22000:2018 introduces the Plan-Do-Check-Act (PDCA) cycle and risk-based thinking. ISO 22000 can help organizations reduce risk exposure and improve safety by combining organizational and operational risk management into one management system. For example, combining PDCA and risk-based thinking to manage business risk with HACCP to identify, prevent and control food safety hazards. Organizationally, this approach provides the opportunity to consider all the different things (both good and bad) that might impact the company [15]. The approach allows for prioritization of the FSMS objectives that it is implemented to accommodate the effects of these risks. On the operational side, risk-based thinking and implementation are based on HACCP principles associated with food safety management. **Figure 2** below shows how they can be seen in the diagram below.

The PDCA Cycle in the food industry.

The PDCA cycle is comprised of the Plan, Do, Check and Act concepts [16], and the cycle is suggested for beginning a new improvement project, implementing changes, continuous process improvements, and planning data collection and analysis (ASQ) [17]. There are four main stages for the PDCA cycle: Plan, Do, Check and Act [18]:

The "Plan Stage": The problem is identified during the planning stage, and data on the intended root causes are collected. Lastly, the intended outcomes are selected, as well as developing a plan to meet the outcomes. The planning stage is performed to assist in evaluating and forecasting problems that might occur during the execution stage and provide alternative modification strategies to prevent possible problems.

The Implementation stage: In this stage, the solution to the problem is developed and implemented, and the results are measured.

The "Check Stage": During this stage, the status and effectiveness of the plan are implemented. For example, checking whether the intended outcome was met and the reasons for not meeting the intended outcome if the outcome was not met.

Act Stage: This is the final stage of the PDCA cycle process and the first stage for the next cycle. In this stage, solutions are reviewed against standards, and actions are taken; information and results about the process and recommended changes are documented.

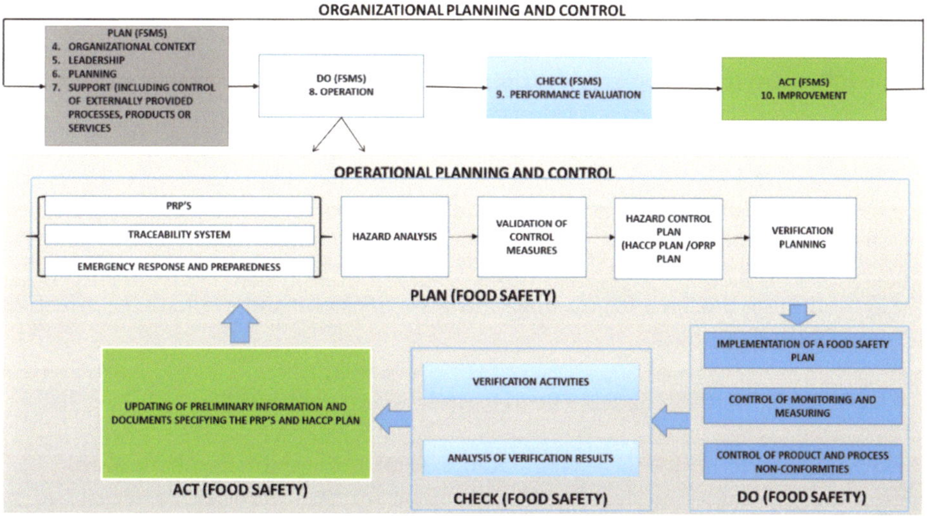

Figure 2.
The food safety PDCA cycle.

2.3 Uses of the PDCA cycle in the food industry

The PDCA cycle is a managerial decision-making strategy that guarantees that process, product, or service goals are accomplished [19]. The plan involves establishing goals and procedures to achieve the set goals. All employees are expected to meet performance standards and behave appropriately hence, contributing to the effective achievement of the goals set by the organizational management.

2.3.1 Corrective actions

Corrective action is a practice whereby management communicates with organizational employees to improve their behavior after other methods such as coaching and performance appraisal failed. Corrective action is also considered an aspect of quality management that aims to remedy a task, process, product/service, or a person's behavior when any deviation from an intended plan is identified. Once the deliverables deviate from the required output, corrective actions can be applied to the entire project whether tangible or service. For example, in Human resources for higher education institutions, corrective action also applies to individual employees and functions to communicate aspects of attendance, unacceptable behavior, or performance that require improvement.

For corrective actions, [20] suggest not using the PDCA cycle as a whole however, it must be broken down into the following seven food safety management system steps for corrective action procedures: **Planning Stage -Step one and two: Understanding the system requirements and planning the process**.

In the course of the planning stage, the managers must understand the FSMS, the nature of the deviations, and a root cause analysis must be conducted to determine the cause of the problem. The risk and consequences of the deviations must be frequently evaluated.

Do Stage-Steps 3, 4, and 5: Develop and Document, Conduct Training, and Implementation.

After the root cause analysis of the problem is determined, planning to correct the deviations can be performed. When developing corrective action, [21] is recommending the following actions by organizations:

- Determining the actual cause for the deviations.
- Developing action plans to ensure the effectiveness of corrective actions and preventative actions.
- Determining the need for training and ways to ensure and evaluate the effectiveness of the training.
- Determine whether there is a need to update the procedures.

Corrective actions may be implemented as soon as the right ones have been determined, procedures updated, and training performed. Implementation of the corrective actions could take account of retesting the food products, confirming and observing procedures, and revising food safety records to make certain employees follow the procedures.

Check Stage-Step 6: Test/check the system.

After a few cycles of corrective actions implementation, ensuring that the corrective actions become a permanent solution is essential. The check stage can be done by gathering employee feedback, employee interviews, reviewing the documentation, and monitoring employee activities.

Act Stage-Step 7: Adjust and improve.

In the last stage, the effectiveness of the corrective actions and preventative actions are reviewed, and the efficiency and effectiveness of the corrective actions are determined. Based on data obtained from the users, the corrective actions may be improved.

2.3.2 Internal audit

Internal audit is a fundamental process in any food safety management system because it helps evaluate its functioning as intended. It enables checking for the process systems and validates processes against their intended result—furthermore, internal audit assists in preparation for third-party audits [22]. This section will review the internal audit process from the PDCA cycle perspective.

The Internal Audit comes into play during the "Check" stage and allows checking of the process put in place during the "Plan" and "Do" Stages. During internal audits data is collected using document reviews, observations, and employee interviews, and used as evidence of the effectiveness of the implementation of the FSMS hence, the process allows for a full systemic review of the FSMS.

The envisioned internal audit purpose is to assure that one finds and resolves the deviations or gaps in the food safety management system before the third-party audit identifies them. The deviations found during the internal audits are documented and further reviewed for immediate corrections and follow the Corrective Actions and Preventative Action procedures. The Internal Audit and Corrective Actions procedures are inter-connected. That is the "Act" stage, where information gathered can be used to improve the organization's food safety management system [23].

2.4 Advantages and disadvantages of PDCA cycle

The PDCA cycle has its advantages and disadvantages. One of its advantages is that it is intended to be repeatable and reused as necessary, thus permitting continuous improvements. It allows for mistake documentation, assessment, and rectifications that can be frequent as needed. Any changes can be tested on a small scale before being implemented on a large scale [24].

According to [25], one of the disadvantages to the PDCA Cycles is that including the actual work only comes in the action plan; it can take very long and even get stuck at the "Plan" stages while being analyzed and not proceeding to the next step.

3. Conclusions

The literature discussed above evidence indicates the importance of implementing a food safety management system. Developing and implementing a food safety management system can assist any type of food production and manufacturing organization to ensure that they provide services or safe food products to their customers. As such, it is apparent that each organization can develop a FSMS relevant and suitable to address the needs of the interested parties. For effective implementation of the FSMS, ISO 22000:2018 was developed on a high-level structure where organizations use a process approach (PDCA) cycle along with risk-based thinking. The high-level structure assists organizations in integrating the FSMS with other management standards. The ISO 22000:2018 standard applies to all types and sizes of organizations in the food industry and supports organizations to control food safety hazards along the food chain. This standard also applies. Therefore, it is critical for food safety and quality management in the food industry.

Author details

Lindiwe Julia Ncube
Sefako Makgatho Health Sciences University, Pretoria, South Africa

*Address all correspondence to: lindiwen@yahoo.com

References

[1] Uçar A, Yilmaz MV, Çakıroğlu FP. Food safety—Problems and solutions. In: Significance, Prevention and Control of Food Related Diseases. London: IntechOpen; 2016. DOI: 10.5772/63176

[2] Gizaw Z. Public health risks related to food safety issues in the food market: A systematic literature review. Environmental Health and Preventive Medicine. 2019;**24**:68. DOI: 10.1186/s12199-019-0825-5

[3] The European Food Information Council (EUFIC). From Farm to Fork: Farming, The Beginning of the Food Chain. Brussels, Belgium: EUFI; 2006. Available from: https://www.eufic.org/en/food-production/article/from-farm-to-fork-the-farm-beginning-the-food-chain [Accessed: December 15, 2021]

[4] Kussaga JB, Jacxsens L, Tiisekwa BP, Luning PA. Food safety management systems performance in African food processing companies: A review of deficiencies and possible improvement strategies. Journal of Science Food Agriculture. 2014;**11**:2154-2169. DOI: 10.1002/jsfa.6575

[5] Jackson L. Food Safety Management and Associated Food Handler Behaviours in a Prominent South African Entertainment Facility. (Thesis). Free State: Central University of Technology;

[6] Pang T, Guindon GE. Globalization and risks to health. EMBO Reports. 2004;**5**(1):S11-S16. DOI: 10.1038/sj.embor.7400226

[7] World Health Organisation. Food Safety. Geneva, Switzerland: WHO; 2020. Available from: https://www.who.int/news-room/fact-sheets/detail/food-safety [Accessed: January 25, 2022]

[8] Scallan E, Hoekstra RM, Angulo FJ, Tauxe RV, Widdowson M, Roy SL, et al. Foodborne illness acquired in the United States—Major pathogens. Emerging Infectious Diseases. 2011;**17**(1):7-15. DOI: 10.3201/eid1701.p11101

[9] Smith AM, Tau NP, Smouse SL, Allam M, Ismail A, Ramalwa NR, et al. Outbreak of listeria monocytogenes in South Africa, 2017-2018: Laboratory activities and experiences associated with whole-genome sequencing analysis of isolates. Foodborne Pathogens and Disease. 2019;**16**(7):524-530. DOI: 10.1089/fpd.2018.2586

[10] Kibret M, Abera B. The sanitary conditions of food service establishments and food safety knowledge and practices of food handlers in Bahir Dar town. Ethiopian Journal of Health Sciences. 2012;**22**(1):27-35

[11] WHO. How Food Gets Contaminated—The Food Production Chain. Geneva, Switzerland: WHO; 2020. Available from: https://www.cdc.gov/foodsafety/production-chain [Accessed: January 27, 2022]

[12] Dubbeling M, Bucutariu C, Santini G, Vogt C, Eisenbeiß K. City Region Food Systems and Food Waste Management- Linking Urban and Rural Areas for Sustainable and Resilient Development. RUAF Foundation and UN FAO. Bonn and Eschborn: GIZ; 2016. Available from: http://www.ruaf.org/projects/city-region-food-systems-and-food-waste-management-linking-urban-and-rural-areas [Accessed: February 12, 2022]

[13] ISO 22000:2018 (EN). Food Safety Management Systems—requirements for any Organisation in the Food

Chain. Geneva, Switzerland: ISO. Available from: https://www.iso.org/obp/ui/#iso:std:iso:22000:ed-2:v1:en [Accessed: February 10, 2022]

[14] Soares NF. ISO 22000:2018—What's New (Part 1). Geneva, Switzerland: ISO; Available from: https://nunofsoares.com/iso-220002018-whats-new-part-1/ [Accessed: February 02, 2022]

[15] ISO. ISO 22000 Implementation Guide. Geneva, Switzerland: ISO. Available from: https://www.nqa.com/medialibraries/NQA/NQA-Media-Library/PDFs/NQA-ISO-22000-Implementation-Guide.pdf [Accessed: February 03, 2022]

[16] Hussain N, Whyman M, Gold JKO, Henshall A. How to use the Deming cycle for continuous quality improvement: Process street: Checklist, Workflow and SOP software? Process Street. San Francisco, California: Process.st; 2020. Available from: https://www.process.st/deming-cycle/ [Accessed: December 22, 2021]

[17] Skhmot N. Using the PDCA Cycle to Support Continuous Improvement (Kaizen). San Francisco, California: The Lean Way; 2017. Available from: https://theleanway.net/the-continuous-improvement-cycle-pdca [Accessed: January 29, 2022]

[18] Patel PM, Deshpande VA. Application of plan-do-check-act cycle for quality and productivity improvement—a review. International Journal for Research in Applied Science & Engineering Technology (IJRASET). 2017;**5**(1):197-201

[19] FoodBio Network. The Deming Wheel (PDCA-Cycle). Johannesburg: FoodBio Network; 2020. Available from: https://foodbionetwork.com/2020/08/23/the-deming-wheel-pdca-cycle/ [Accessed: February 05, 2022]

[20] Brooks B. 7 key Steps to Plan and Implement an Effective Corrective Action System. Santa Barbara, California: QAD Blog; 2021. Available from: https://www.qad.com/blog/2020/01/effective-corrective-action-system [Accessed: January 05, 2022]

[21] Gemechu ED, Sonnemann G, Remmen A, Frydendal J, Jensen AA. How to implement life cycle management in business. In: Sonnemann G, Margni M, editors. Life Cycle Management. LCA Compendium—The Complete World of Life Cycle Assessment. Dordrecht: Springer; 2015. DOI: 10.1007/978-94-017-7221-1_4

[22] Safe Food Alliance. The Benefits of an Internal Auditing System. Sacramento, California: Safe Food Alliance; 2020. Available from: https://safefoodalliance.com/food-safety/the-benefits-of-an-internal-auditing-system/ [Accessed: February 02, 2022]

[23] Clarke I. What is an Internal Audit? Answers to Common Questions. Denver, CO: Linford & Co LLP; 2018. Available from: https://linfordco.com/blog/what-is-internal-audit/ [Accessed: February 03, 2022]

[24] Paiva CL. Quality Management: Important Aspects for the Food Industry. London: IntechOpen; 2013 . Available from: https://www.intechopen.com/chapters/41652 [Accessed: March 03, 2022]

[25] Chapman L. PDCA Benefits. St, San Francisco, California: Bizfluent; 2019. Available from: https://bizfluent.com/info-8605815-pdca-benefits.html [Accessed: February 04, 2022]

Chapter 8

Updates on Software Usability

Ahmed Mateen Buttar and Muhammad Majid

Abstract

Network security ensures that essential and accessible network assets are protected from viruses, key loggers, hackers, and unauthorized gain. Interruption detection system (IDS) is one of the most widespread significant network tools for network security management. However, it has been shown that the current IDS is challenging for network professionals to use. The interface, which assists users in evaluating the software usability, is a crucial aspect that influences the effectiveness of IDS, while security software such as IDS is effective. Usability testing is essential for supporting users in successful interaction and IDS utilization because the user finds it difficult to assess and use the output quality. Usability engineers are responsible for the majority of usability evaluations. Small and large business software engineers must master multiple usability paradigms. This is more difficult than teaching usability engineers how to write software. The Cognitive Analysis of Software Interface (CASI) technology was created as a solution for software engineers. Furthermore, this system aids software engineers in evaluating IDS based on user perception and evaluation perspectives. This study also discusses a large body of research on software interfaces and assessment procedures to evaluate novel heuristics for IDS. Finally, additional interface challenges and new ways for evaluating programmed usability are discussed. Topic Subject Areas: Intrusion Detection System (IDS) Usability.

Keywords: IDS, usable security, heuristics evaluation, cognitive analysis, SDLC

1. Introduction

The Internet has evolved recently, and users have been confronted with network security issues. Many firms are concerned about protecting their valuable and private data from dangers inside and outside society. Human and organizational variables, according to research, have an impact on network security. Security is a challenge for network practitioners. As a result, they employ specific tools, such as intrusion detection systems, firewalls, antivirus software, and Nmap, among others, to reduce or completely eradicate incursion. Interruption detection system (IDS) is critical in detecting malevolent behavior quickly and supporting real-time attack response. But many intrusion detection systems are challenging to use, and users cannot take advantage of all of their functions. These issues must be to boost IDS efficiency. One option is to create an effective solution that may assist network administrators in controlling security. Usability is a critical factor that has a significant impact on security management. Software developers acknowledge that the software interface is critical to its success. In terms of software usability, this success can be measured. Usability discusses the quality of a user's know-how when interacting with products or systems, including websites, software, devices, or applications. Usability is an essential term in the human-computer interaction (HCI) discipline. One option to overcome the issues of IDS is to create a user-friendly interface to assist network experts in effectively managing security.

2. Usability

The way businesses and people interact has altered due to Twitter, which was created in 2006. Therefore, usability is a crucial aspect of software quality. It is described by ISO 9241 as the degree to which confident clients can use a product to succeed in preset goals with sufficiency, competency, and fulfillment in an exact set of users. The capacity of the product item is to be perceived, learned, and enjoyed by the client when used in endorsed settings [1]. Definitions emphasize convenience as a key component of programming that enables users to do tasks quickly and without any issues. Nielson lists the five characteristics of learnability, memorability, and adaptability essential to usability.

According to the client's point of view, ease of use ensures that the result produced is easy to measure, use, and recollect. The objective of effectiveness, adequacy, security, utility, learnability, and memorability is reached. HCI's center has grown, and the errand-focused convenience worldview has expanded to include a refined and epicurean client experience UX worldview.

Various methodologies assess the convenience of programming in ease of use. There are two convenience testing techniques: ease-of-use assessment and ease-of-use testing strategies. Convenience issues are perceived by ease-of-use professionals in convenience assessment. However, ease-of-use issues are found in clients' perceptions of how they utilize the framework and connect with the product interface in ease-of-use testing strategies.

3. Heuristic evaluation

Users believe that testing applications are an essential step in making them better. Heuristic evaluation is a well-known low-cost approach to usability testing. According to some authors [2], heuristics and recommendations can be used interchangeably. Up to 60% of the usability flaws were identified [2]. However, a collection of heuristics has never been designed expressly for evaluating security-related applications. The project's objective at this stage is to create criteria for assessing usability for this particular problem space. These strategies are used to evaluate the quality of existing products and to discover demands that products can meet. For the heuristic evaluation, users selected snort as a candidate application. Snort is a simple yet popular intrusion detection system. It can track and record IP traffic. Because it is a command line-based tool, users decided to use a web-based application. Silicon defense has created a user-interface front end.

Usability testing can be done in various ways, including cognitive walkthroughs, formal usability inspections, heuristic evaluations, and pluralistic walkthroughs. Heuristic evaluation was used to assess the usefulness of IDS additionally, and heuristics are specifically developed for IDS. Heuristic evaluation entails a small group of convenience specialists looking through the framework and comparing it with usage standards. Customers can assess the ease of use of IDS and identify and address usability matters more successfully by employing new heuristics.

However, given that assessment can be expensive in terms of time, money, and human exertion, semi-mechanized or fully robotic evaluation is a viable option to improve current assessment approaches. Additionally, research reveals [2] the significance of a particular framework in facilitating convenience assessment.

Regarding programming projects, utilizing a computerized or self-loader audit framework is basic to guarantee the venture's adequacy, mainly when the cutoff time is tight. To guarantee project achievement, one choice is to develop further manual evaluation utilizing robotization or semi-mechanization. This will help assessors follow guide cycles and catch more mistakes significantly quicker. Finally, the assessment's discoveries are summed up and introduced to the planning group, alongside ideas for development.

4. Intrusion detection system

IDS is continuously monitoring and evaluating events within a computer system or organization for precursors to upcoming events, such as infringement or dangers of violation of PC security guidelines, acceptable use approaches, or usual security rehearses. The interruption detection system (IDS) is an organization-specific security arrangement that screens the organization for unapproved access. In IDS, users deal with two essential issues: The first is related to best-in-class, and the second is related to the state of training; the strategies or calculation used to recognize the assault, and the human point of interaction that permits security overseers or organization specialists to identify and answer the assault rapidly. Different techniques and calculations are being created to expand IDS's capacity to distinguish unapproved network access as depicted [3] in **Figure 1**. On the other hand, when the UI is not good, functional programming frequently fails.

Traditionally, IDS users have been network officers; however, the benefits of employing IDS have turned out to be so well-known that users today range from PC users who need to monitor network traffic passing through their business. There are three different types of clients: network administrators, security-trained professionals, and software engineers. An organization developer's skill is the ability to design networks with traffic in mind. While LAN professionals manage and support an organization's LAN, security professionals have a comprehensive understanding of

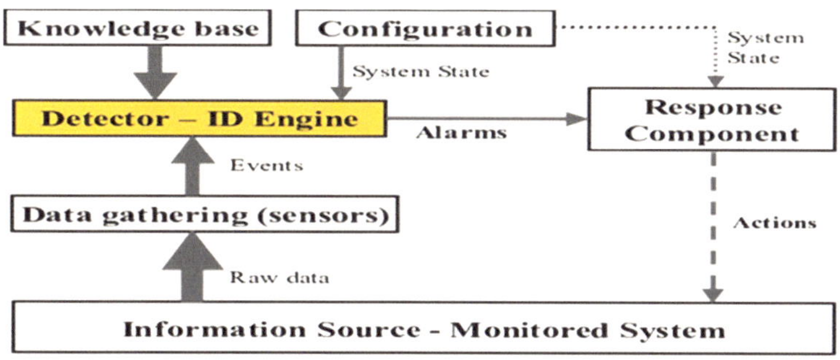

Figure 1.
IDS architectural data flow diagram.

technology, including anti-infection, strong validation, interruption discovery, and biometrics. While interruption discovery frameworks watch out for networks for possible antagonistic exercises, they are inclined to deception. Thus, when ventures first carry out IDS items, they should twist them. It involves properly arranging interruption documentation frameworks to recognize genuine organization traffic and noxious exercises.

The interruption counteraction frameworks screen network parcels entering the frameworks to search for pernicious action and immediately give cautioning signals.

4.1 Intrusion detection system classification

4.1.1 NIDs (network intrusion detection system)

NIDs are implemented at a prearranged point in the organization to examine traffic from all associated devices. It inspects completely subnetwork correspondence and looks at it as an information base of perceived dangers. An alert can be given to the chairman at whatever point an attack has been identified or bizarre conduct has been found. To determine whether someone is attempting to breach the firewall, NIDs are introduced on the subnet where firewalls are installed.

4.1.2 HIDs (host intrusion detection system)

HIDs are network interruption recognition frameworks that suddenly spike demand for independent hosts or gadgets. HIDs just screen the gadget's approaching and active bundles, alarming the manager by assuming that dubious or malignant action is found. It thinks about the ongoing depiction of the past preview of existing framework records. An alarm is given to the director of the insightful framework records that have been adjusted or eliminated. HIDs should be visible in real life on tactical equipment that is not designed to change its format.

4.1.3 PIDS (protocol-based intrusion detection system)

PIDS is a structure that is frequently seen at the front end of the server, supervising and deciphering the communication between the client/contraption and the server. By consistently examining the HTTPS show stream and enduring the connected HTTP show, it attempts to connect to the web server. This system would need to remain in collaboration for HTTPS to be used because HTTPS is not mixed until it manifests at the web show layer.

4.1.4 APIDS (application protocol-based intrusion detection system)

A framework that exists inside an assortment of servers is called APIDS. It identifies interruptions by checking and investigating application-explicit convention traffic, for instance, the way the SQL convention the work communicates with the information base on the web server.

4.1.5 HIDS (hybrid intrusion detection system)

HIDS is made by joining at least two interruption identification advancements. First, the hosting specialist or framework data are converged with network data to get

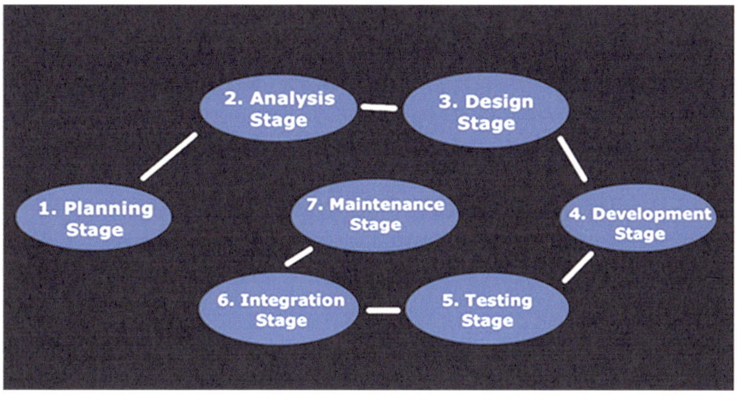

Figure 2.
Life cycle or system flow diagram.

an entire viewpoint on the organizational frameworks in the crossover interruption identification framework. The crossover interruption discovery framework shown in **Figure 2** is more powerful [4].

5. Detection method of IDS

5.1 Signature-based method

Signature-put-together IDS recognizes attacks based on specific examples in network traffic, that is, the quantity of 1 s or 0 s. It additionally identifies malware given the infection's recently realized hazardous guidance arrangement. Marks are models that IDS perceives.

While fresh malware attacks are attempting to be recognized because their model signature is dark, signature-based IDS can quickly identify attacks whose model signature already exists in the system.

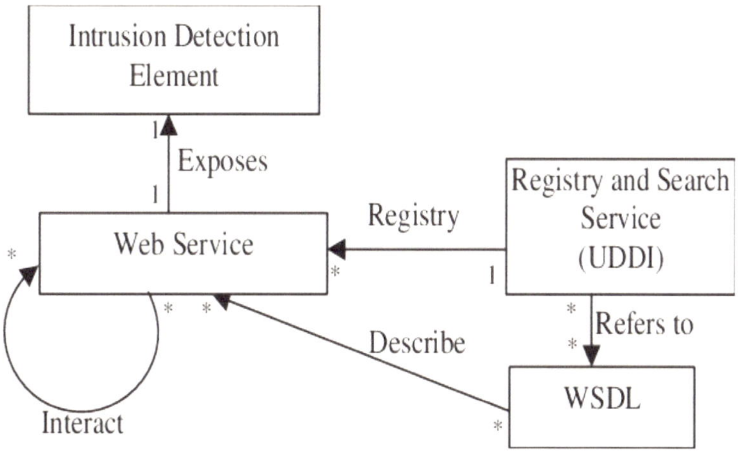

Figure 3.
Internal life cycle model.

5.2 Anomaly-based methods

A peculiarity-based IDS was designed to identify the hazards posed by dark malware because new malware is being produced at a rapid rate. A dependable development model is fostered by computer-based intelligence in characteristic-based intrusion detection systems IDS, and anything that enters is diverged from that model and stepped suspect if it is not detected. In contrast, because these models may be prepared by the applications and equipment plans, **Figure 3** represents AI-based IDS that has a prominent regular property [5].

6. Software interface cognitive analysis

According to studies [6], the software is currently being developed by businesses that can test ease of use completely on their own or with very little assistance from humans. This is because many businesses dislike hiring convenience specialists, because it examines and evaluates customer discernments, such as what clients think of the connection point, how they associate with it, and how they believe it should be. CASI is a strategy that helps programmers and IT clients assess user interface without needing to enlist convenience specialists. CASI does not just recognize convenience shortcomings in a framework's connection point, yet, in addition, makes suggestions to further develop it and make it more intelligent for the client.

The product connection point is essential in deciding the ease of use of programming. Every IDS interface in CASI is evaluated for usability, and flaws are identified. To show this test, the proposed IDS heuristics are executed on the IDS connection point. Proposed heuristics are installed in CASI and run on each ID connection point to recognize and suggest ease-of-use issues. The IDS connection point is picked and organized by the client's prerequisites at the primary level. Those IDSs that are as yet being created can be used to work on their helpfulness during the advancement stage. The authors believe that users should choose a single IDS point of interaction and then execute the suggested heuristics at the following level.

7. IDPS methodologies

IDPs utilize various ways to deal with distinguishing changes in the frameworks they screen. Outside assaults or interior staff abuse can cause these changes. Four procedures stand apart among the numerous others and are ordinarily utilized. The four options are as follows:

- Signature-based,
- Oddity-based,
- State full convention examination-based, and
- Half-and-half-based.

The half-breed strategy, which consolidates various approaches to give prevalent location and avoidance capacities, is utilized by most of the current IDPS frameworks.

Each methodology follows a similar broad framework; the main variations lie in how they analyze data from the observed environment to determine whether an agreement violation has happened as explained in **Table 1** [7].

7.1 Anomaly-based methodology

The system of irregularity-based procedure analyzes noticed action to a gauge profile. The gauge profile is the practical framework's learned typical way of behaving that is created during the learning time frame when the IDPS learns the climate and produces an ordinary profile. This climate can incorporate organizations, clients, frameworks, and other things. Fixed or dynamic profiles are accessible. A decent profile stays consistent over time, yet a robust profile differs when the practical frameworks change. A robust profile adds critical upward to the framework because the IDPs continue refreshing it, making it defenseless against avoidance. By spreading the assault throughout an extensive period, an aggressor can sidestep the IDPS that utilizes a powerful profile.

7.2 Signature-based methodology

Signature-based approach thinks about noticed marks to marks put away on record. An information base or a rundown of known assault marks may be remembered for this record. Any signature that matches the marks on a document in the checked climate is set apart as a security strategy infringement or an assault. Since it does not assess each activity or organization traffic on the observed climate, the mark-based IDPS has a low upward. It simply looks at the information base or document for perceived marks. Unlike irregularity-based approaches, signature-based systems are simple to apply since they do not require studying the climate. This technique looks, investigates, and analyzes the items in caught network bundles for known danger marks. It likewise thinks about conduct marks to those that are allowed. The frameworks' known hazards payload is also broken down using a mark-based approach. Signature-based systems are very effective against known attacks and infringements, but they cannot identify fresh attacks unless new marks are introduced.

Solar winds security	Top features	Common features
Kismet	Risk assessment report trend	Export to the PDF
Zeek	Various plugins available	Monitor SNMP Traffic
Open DLP	Customizable policy scripts	Agents
Sagan	Identifies at rest data across thousands of system	Snort-like design
Suricata	Compatible with rule management software	Detects complex threats
Security Onion	Supported standard output and input formats	Traffic pattern insight
Security Level	NIDS/HIDS Hybrid	Automated asset discovery

Table 1.
Best intrusion detection software tools and features.

Signature-based IDPSs are not difficult to overcome because they depend on existing assaults and require the utilization of new marks before they can identify new ones. Attackers can easily lose signature-based identification frameworks if they modify known attacks and target frameworks that have not been updated with new marks that identify the alteration. Signature-based procedures demand significant resources to maintain awareness of the potentially endless number of changes to known risks. Systems based on signatures are easier to modify and enhance since the markings or rules used to display them can still change.

7.3 Hybrid-based methodology

1. With the advancing assortment of assaults, the two old-style IDSs referenced above can safeguard our data frameworks. New strategies for joining different interruption location frameworks to further develop their adequacy have been planned. The inquiry has shown that consolidated calculations perform well compared with only calculations [8].

2. The objective of half-and-half interruption identification frameworks is to join a few discovery models to accomplish improved results. A crossbreed interruption location framework comprises two parts. The main part processes the unorganized information. The subsequent part takes the handled information and sweeps it to sleet available interruption exercises [9].

3. Crossbreed interruption location frameworks depend on consolidating two learning calculations. Each learning calculation has novel highlights, it helps to work on half-breed offering IDS mixture, and it can be broadly classified as fluid half-breed, coordinated-based crossover over the bunch single, and half-breed.

4. A crossover interruption discovery framework in light of mark-based and irregularity location parts. In the principal phase of the model, an abuse discovery part was applied to recognize realized assaults in light of the caught designs. The next phase included an irregularity recognition component to capitalize on the flaws of the abuse discovery component. Various one-class SVM calculations were used to support the model's second component. The KDD Cup 99 dataset was used to test the model's presentation. When compared with a single traditional IDS, the model outperformed it [10].

5. Experts combine highlight extraction strategies and management methods to increase detection rates as well as reduce the amount of fraud. The crossover's initial phase employed chi-square to identify the highlights. The goal of this stage was to reduce the number of entries in the dataset while maintaining the important highlights that detect attacks. A multiclass support vector machine (SVM) calculation was used for grouping in the following stage. To improve the characterization rate of this model, a multiclass support vector machine was used. The NSL-KDD dataset [11] was used to evaluate the model, and the results showed that the model had a high discovery rate and a low misleading problem rate.

6. In light of a C5 choice tree classifier and a one-class support vector machine, scientists developed a mixed location model OC-SVM. Two key components made up the model [12]. The primary component of the abuse identification model

was developed using a C5.0 decision tree classifier. The next section was developed with OC-SVM for irregularity discovery. The NSL-KDD and Australian Defense Force Academy (ADFA) datasets were used by the experts to demonstrate the model, and the results revealed that the half-and-half model performed better than single-based models.

7. In light of repetitive brain architecture and convolutional brain organization CNN, the author [13] developed a crossover interruption discovery model RNN. The investigation is anticipated to advance highlight extraction, in the presentation of interruption finding frameworks, which is crucial. The RNN was used in the second stage to extract transient elements from the dataset, while CNN was used in the main stage to differentiate neighboring highlights in the dataset. The information irregularity on the accessible dataset was resolved by this tactic. The CSE-CIC-DS2018 dataset, which is the updated dataset, was used to test the model's presentation. With an interruption identification exactness of 97.75%, the model outperformed other interruption location models.

8. For smarter home security, the experts [14] suggested a half-and-half model interruption identification model. The model was divided into two pieces. The majority of the section used AI calculations to recognize continuous interruptions. In this section, calculations using irregular forests, XG Boost, choice trees, and K-closest neighbors were used. The abuse interruption identification approach was used in the next section to find known assaults. Both the CSE-CIC-IDS2018 and NSL-KDD datasets were used to test the model's presentation. For the location of both organizational disruption and client-based anomalies in cunning homes, the model captured an amazing display.

9. A mixture location model given Catalyst ML and the convolutional_LSTM Conv-LSTM network was planned. The model comprises two parts: The principal part utilizes Catalyst ML to identify inconsistency interruption, while the subsequent part sends Conv-LSTM for abuse discovery. To explore the exhibition of the perfect, the specialists utilized the ISCX_UNB datasets [15]. The model kept a remarkable presentation of 97.29% precision in identification. The specialists suggested that the ideal can be assessed further utilizing an alternate dataset as an approach to endeavoring to replicate the outcomes.

10. The creators [16] fostered an interruption location framework by joining firefly and Hopfield brain organization HNN calculations. The analysts utilized Firefly calculation to identify refusal-of-rest assaults through hub grouping and verification.

11. The scientists [17] proposed a crossbreed recognition framework for VANET vehicular impromptu organization. The model comprises two parts. The scientists conveyed an order calculation on the main part and a grouping calculation on the subsequent part. In the main stage, they utilized irregular woodland to identify known assaults through the order. They sent a weighted K-implies computation for the next step, which was the finding of an odd interruption. The most recent dataset, the CICIDS 2017 dataset, was used to evaluate the model. The experts suggested conducting additional testing on the model under verifiable circumstances. They also combined arbitrary woods computation with unsupported bunching calculation in light of corsets in another work. This model was used to

identify persistent VANET disruptions. In comparison with other models, this maintained a better presentation in terms of accuracy, computational efficiency, and identification rate.

12. The author [18] projected a mixture location perfect given hereditary calculation and fake-resistant framework AIS-GAAIS for interruption identification on impromptu on-request distance vector-based versatile impromptu organization AODV-based MAN, ET. The model was assessed utilizing different steering assaults. In contrasted and different models, the model had superior recognition rates and diminished the deception rates.

13. The scientists [19] involved incorporated firefly calculation with a hereditary calculation to include determination MANET. To group the chosen highlights in the main phase of the model as one or the other interruption or typical, the specialists utilized a replicator brain system for arrangement. The models' exhibition was contrasted with that of fluffy-based IDS. The model beat fluffy-based IDS in exactness as well as accuracy.

7.4 Literary analysis

The objective of the literary investigations is to look into IDS and convenience to track down replies to explore issues. Users will do an abstract analysis of IDS' usability to identify any usability challenges and determine the best course of action. To advance the usability of IDS, **Figure 4** shows the users will also need to ascertain the present state of craftsmanship and methods [20].

To improve convenience, users want to identify and study the IDSs that are used the most frequently. Users have Grunt, KF Sensor, and Easy *KF Sensor* is a viable host *based*-intrusion detection system (IDS) that acts as a honeypot to invite and detect hackers by pretending weak systems. A few fundamental highlights of IDS are seen during the examination, including client sorts, ease of use issues, and client collaboration with IDS.

7.5 Selection of IDS practitioners

It is critical to understand who the actual IDS users are to gain meaningful user input in defining the heuristics for IDS. In addition, this will aid in

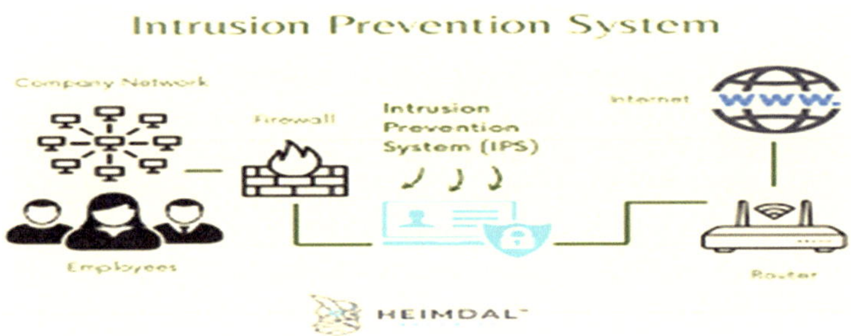

Figure 4.
Selection-and-study of IDS.

identifying IDS usability issues and determining ways to improve IDS usability based on user perceptions.

7.6 Survey questionnaire

Before working into the specifics of IDS convenience difficulties, an overview survey is intended to provide more insight into how ease-of-use and IDS are handled practically. Since users oversee various individuals with various backgrounds and levels of expertise, information, and aptitude, this system was picked. While utilizing IDS, it will help with getting what applies to these clients.

7.7 Designing of heuristics for IDS

Based on the responses to the review questionnaire, users determine the problems users have using IDS. This will support the creation of fresh IDS heuristics. The heuristics are broken down into various groups, including:

a. Installation heuristics.

b. Interface heuristics.

c. Output heuristics.

d. Customization heuristics.

e. Help heuristics.

7.8 Lab-based testing

After the heuristics have been planned, now is the right time to scrutinize them in the lab. The good thing about CASI is that the user may use the provided calculations at any point in the IDS process, including the result and customization phases. This study aims to evaluate CASI's performance in identifying and fixing ease of use flaws compared with conventional heuristics.

7.9 Experts-based testing

Following lab testing, the wished-for heuristics are currently prepared for exact testing, in which network experts can participate in IDS interface assessment challenges and receive the outcomes. At the same time, another IDS interface mock-up is assembled and tried for assessment relying upon the experimental outcomes. Assuming that network experts find the point of interaction engaging and easy to utilize, it will ultimately supplant the past IDS interface.

8. Evaluation of intrusion detection system (IDS)

To observer-assess IDS, this can be achieved *via* the CASI and (Nielson) [2] Usability on IDS to decide the number of ease-of-use flaws found and eliminated from the IDS interface. The researcher's ease-of-use was picked because they are the most routinely used. The objective of contrasting the convenience of how CASI functions contrast with scientists' ease of use. A few elements should be considered while contrasting including the quantity of ease-of-use defects distinguished, time, dependability, proficiency, and accuracy.

8.1 Challenges in intrusion detection for web-based applications

In the web application security field, the interruption identification system is still at its outset. The identifying frameworks are mostly used as an organization security gadget. In contrast to standard network IDS design, tackling the intricacies associated with online applications necessitates a novel methodology in this segment. One should outline some of the characteristics of online apps and web traffic that make designing the IDS challenging. The elements depicted in the accompanying subsets structure the theoretical starting point for fostering the web's IDS. This will aid in understanding the essential knowledge needed to create a solid engineering framework.

8.2 Communication protocol (HTTP//HTTPS)

To take advantage of online application weaknesses, aggressors solely use HTTP/HTTPS conventions. HTTPS guarantees a protected and encoded association. Hypertext transfer protocol (HTTP) is a solicitation reaction convention intended to ease correspondence between the client and server. One major disadvantage of noticing HTTPS traffic from an IDS stance is that encryption blinds network-based location frameworks. Based on their work on the application layers or the Internet layer of the TCP/IP worldview, IDS can be delegated host-based intrusion detection systems (HIDs) or networks-based intrusion detection systems (NIDs).

NIDS observes the organization bundles, and in HTTPS association, the parcel information is scrambled, which the framework neglects to check. If these frameworks approach the SSL testament's private key, they can examine HTTPS traffic. HIDS, then again, experiences no difficulty managing HTTPS traffic since it safeguards endpoints where the encoded information is unscrambled once more into its unique structure.

8.3 Internet request

Information is sent from the client to the server through a web demand. The data is sent utilizing HTTP demand header fields or solicitation boundaries. The solicitation header fields contain client demand control data, while the solicitation boundaries contain extra client data required by server-side projects to play out a movement. GET and POST are the two standard strategies for passing boundaries to the server. Boundary values are provided in the inquiry line of the URL in the GET demand, and these qualities are conveyed in the solicitation body in the POST demand. The client program typically characterizes the header fields. However, the boundary values are either given by the client or recently arranged by waiter side projects, for example, treats, stowed away fields. The hidden test with electronic application security is that client information can be truly a factor and similarly mind boggling, making it hard to interface them along with a legitimate arrangement of values.

The primary function of identification frameworks is to scrutinize the attributes listed in header fields and solicitation borders. Positive or negative methodologies may be utilized to approve these qualities. The positive approval procedure indicates what information the program anticipates. It incorporates information type (string, the negative strategy, then again, involves filtration of values that contain attack designs). Positive (whitelisting) and negative boycotting approval are remembered for metaphysics and mark-based frameworks, while inconsistency-based frameworks are concerned mostly with certain approvals. The information sent in a web solicitation could contain a wide scope of values, and the methodology to utilize (whitelist or boycott) is profoundly reliant upon the kind of significant worth set. The accompanying classes have been created from the worth set.

8.4 Finite values

These characteristics exist in a restricted reach and can be free, that is, general to all or tweaked to the application's business rationale. The main gathering contains an assortment of normal qualities, for example, header fields, Accepts, Accept Charest, Accept-Language. Since these qualities are regularly something similar across applications, they can be checked against a SIDS allow list. The last gathering of boundaries contains values for HTML controls, such as dropdown records, checkboxes. These controls assist clients with choosing values from a restricted determination of choices. However, the business case for the application leaves the value arrangement of these uncertain. Because of an assortment of elements, keeping up with the whitelist to assess such boundary values can become a tedious activity for SIDS. First, the whitelist has become excessively intended for the assortment of values that match the business rationale. Second, this rundown may be huge dependent upon how much an application controls. Third, staying up with the latest is troublesome since the passable arrangement of values could shift rapidly as business rationale changes. However, assistance can be beneficial in this situation as it allows one to become familiar with the benefits of boundaries.

9. Application values

This class provides values given by server-side projects that should not be changed on the client side. Treats are stowed away fields, and designers utilize question strings to store a scope of significant information, for example, item cost and amount, meeting ID. IDS should check that these qualities match those set by the application. Signature-based IDS cannot detect changed values because they need an attack strategy and changed values frequently resemble real information. Inconsistency-based frameworks, then again, can be utilized to realize which boundaries should not be changed on the client side. Boundary-altering assaults were found in the exploration portrayed.

9.1 Multiple users with multiple roles

Web applications typically have a large number of clients with varying levels of honors. These honors are supervised by the approval interaction, which ensures that the client is only leading legal activities. Applications follow each client-server connection and direct each solicitation to a specific client before deciding whether to handle it. Every time a user logs in to the program, a meeting ID is assigned the responsibility of identifying the solicitations from the solicitation pool and appending them to the user.

Utilizing discovery frameworks allows the user to provide various clients with unique honors arrangements. IDS should initially have the option to follow client meetings to relate client solicitations to the suitable meeting. IDS should also observe asset utilization and client actions during a meeting. Unapproved access can be acquired with an all-around created honor heightening attack. This element helps the IDS in monitoring the situation with a solitary meeting. Finally, the full state strategy can associate the grouping of solicitations to a given client, while stateless IDS treats each solicitation freely and does not monitor them. Frameworks that come up short on means to connect the current solicitation to recently got demands will probably not recognize state support and authorization infringement.

10. Conclusion

Interruption identification frameworks are confounded and present various obstacles to security experts. Earlier IDS research has generally centered on expanding the precision of these frameworks and giving help to experts and dissecting potential security issues. Further developed IDS convenience is one region that has received insignificant consideration. Yet, present heuristics are not laid out for IDS frameworks and can go about as a hindrance to utilization. An overview of the ease-of-use assessment was provided. This project includes convenience evaluations and difficulties with usability. In terms of computer programming, organization and programming connection points, and the proximity of the correlation of ease of use assessment, the analysis further added to the categorization of convenience issues, which check to take the issues and inadequacies in this field into account. Moreover, the suggested heuristics for clients and IDS give the principal standards for creating and developing IDS connection points to opposing security breaks.

Abbreviations

ADFA	Australian Defense Force Academy
APIDS	Application protocol-based intrusion detection system
CASI	Cognitive analysis of software interface
HIDS	Host intrusion detection system
HIDS	Hybrid intrusion detection system
HTTP	Hypertext transfer protocol
IDS	Intrusion detection system
NIDS	Network intrusion detection system
PIDS	Protocol-based intrusion detection system

Updates on Software Usability
DOI: http://dx.doi.org/10.5772/intechopen.107423

Author details

Ahmed Mateen Buttar* and Muhammad Majid
Department of Computer Science, University of Agriculture, Faisalabad, Pakistan

*Address all correspondence to: ahmedmatin@hotmail.com

References

[1] Usability Evaluation. The Encyclopedia of Human-Computer Interaction. 2nd ed. Available from: https://www.interaction-design.org/literature/book/the-encyclopedia-of-human-computer-interaction-2nd-ed/usability-evaluation. [Accessed: August 5, 2022]

[2] Becker FG, et al. Available from: https://www.researchgate.net/publication/269107473_What_is_governance/link/548173090cf22525dcb61443/download%0A. http://www.econ.upf.edu/~reynal/Civilwars_12December2010.pdf%0A. https://think-asia.org/handle/11540/8282%0Ahttps://www.jstor.org/stable/41857625

[3] Naqvi I, Chaudhary A, Kumar A. A systematic review of the intrusion detection techniques in VANETS. TEM Journal. 2022;**11**(2):900-907. DOI: 10.18421/tem112-51

[4] 7 Phases of the System Development Life Cycle Guide. Available from: https://www.clouddefense.ai/blog/system-development-life-cycle. [Accessed: August 5, 2022]

[5] Lazarevic A, Kumar V, Srivastava J. Intrusion detection: A survey. Managing Cyber Threats, Massive Computing. 2005;**5**:19-78. DOI: 10.1007/0-387-24230-9_2

[6] Masood Butt S, Majid MA, Marjudi S, Butt SM, Onn A, Masood Butt M. Casi method for improving the usability of IDS. Science International (Lahore). 2015;**27**(1):275-286

[7] Best Intrusion Detection Software - IDS Systems - DNSstuff. Available from: https://www.dnsstuff.com/network-intrusion-detection-software. [Accessed: August 5, 2022]

[8] Preparing Simple Consolidated Financial Statements | F3 Financial Accounting | ACCA Qualification | Students | ACCA Global. Available from: https://www.accaglobal.com/my/en/student/exam-support-resources/fundamentals-exams-study-resources/f3/technical-articles/preparing-simple-consolidated-financial-statements.html. [Accessed: August 5, 2022]

[9] Son LH, Pritam N, Khari M, Kumar R, Phuong PTM, Thong PH. Empirical study of software defect prediction: A systematic mapping. Symmetry. 2019;**11**:212. DOI: 10.3390/SYM11020212

[10] Available from: https://ieeexplore.ieee.org/stamp/stamp.jsp?arnumber=7937814. [Accessed: August 5, 2022]

[11] Sumaiya Thaseen I, Aswani Kumar C. Intrusion detection model using fusion of chi-square feature selection and multi class SVM. Journal of King Saud University - Computer and Information Science. 2017;**29**(4):462-472. DOI: 10.1016/J.JKSUCI.2015.12.004

[12] Khraisat A, Gondal I, Vamplew P, Kamruzzaman J. Survey of intrusion detection systems: Techniques, datasets and challenges. Cybersecurity. 2019;**2**(1):1-22. DOI: 10.1186/S42400-019-0038-7

[13] Khan A, Sohail A, Zahoora U, Qureshi AS. A survey of the recent architectures of deep convolutional neural networks. Artificial Intelligence Review. 2020;**53**:5455-5516. DOI: 10.1007/s10462-020-09825-6

[14] Alghayadh F, Debnath D. A hybrid intrusion detection system for smart home security based on machine learning and user behavior. Advanced Internet of Things. 2021;**11**:10-25. DOI: 10.4236/ait.2021.111002

[15] Kim S et al. A critical function for the actin cytoskeleton in targeted exocytosis of prefusion vesicles during myoblast fusion. Developmental Cell. 2007;**12**(4):571-586. DOI: 10.1016/J.DEVCEL.2007.02.019

[16] Maseno EM, Wang Z, Xing H. A systematic review on hybrid intrusion detection system. Security and Communication Networks. 2022;**2022**:42-61. DOI: 10.1155/2022/9663052

[17] Bangui H, Ge M, Buhnova B. Exploring big data clustering algorithms for internet of things applications, IoTBDS 2018—Proc. 3rd Int. Conf. Internet Things. Big Data Security. 2018;**2**:269-276. DOI: 10.5220/000677 3402690276

[18] Amiri E, Keshavarz H, Heidari H, Mohamadi E, Moradzadeh H. Intrusion detection systems in MANET: A review. Procedia—Social and Behavioral Sciences. 2014;**129**:453-459. DOI: 10.1016/J.SBSPRO.2014.03.700

[19] Shona D, Kumar MS. Efficient IDs for MANET Using hybrid firefly with a genetic algorithm. Proceedings of International Conference on Inventive Research in Computing Applications ICIRCA. 2018;**2018**:191-194. DOI: 10.1109/ICIRCA.2018.8597268

[20] What Is an Intrusion Prevention System (IPS)?. Available from: https://heimdalsecurity.com/blog/intrusion-prevention-system-ips/. [Accessed: August 5, 2022]

Chapter 9

Design of Low-Cost Reliable and Fault-Tolerant 32-Bit One Instruction Core for Multi-Core Systems

Shashikiran Venkatesha and Ranjani Parthasarathi

Abstract

Billions of transistors on a chip have led to integration of many cores leading to many challenges such as increased power dissipation, thermal dissipation, occurrence of faults in the circuits, and reliability issues. Existing approaches explore the usage of redundancy-based solutions for fault tolerance at core level, thread level, micro-architectural level, and software level. Core-level techniques improve the lifetime reliability of multi-core systems with asymmetric cores (large and small cores), which have gained momentum and focus among a large number of researchers. Based on the above implications, multi-core system using one instruction cores (MCS-OIC) factoring its features are proposed in this chapter. The MCS-OIC is an asymmetric multi-core architecture with MIPS core as the conventional core and OICs as the warm standby-redundant core. OIC executes only one instruction named 'subleq _ subtract if less than or equal to zero'. When there is one of the functional units (i.e., ALU) of any conventional core fails, the opcode of the instruction is sent to the OIC. The OIC decodes the instruction opcode and emulates the faulty instruction by repeated execution of the 'subleq' instruction, thus providing fault tolerance. To evaluate the idea, the OIC is synthesized using ASIC and FPGA. Performance implications due to OICs at instruction and application level are evaluated. Yield analysis is estimated for various configurations of multi-core system using OICs.

Keywords: fault tolerance, reliability, one instruction core, multi-core, yield

1. Introduction

Researchers have predicted about an eight percent increase in soft-error rate per logic state bit in each technology generation [1]. According to the International Telecommunication Roadmap for Semiconductors (ITRS) 2005 and 2011, reduction in dynamic power, increase in resilience to faults and heterogeneity in computing architecture pose a challenge for researchers. According to the International Roadmap for

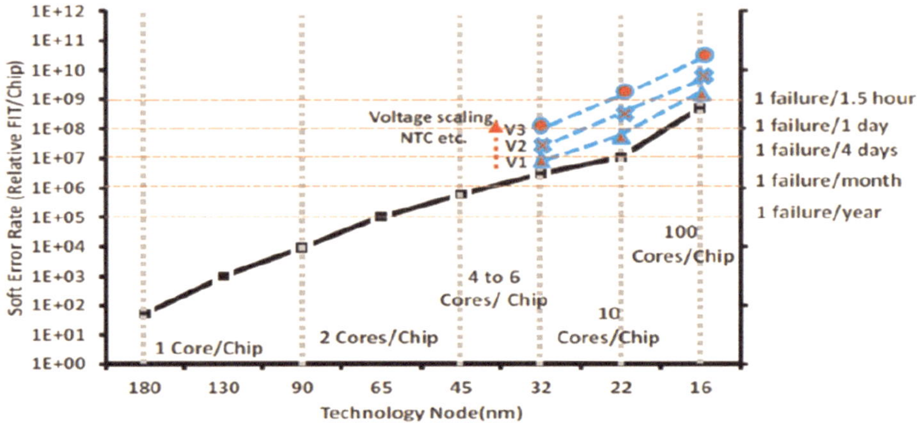

Figure 1.
SERs at various technology node.

Device and System (IRDS) roadmap 2017, device scaling will touch the physical limits with failures reaching one failure per hour as shown in **Figure 1**. The soft error rate (SER) is the rate at which a device or system encounters or is predicted to encounter soft errors per unit of time, and is typically expressed as failures-in-time (FIT). It can be seen, from **Figure 1** [2–4] that, at 16 nm process node size, a chip with 100 cores could come across one failure every hour due to soft errors.

This decrease in process node size and increase in integration density as seen in **Figure 1**, has the following effects.

1. Number of cores per chip has increased. Due to increase in number of cores, size of the last level cache (LLC) has increased. For example, NVIDIA's GT200 architecture GPU did not have an L2 cache, the Fermi GPU, Kepler GPU, Maxwell GPU has 768KB LLC, 1536KB LLC and 2048KB LLC respectively [5]. Similarly, Intel's 22 nm Ivytown processor has a 37.5 MB static random-access memory (SRAM) LLC (Rusu 2014) [6] and 32 nm Itanium processor had a 32 MB SRAM LLC (Zyuban 2013) [7]. Consequence of larger cache size has led to exponential increase in SER.

2. Low swing interconnect circuits are being used in CMOS transmission system. This has proved to be an energy efficient signalling system compared to conventional full swing interconnects circuits. However, incorrect sampling of the signals in low swing interconnect circuits together with interference and noise sources can induce transient voltages in wires or internal receiver nodes resulting in incorrect value being stored at receiver output latch [8].

This scenario can be envisaged as a "fault wall". In order to surmount the fault wall scenario, reliability has been identified as a primary parameter for future multi-core processor design [9, 10]. Similarly, ITRS 2005 and 2011, have also identified increase in resilience to faults as a major challenge for researchers. Hence, a number of researchers have started focusing on resilience to faults and reliability enhancement in multi-core processors. The chapter focuses on providing fault tolerance solutions for processor cores in multi-core systems.

2. Motivation

As seen in **Figure 1**, the total FIT per chip increases with number of cores per chip increasing. In order to accommodate higher number of cores per chip, (1) total FIT per chip has to be maintained constant (or no change), and (2) SER per core needs to be reduced. In the present-day processor cores, the frontend of the core comprises of decode queue, instruction translation lookaside buffer, and latches. The backend of the core comprises of arithmetic logic unit, register files, data translation lookaside buffer, reorder buffers, memory order buffer, and issue queue. SER from backend and the frontend of the core is 74.48% and 25.22% respectively. In the present processor cores, latches are hardened [11, 12] cache and large memory arrays are protected using error correcting codes (ECC) [13, 14]. The SER from backend of the processor is more when compared to front end and is mainly due to arithmetic logic unit. The FIT from the arithmetic logic unit of the processor core has started reaching higher levels which needs robust fault mitigation approaches for present and future processors. Hence addressing the reliability issues of the core (arithmetic logic unit in backend) is more significant in improving the reliability of the multi-core system [15, 16]. Conventional approaches to handle soft errors consumes more power and area. Hence, the chapter focuses on using heterogeneous model with low cost ("low cost" denote low power and lesser area of OICs) fault tolerant cores to improve reliability of multi-core systems.

2.1 Chapter contributions

Contributions of the chapter are briefly presented below.

1. The microarchitecture consisting of control and data path for OIC is designed. Four modes of operation in 32-bit OIC namely (a) baseline mode (b) DMR mode (c) TMR mode and (d) TMR with self-checking subtractor (TMR + SCS) are introduced.

2. The microarchitecture of 32-bit OIC and multi-core system integrated with 32-bit OIC are implemented using Verilog HDL. The design is synthesized in Cadence Encounter (R) RTL Compiler RC14.28 –V14.20 (Cadence design systems 2004) using TSMC 90nm technology library (tcbn90lphptc 150).

3. Dynamic power, area, critical path and leakage power for four modes of OIC are estimated and compared.

4. Dynamic power and area of OIC and URISC++ are compared.

5. Area and power are estimated for multi-core system consisting of 32-bit OIC.

6. The OIC is synthesized using Quartus prime Cyclone IVE (Intel, Santa Clara, CA) with device EP4CE115FE29C7. Number of logical elements and registers are estimated.

7. Number of logical elements and registers in OIC and URISC++ are compared.

8. Using Weibull distribution, the reliability for the four modes of OIC are evaluated and compared.

9. Using Weibull distribution, the reliability for OIC and URISC++ are evaluated and compared.

10. Performance overhead at instruction level and application level is estimated.

11. Yield analysis for proposed multi-core system with OICs is presented.

2.2 Chapter organization

The remaining portion of the chapter is organized as follows as: Section titled “3. An Overview on 32-bit OIC” presents (a) an outline of 32-bit OIC (b) one instruction set of OIC (c) modes of operation of OIC (d) microarchitecture of OIC (e) microarchitecture of multi-core system consisting of OIC (f) instruction execution flow in multi-core system using one instruction cores (MCS-OIC); Section titled “4. Experimental results and discussion” presents power, area, register and logical elements estimation for OIC, and power, area estimation for MCS-OIC; Section titled “5. Performance implications in multi-core systems” presents performance implications at instruction level and application level; Section titled “6. Yield analysis for MCS-OIC” presents yield estimates for the proposed MCS-OIC; Section titled “7. Reliability analysis of 32-bit OIC” presents reliability modelling of OIC and its estimate in different operational modes; the conclusion of the chapter is presented in the Section titled “8. Conclusion”; the relevant references are citated in the Section titled “References”.

3. An overview on 32-bit one instruction core

A 32-bit OIC [17] is designed to provide fault tolerance to a multi-core system with 32-bit integer instructions of conventional MIPS cores. OIC is an integer processor. The terms “32-bit OIC” and “OIC” are interchangeably used in this thesis. OIC executes only one instruction, namely, “subleq – subtract if less than or equal”. The OIC has three conventional subtractors and an additional self-checking subtractor. A conventional core that detects faults in one of the functional units (i.e., ALU) sends the opcode with operands to the OIC. In this thesis, the OIC is designed to support the instruction set of 32-bit MIPS core. However, it can be designed to support 32 bit ×86/ARM instruction set by making necessary changes in the instruction decoder. The OIC emulates the instruction by repetitively executing the subleq instruction in a predetermined manner. There are four modes of operation in OIC and they are (a) baseline mode (b) DMR mode (c) TMR mode and (d) TMR + Self Checking Subtractor (SCS) or TMR + SCS mode. TMR + SCS is the “high resilience mode” of OIC. Baseline mode is invoked only when soft error detection and correction alone are required.

3.1 One instruction set

“Subleq – subtract if less than or equal” is the only instruction executed by the OIC. The syntactic construct of the subleq instruction is given below.

Subleq A, B, C; Mem [B] = Mem [B] – Mem [A]

ADD a,a,b	INC a	MOV a, b	RSB b, a, b
1.Subleq a, z,2	1.Subleq One, z,2	1.Subleq a, a,2	1.Subleq a, b,2
2.Subleq z, b,3	2.Subleq z, a,3	2.Subleq b, z,3	2 ret
3 Subleq z, z,4	3.Subleq z, z,4	3.Subleq z, a,4	**DEC a**
4 ret	4.ret	4.Subleq z, z,5	Subleq one, a
		5.ret	ret

Table 1.
Sequence of synthesized Subleq instruction.

; If (Mem [B] ≤ 0) go to C;

It is interpreted as: "subtract the value at the memory location A from the value at the memory location B; store the result at the memory location B; If the value at the memory location B is less than or equal to zero, then jump to C." The subleq instruction is Turing complete. The instruction set of a core or processor is said to be Turing complete, if in principle, it can perform any calculation that any other programmable computer can. As an illustration, the equivalent synthesized subleq instructions for ADD, INC, MOV, DEC and RSB (Reverse subtract) instructions are given in the **Table 1**.

3.2 Modes of operation

The OIC operates in four modes as mentioned above. They are (a) baseline mode (b) DMR mode (c) TMR mode and (d) TMR + Self Checking Subtractor (SCS) or TMR + SCS mode.

a. *Baseline mode*: In this mode, only the self-checking subtractor is operational. The results from the subtractor are verified by the self-checker. If the results differ, the subtraction operation is repeated to correct the transient faults. Transient faults are detected and corrected in this mode. If the results do not match again, a permanent fault is detected.

b. *DMR mode*: In this mode, only two subtractors are operational. The results of the two subtractors are compared using a comparator. If the results differ, the subtraction operation is repeated to correct the transient faults. The transient faults are detected and corrected in this mode. If one of the two subtractors fails, a permanent fault is detected, and the OIC switches to baseline mode.

c. *TMR mode*: In this mode, all three subtractors are operational. The results from the three subtractors are compared using three comparators. The voters check the results from the comparators and perform majority voting. To correct the transient faults, the operations are repeated. If anyone subtractor fails, the faulty subtractor is disabled. In this mode, results from the redundant subtractors are fed back on special interconnects to the inputs of the multiplexer. OIC then switches to DMR mode. It is assumed that two subtractors do not fail simultaneously. Occurrence of one permanent fault is detected and tolerated in this mode.

d. *TMR + SCS mode:* TMR + SCS mode is the initial mode of operation in OIC. In this mode, all three subtractors and SCS are operational. Both permanent and transient faults are detected and corrected. The results of three subtractors and SCS are compared using a comparator. If the results differ, then entire operation is repeated to correct the transient faults. If results continue to differ, then OIC switches to TMR mode.

3.3 Micro-architecture of OIC

The micro-architecture of the OIC is given in **Figure 2**. The micro-architecture of the OIC can be divided into two parts: the control unit and data-path unit. The control unit consists of a 12-bit program counter (PC), an instruction decoder, a 12-bit control word memory register and control word memory. The control memory is safeguarded by (12, 4) Hamming codes [18]. All single-bit errors are detected and corrected by Hamming codes. The data-path unit consists of four multiplexers, one demultiplexer, three subtractors, one self-checking subtractor (SCS), three comparators and one voter unit. Normally, the register files occupy a large die area in a core and are exposed to high energy particles. In the spheres of replication, the register files also have high access latency and power overhead due to their fortification from ECC. The OIC does not have large register files that are likely to propagate transient faults or soft errors to other subsystems. The OIC uses very few registers. Once the operands from faulty core are admitted, they are stored in the registers. The results computed by the subtractors are compared and fed back on a separate interconnect line to the respective multiplexers. The intermediate results are not stored in the registers.

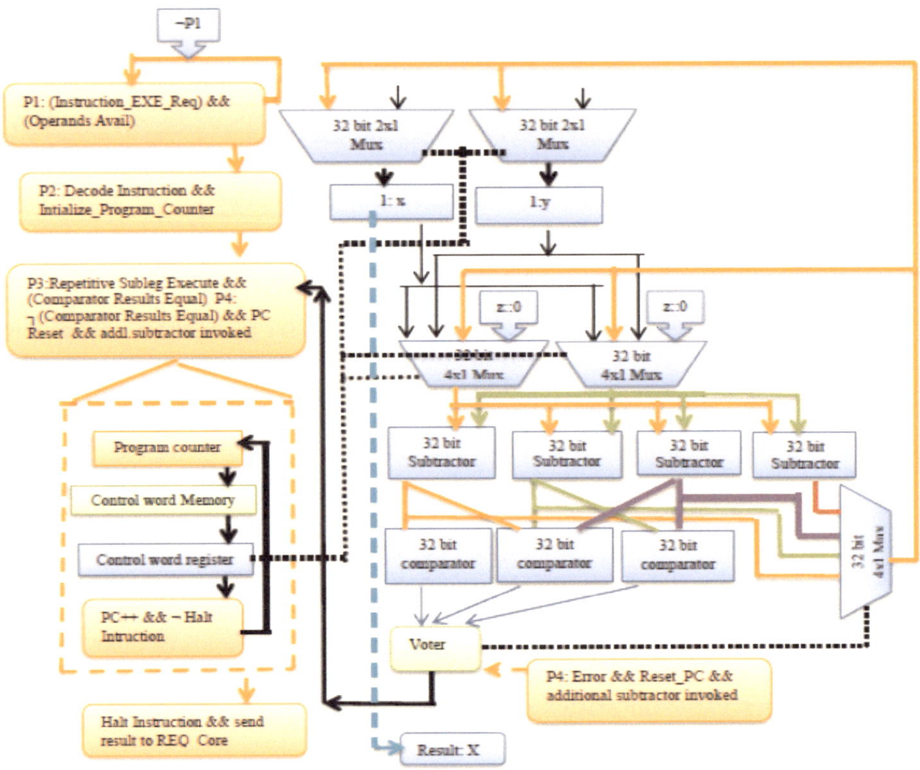

Figure 2.
Control unit and data path unit of 32-bit OIC.

3.4 Microarchitecture and instruction execution flow in MCS-OIC

A Multicore system comprising one 32-bit MIPS core and one 32 bit OIC occupying the upper half and lower half portions respectively in the micro-architecture, is shown in **Figure 3**. The MIPS core is a five-stage pipelined scalar processor. Instruction Fetch (IF), Instruction Decode (ID), Execution (EXE), Memory access (MEM) and Write Back (WB) are the five stages in the MIPS pipeline. IF/ID, ID/EXE, EXE/MEM, and MEM/WB are the pipeline registers. PC is a program counter and LMD, Imm, A, B, IR, NPC, Aluoutput, and Cond are temporary registers that hold state values between clock cycles of one instruction. The fault detection logic (FDL) detects faults in all the arithmetic instructions (except logical instructions) by concurrently executing the instructions. The results of ID/EXE.Aluoutput and FDL are compared to detect the fault. If a fault is found then the pipeline is stalled. The IF/ID.opcode (in IR) and operands ID/EXE.A and ID/EXE.B are transferred to OIC as shown in **Figure 4**. The IF/ID.opcode is decoded and concurrently ID/EXE.A and ID/EXE.B values are loaded into the OIC registers (X & Y). The OIC.PC is initialized and simultaneously first control word from memory is loaded into its control word register. During every clock cycle, the control bits from control word register are sent to the selection lines of the multiplexer that control the input lines to the subtractors. At every clock cycle, subtraction is performed to emulate the instruction sent from the MIPS core. Finally,

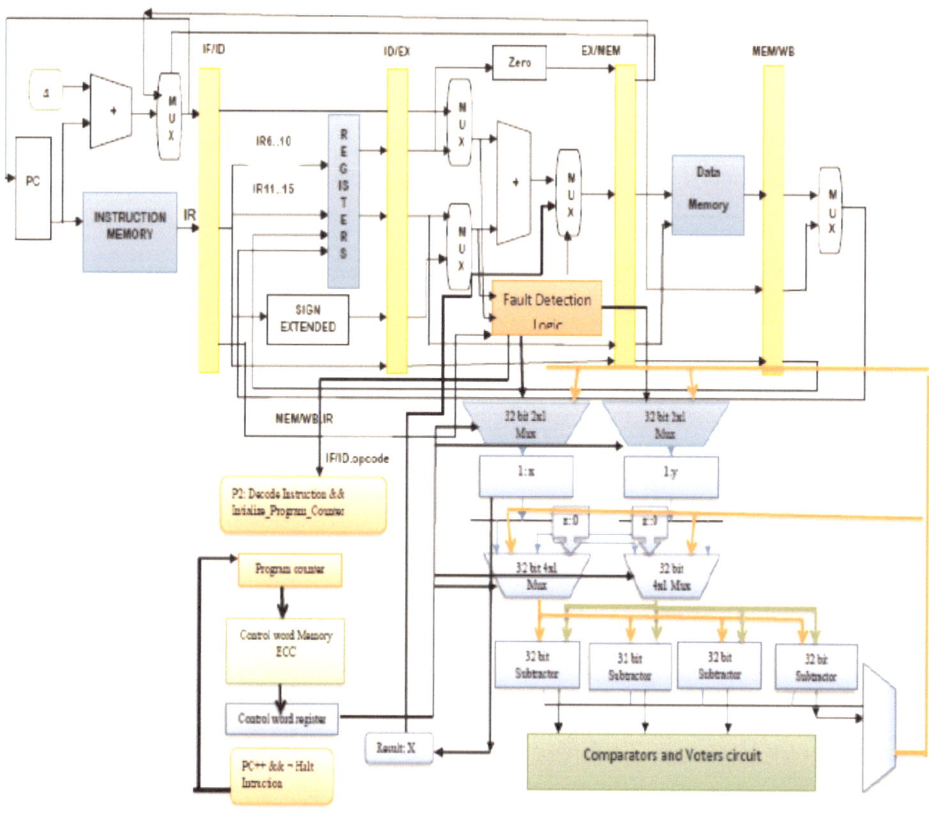

Figure 3.
Multi-core system consisting of one 32-bit MIPS core and one 32-bit OIC.

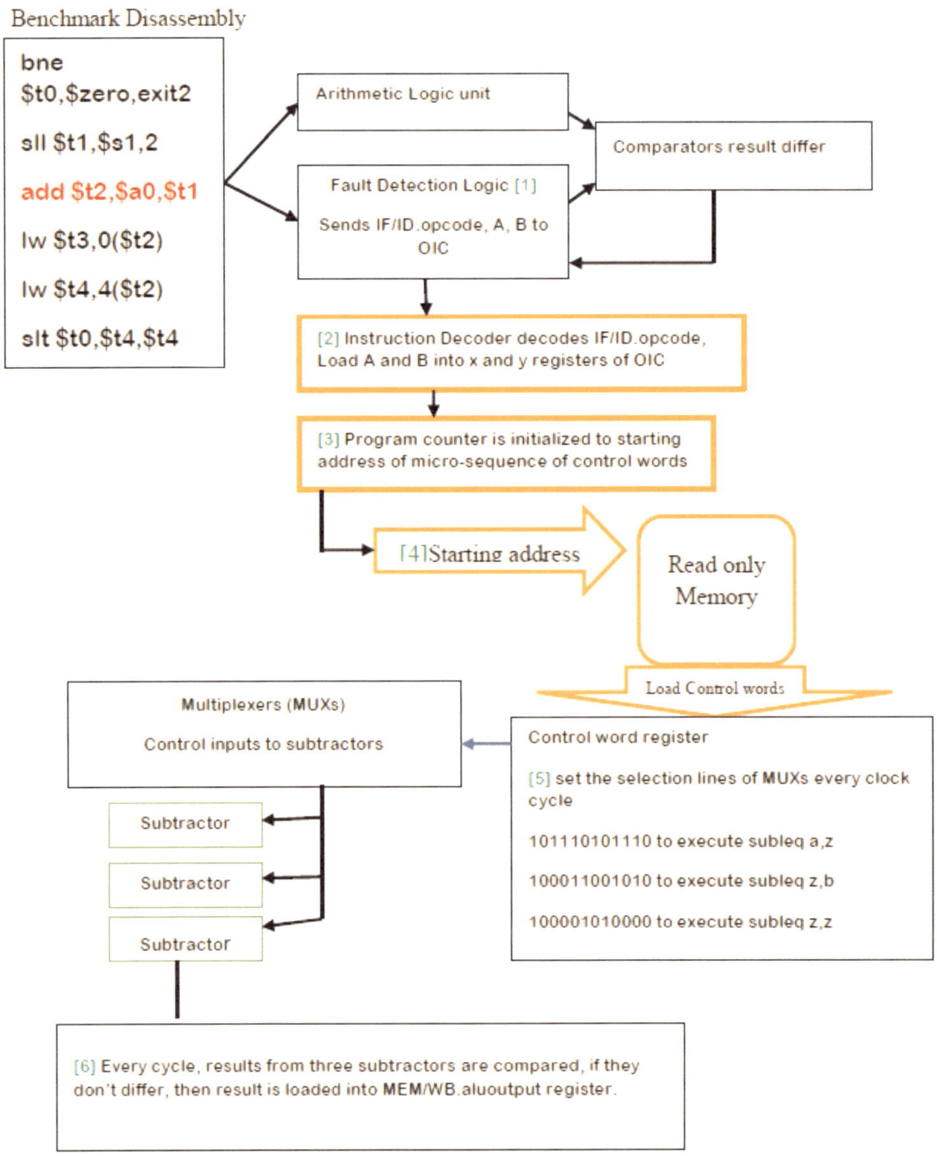

Figure 4.
Sequence of events from fault detection to loading of results into Mem/WB.Aluoutput register of MIPS core.

the computed result is loaded into MEM/WB.Aluoutput and the MIPS pipeline operation is resumed. The sequence of events from fault detection to results loaded into MEM/WB.Aluoutput register of the MIPS core is shown in **Figure 4**.

4. Experimental results and discussion

The micro-architecture of the OIC is implemented using Verilog HDL and synthesised in ASIC and FPGA platforms to estimate hardware parameters (area, critical path delay, leakage power, dynamic power) and number of logical elements, register usage respectively. In the Section 4.1, comparison of area, power, registers and number of logical elements of OIC with an approach named URISC proposed by [19] and URISC ++ proposed by [20] is presented. Notably, URISC/URISC++ implement one instruction set. The URISC/URISC++, a co-processor for TigerMIPS, emulates instructions through the execution of subleq instruction. TigerMIPS performs static code insertion in both control flow and data flow invariants so as to detect faults by performing repeated executions of subleq within the co-processor. Comparative analysis on hardware parameters for different modes of OIC are discussed in Section 4.2.

ASIC simulation: The OIC given in **Figure 2** and multi-core system in **Figure 3** has been implemented using Verilog HDL and then synthesized in Cadence Encounter (R) RTL Compiler RC14.28 –V14.20 (Cadence design systems 2004) using TSMC 90 nm technology library (tcbn90lphptc 150). The area, power (dynamic, leakage, net, internal) and critical path delay are estimated for the OIC and tabulated in **Table 2**.

FPGA synthesis: The OIC is synthesized using Quartus prime Cyclone IVE with device EP4CE115FE29C7 and the results are illustrated in **Tables 3** and **4**.

Leakage power and dynamic power: Power dissipation shown in **Table 2** is understood as sum of dynamic power and static power (or cell leakage). Static power is consumed when gates are not switching. It is caused by current flowing through transistors when they are turned off and is proportional to the size of the circuit. Dynamic power is a sum of net switching power and cell internal power. The net switching power is the power dissipated in the interconnects and the gate capacitance.

Block name	Area (μm^2)	Leakage power (nW)	Internal (nW)	Net (nW)	Dynamic power (nW)	Critical path delay (ps)
Control path	590	39.87	79,498.48	21,881.40	10,1379.88	
(Control path + data path)	8122	704.08	10,51,631.88	346,487.45	13,98,115.34	8608
Sub blocks						
Subtractor	581	67.98	41,676.83	6711	48,387.83	
Comparator	615	67.04	42,457.83	9954.38	52,411.44	

Table 2.
Implementation of 32 bit OIC results using 90 nm TSMC technology.

(A) blocks	Logical elements	Dedicated registers
OIC (TMR + SCS)	530	160
Subtractor (1)	33	
Comparator (1)	43	
(B) modes	Logical elements	
Baseline	100	
DMR	303	
TMR	486	

Table 3.
FPGA synthesis results for OIC.

Cores	Logical elements	Dedicated registers
OIC	530	160
URISC	15,019	5232
URISC++	15,081	5233

Table 4.
FPGA synthesis results comparison.

The cell internal power is the power consumed within a cell by charging and discharging cell internal capacitances. The total power is a sum of the dynamic power and the leakage power.

Multi2sim (version 5.0): Multi2sim supports emulation for 32 bit MIPS/ARM binaries and simulation for 32-bit ×86 architectures. It performs both functional and timing simulations. The performance loss is estimated for compute intensive and memory intensive micro-benchmarks using a Multi2sim simulator. Performance loss for micro-benchmarks listed in **Table 6** are illustrated in **Figures 6–11**.

4.1 Comparative analysis: power, area, registers and logical elements

With the critical path delay at 8608 ps, the operating frequency of the circuit is 115 MHz with power supply at 1.2v. OIC is a low power core consuming 1.3 mW, with die area of 8122 μm^2. The die area of conventional MIPS core is 98,558 μm^2 which is 14.2× larger than OIC core. The MIPS core consumes a total power of 1.153 W and the 32-bit OIC consumes 1.39 mW; order of difference in powers of 10 is three. The registers in OIC are PC and temporary registers which hold the operands. But they are not designed and managed as a register file. **Tables 3** and **4** provide the register count and logical elements count for OIC and URISC++. The number of logical elements in OIC is 3.51% and 3.52% of the logical elements in URISC and URISC++ respectively. The number of registers in OIC is 3.05% of URISC++. URISC++ adds 62 logical elements and one additional register to the architecture of URISC. The logical elements in URISC++ consume 6.6 mW. URISC++ has 650 registers or 14.3% of registers in TigerMIPS. URISC++ has two large register files. URISC++ altogether consumes 1.96 W. Thus, OIC consumes less power than URISC++.

4.2 Comparative analysis: four modes of OIC

The critical path delay, area, dynamic power and leakage power for the four modes of OIC namely baseline mode, DMR mode, TMR mode and TMR + SCS mode are normalized to baseline mode and shown in **Figure 5**. The area overhead of TMR + SCS mode is 68.43% of the baseline, area overhead of TMR mode is 65.37% of the baseline and for DMR mode it is 51.4%. The comparators and subtractors occupy 22.71% and 28.6% of TMR + SCS mode area respectively. The size of the voter is negligible in TMR + SCS mode and TMR mode. In the critical path delay, 10% increase is noticed from the baseline to TMR + SCS mode. The critical path traverses from the subtractor input to the comparator, and then to the voter, passing through select logic and ends at an input line. Delay would not differ much between TMR mode and TMR + SCS mode.

Both the dynamic power and leakage power for TMR mode and DMR mode increase significantly due to redundant subtractors and comparators which are not in the baseline. The dynamic power overhead of TMR mode and DMR mode is 60% and

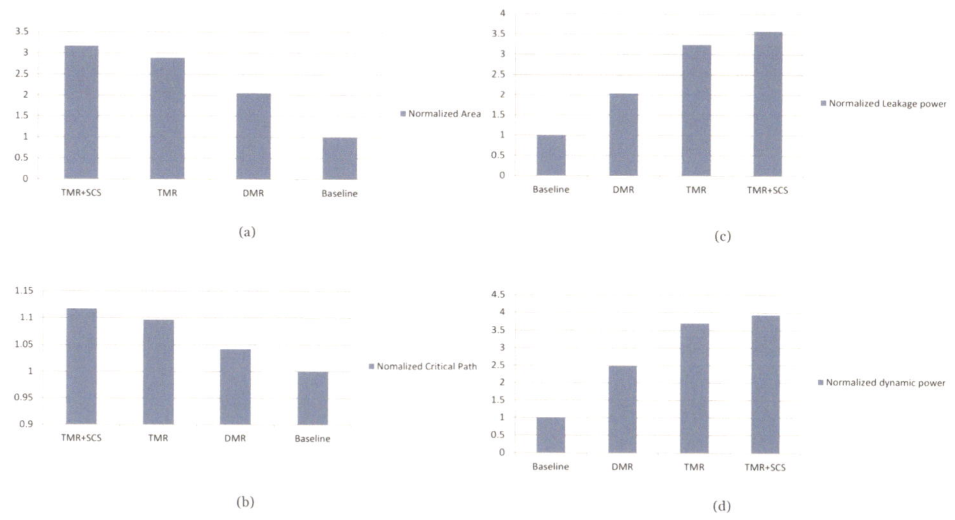

Figure 5.
(a) Area, (b) critical path delay, (c) leakage power and (d) dynamic power (y-axis—normalized values to baseline).

73% of the baseline. It is 75% for TMR + SCS mode. The static power or leakage power is proportional to the size of the circuit. The TMR + SCS mode has leakage power which is 76% more than the baseline. The TMR and DMR mode have leakage power which is 72% and 50% more than the baseline. In **Table 4** which depicts FPGA synthesis results, it is observed that the number of logical elements in TMR + SCS mode and DMR mode is 79% and 66% more than the baseline. From **Tables 2** and **4**, it is observed that TMR mode with additional self-checking subtractor in TMR + SCS mode costs more than the baseline, but still TMR + SCS/OIC will be a suitable fault tolerant core for a low power embedded system.

4.3 Power and area estimation for MCS-OIC

The area and power for the micro-architecture of multi-core system (one MIPS core with one OIC) shown in **Figure 3**, are estimated using ASIC simulation. The multi-core system occupies a total area of 306,283 μm^2 and consumes a total power of 1.1554 W. The FDL occupies an area of 6203 μm^2 which is 2% of the total area occupied by the system. The OIC occupies an area of 8122 μm^2 which is 2.6% of the total area occupied by the system. The FDL consumes a power of 1.2 mW and OIC consumes a power of 1.4 mW which are negligible when compared to the total power. The redundancy-based core level fault mitigation techniques/approaches such as Slipstream [21], dynamic core coupling (DCC) approach proposed by [22], configurable isolation [23], reunion is a fingerprinting technique proposed by Smolens et al. [24] have nearly 100% area overhead and obviously larger power overhead.

5. Performance implications in MCS-OIC

For every instruction emulated on OIC, an additional three clock cycles are needed for transfer of opcodes and operands, and two clock cycles are needed to resume the pipeline in the MIPS processor. The two terms defined below highlight the latency that incur in instruction execution, presented in the following subsection.

5.1 Performance overhead at instruction level

Definitions: (a) The **instruction execution time by emulation (IETE)** is defined as the number of cycles needed to execute the instruction on OIC. (b) **Total execution time (TET)** is defined as the sum of IETE and time (in clock cycles) to transfer opcodes, operands (from MIPS to OIC) and results (from OIC to MIPS). In other words, this indicates that time in clock cycles between pipeline stall and resumption of pipeline. The TET and IETE for instructions are tabulated in **Table 5**.

5.2 Performance overhead at application level

In the previous section, performance loss at instruction level caused due to transfer of operands and results back to host core is discussed. This would cause an accumulative loss in performance of application and the same is discussed in this section. The OIC supports 32 bit ISA of MIPS R3000/4000 processor operating at a frequency of 250 MHz. OIC operates at a frequency of 115 MHz, thereby incurring a performance loss while emulating the instructions from a faulty functional unit in MIPS core. The Multi2sim, a cycle accurate simulator together with a cross compiler, *mips-linux-gnu-gcc/mips-unknown-gnu-gcc* is used to estimate the simulation execution time for a set of micro-benchmarks. The emulation of only arithmetic instructions on OIC is considered for estimating the performance loss as they constitute nearly 60% of total instructions in integer application programs. The compute intensive and memory intensive micro-benchmark programs considered are listed in **Table 6**.

5.2.1. Memory intensive micro-benchmarks

The performance loss for memory intensive micro-benchmark programs, namely, binary search, quicksort (using recursion), and radix sort, are given in **Figures 6–8** respectively. The performance loss for CPU intensive micro-benchmark programs, namely, matrix multiplication, CPU scheduling, and sieve of Eratosthenes, are given in the **Figures 9–11** respectively. The baseline indicates the simulated execution time

Instructions	IETE	TET	Clock cycle in MIPS/LEON 2FT/3FT
ADD	4	9	1
MOV	5	10	1
INC	4	9	1
DEC	1	5	1
SUB	1	5	1
MUL	7 (per iteration)	Min 12	6
DIV	5 (per iteration)	Min 10	34

Table 5.
IETE and TET for instructions.

S. no	Micro-benchmark	CPU/memory intensive	Input form	Input size
1	Matrix multiplication (single/multithreaded)	CPU	Matrix	[10 × 10], [100 × 100], [1000 × 1000] elements
2	Binary search (single/multithreaded)	Memory	Array	3000, 30,000, 300,000 elements
3	Sieve of Eratosthenes	CPU	Array	1000, 10,000, 100,000 prime number limit
4	CPU-scheduling	CPU	Array	1000, 10,000, 100,000 processes
5	Quicksort (recursion)	Memory	Array	Sorted 100, 1000, 10,000 elements for worst case analysis
6	Radix sort	Memory	Array	1000, 10,000, 100,000 elements

Table 6.
CPU intensive and memory intensive micro-benchmarks.

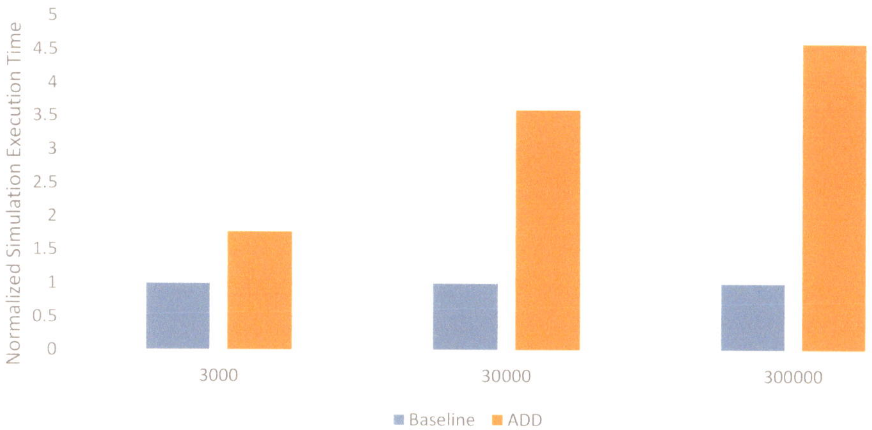

Figure 6.
Performance overhead in binary search by emulating ADD using subleq instruction.

of micro-benchmarks with no arithmetic instructions emulated on OIC. The performance loss is quantified for micro-benchmarks with respect to simulated execution time of the baseline (with varying input data sets/size).

As shown in **Figure 6**, Binary search with emulation of ADD instructions incurs performance loss of 1.77×, 3.59× and 4.59× with input size of 3000, 30,000 and 300,000 respectively, when compared to baselines. Significant proportion of ADD instructions is associated with incrementing or decrementing counters and effective addresses. OIC do not fetch operands or store results directly to main memory. Main memory latency is not taken into account for performance loss estimation. The number of ADD instructions executing as a part of the algorithmic phase of program execution does not increase exponentially with increase in input data sets. Hence, performance loss impact is minimal in algorithmic phase and is higher during fetching and storing of the input data sets. In case of multithreaded binary search, multi-core setup consisting of two cores core-0 and core-1 each with single thread is used to estimate the performance loss. The performance loss is similar to that of single

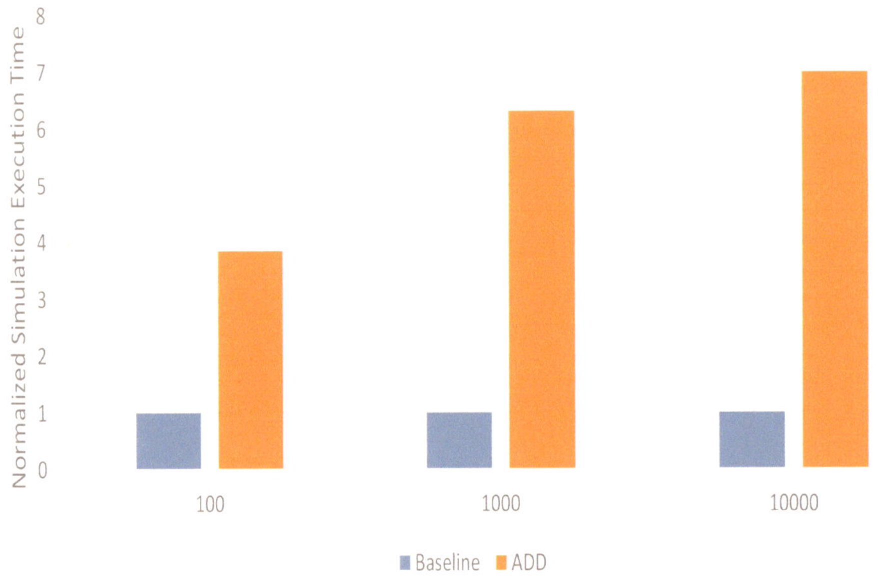

Figure 7.
Performance overhead in Quicksort by emulating ADD using subleq instruction.

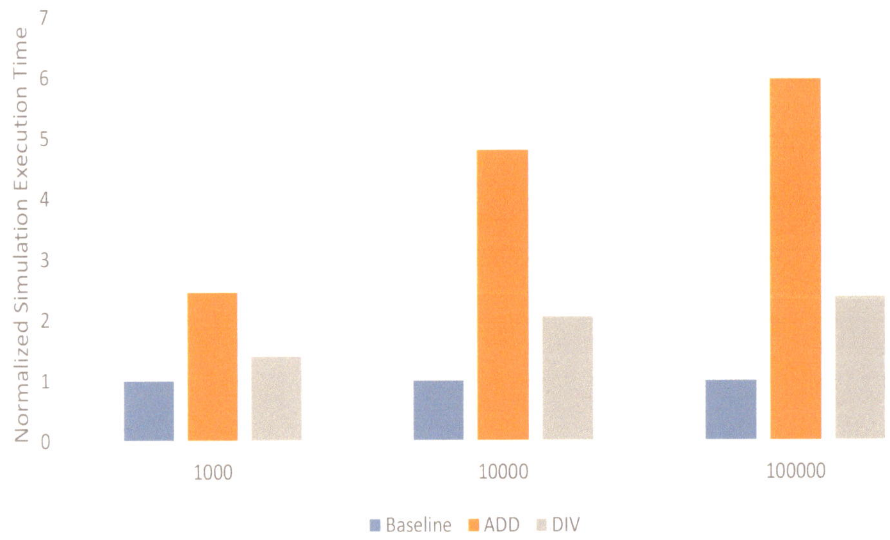

Figure 8.
Performance overhead in Radix sort by emulating ADD and DIV using subleq instruction.

threaded binary search. It is due to the fact that majority of the ADD instructions are associated with LOAD and STORE instructions.

Quicksort (with emulation of ADD instruction), implemented using recursion for sorted data elements (worst case analysis) incurs performance loss of 3.85×, 6.31×, and 6.99× for data size of 100, 1000 and 10,000 respectively as shown in **Figure 7**. For best case analysis of quicksort for 10,000 elements, performance loss reduces to 1.008×. Due to recursion, majority of ADD instructions are associated with LOAD/STORE instructions. In radix sort, occurrence of ADD instructions is more compared to DIV instructions. Since it is memory intensive method of sorting, large number of

ADD instructions is used to increment counters and constants associated with LOAD/ STORE instructions. The performance loss due to emulation of ADD instructions for radix sort is 2.45×, 4.79×, and 5.96× for input sizes of 1000, 10,000 and 10,000 as shown in **Figure 8**. For DIV instructions, performance loss is 1.4×, 2.05×, and 2.37× for input sizes of 1000, 10,000 and 10,000 elements.

As shown in **Figure 9**, matrix multiplication with emulation of ADD and MUL instructions executing in the algorithmic phase of the program, incurs a performance loss of <1.56×, 4.09×, 4.0×> (for ADD) and <1.632×, 7.62×, 7.99×> (for MUL), for input matrix sizes of 10 × 10, 100 × 100, and 1000 × 1000 respectively. In CPU scheduling, ADD and SUB instructions emulation incur a performance loss of <2.45×, 4.79×, 5.96×> and <1.4×, 2.05×, 2.3×> for input data set of 1000, 10,000 and 100,000 processes respectively as shown in **Figure 10**. In sieve of Eratosthenes, emulation of MUL and ADD instructions incur a performance loss of <1.89×, 5.03×, 7.63×> and <1.48×, 2.9×, 3.8×> for input data set size of 1000, 10,000 and 100,000 respectively as shown in **Figure 11**.

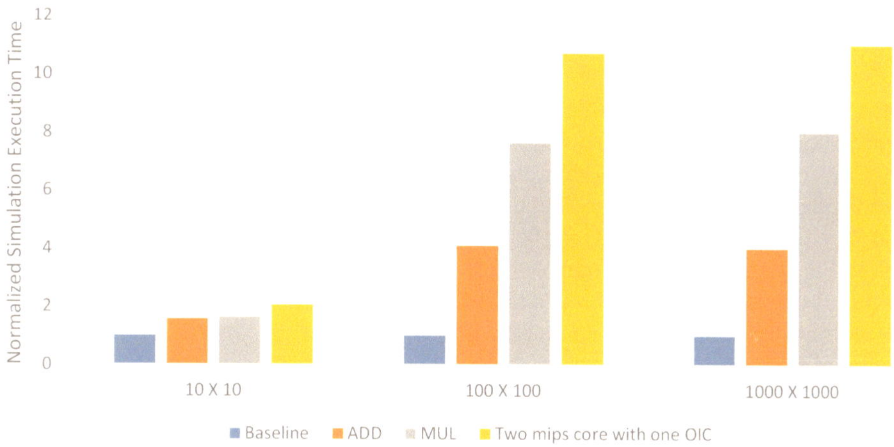

Figure 9.
Performance overhead in matrix multiplication by emulating ADD and MUL using subleq instruction.

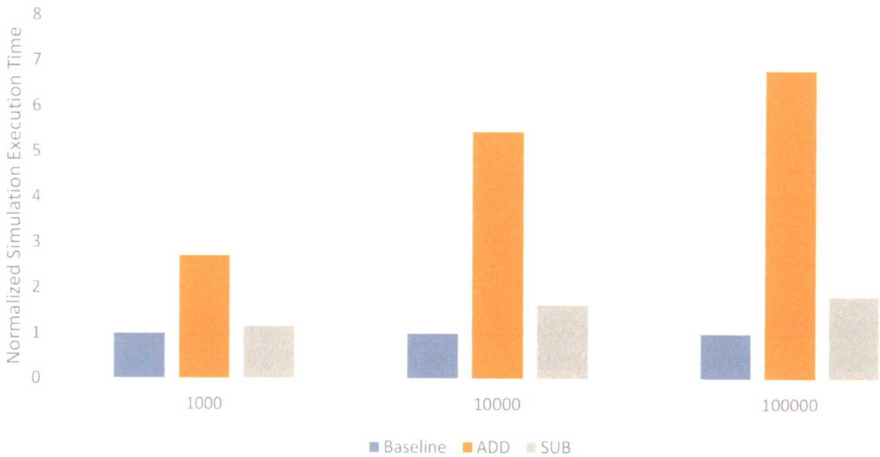

Figure 10.
Performance overhead in CPU scheduling by emulating ADD and SUB using subleq instruction.

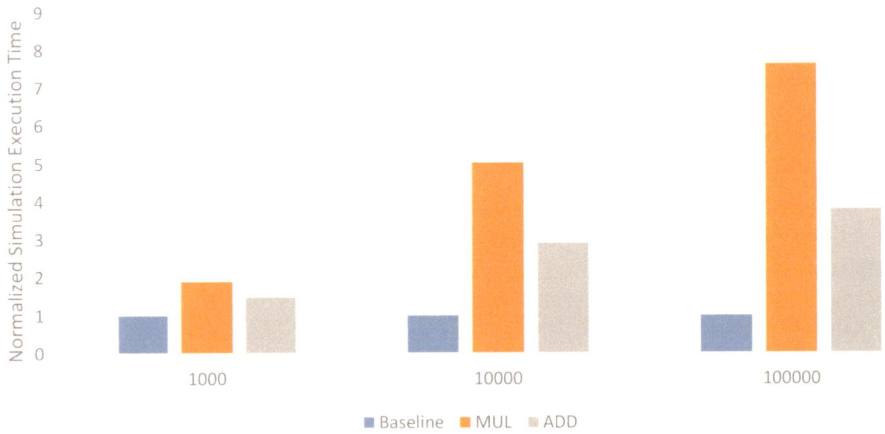

Figure 11.
Performance overhead in Sieve of Eratosthenes by emulating MUL and ADD using subleq instruction.

For multithreaded matrix multiplication, multi-core configuration consisting of two cores: core-0 and core-1 with single thread each is considered. The ADD and MUL instructions of core-0 and core-1 are emulated on single OIC due to failures in adder and multiplier units respectively. The performance loss is estimated as 2.04×, 10.07×, and 10.99× for matrix size of 10 × 10, 100 × 100, and 1000 × 1000 respectively as shown in **Figure 9**. Since, simultaneous access to single OIC from two cores is not permitted, performance loss includes the waiting time between subsequent ADD and MUL instructions emanating from core-0 and core-1. Waiting time alone is greater than 45% of the performance loss. In this multi-core configuration, consisting of two MIPS cores with single OIC, it bears the brunt of multiple functional unit failures in two cores. An Additional OIC would bring down the performance loss by 1.5× (for matrix size of 10 × 10) and 7× (for matrix size of 100 × 100/1000 × 1000) and eliminate the need for instructions to wait for execution on OIC. On 1:1 and 1: N basis i.e., one MIPS core with one or more OICs can scale to 100 MIPS core with 100 or more OICs.

It may be noticed that the performance loss does not vary when there is change of mode in OIC from TMR + SCR to TMR, or TMR to DMR as the number of instructions executed remains the same.

6. Yield analysis for MCS-OIC

This section examines the effect of fault tolerance provided in MCS-OIC on the yield. As discussed in the section which presents design of OIC, it is assumed that two subtractors do not fail simultaneously. In the TMR + SCR, TMR, and DMR modes, OIC repeats the instruction execution if the results differ, to avoid transient failures. The spatial and temporal redundancy to avoid permanent and transient faults in OIC makes it defect tolerant. The arithmetic logic unit in MIPS is protected by functional support provided by OIC. The remaining portion of MIPS are hardened and protected by ECC. The die yield for proposed different configurations of MCS-OIC is estimated using the equations presented below.

6.1 Terms and parameters

a. *Original die*: It is the die consisting of MIPS cores only.

b. *Fault tolerant die*: It is the die *consisting* of MIPS cores and OICs.

c. *Regular dies per wafer*: It is the number of original dies per wafer. The number of regular dies per wafer is estimated using the Eq. (1).

$$\text{Regular dies per wafer} = \frac{\pi(\text{diameter}/2)^2}{\text{Area}} - \frac{\pi \times \text{diameter}}{\sqrt{2} \times \text{Area}} \tag{1}$$

Where diameter refers to the diameter of the wafer, Area refers to the area of the die.

d. *Die yield*: Ignoring full wafer damage, the yield for single die is approximated using negative binomial approximation as given in the Eq. (2).

$$\textit{Die yield} = \left(1 + \frac{\text{defect density} \times \text{Area}}{\text{cp}}\right)^{-\text{cp}} \tag{2}$$

Where cp denotes cluster parameter or manufacturing complexity, defect density denotes number of defects per unit area.

e. *Regular working dies per wafer*: It is die yield times the regular dies per wafer. It is estimated using the Eq. (3).

$$\text{Regular working dies per wafer} = \left(1 + \frac{\text{defect density} \times \text{Area}}{\text{cp}}\right)^{-\text{cp}} \cdot \left(\frac{\pi(\text{diameter}/2)^2}{\text{Area}} - \frac{\pi \times \text{diameter}}{\sqrt{2} \times \text{Area}}\right) \tag{3}$$

f. *Regular fault tolerant dies per wafer*:

The area of fault tolerant core is expressed as summation of area of original die and area of OIC. If the area of OIC is expressed as $\delta(0 < \delta < 1)$ times the area of original design then $((1 + \delta) \times$ area of the original design$))$ denotes the area of the fault tolerant die. By substituting $(1 + \delta) \times$ area in the number of regular fault tolerant cores per wafer can be estimated and is given in the Eq. (4).

$$\text{Regular fault tolerant dies per wafer} = \frac{\pi(\text{diameter}/2)^2}{(1+\delta)\text{Area}} - \frac{\pi \times \text{diameter}}{\sqrt{(2\ (1+\delta)\ \text{Area})}} \tag{4}$$

g. *Regular working fault tolerant dies per wafer*: It is die yield times the regular fault tolerant die. It is estimated using the Eq. (5).

$$\text{Regular working fault tolerant dies per wafer} = \left(1 + \frac{\text{defect density} \times \text{Area}}{\text{cp}}\right)^{-\text{cp}} \cdot \left(\frac{\pi(\text{diameter}/2)^2}{(1+\delta)\text{Area}} - \frac{\pi \times \text{diameter}}{\sqrt{(2(1+\delta)\text{Area})}}\right) \quad (5)$$

6.2 Parametric evaluation and discussion

The die yield for the original die and fault tolerant die estimated for one MIPS core with one/two/four/ OICs, two MIPS core with one/two/four/ OICs, four MIPS core with one/two/four/ OICs, and eight MIPS core with one/two/four/six OICs is tabulated in **Tables 7–10** respectively. The defect density is varied from 9.5, 5.0, to 1.0 and wafer diameters varied from 300 mm, 200 mm to 100 mm to estimate die yield of the original die and fault tolerant die. The cp is fixed at 4.0. The die yield of the original die at defect densities are 1.0, 5.0, and 9.5 are 0.9971, 0.9855, and 0.9727 respectively. The die yields for three fault tolerant dies each consisting of one MIPS core with first die with one OIC, second with two OICs, third with four OICs for 300 mm wafer with defect density at 1.0 is <0.9970/0.9969/0.9967> respectively as shown in the **Table 7**, which is slightly lesser than the yield of the original die. The average of the differences between yield of original die and fault tolerant dies with defect density 1.0 is 0.0002 which is negligible value. Similarly, the average of the differences between yield of original die and fault tolerant dies at defect density 5.0 and 9.5 are 0.0009 and 0.0017 respectively. It is observed that an increase in the defect density decreases the yield.

Wafer diameter		**100 mm**			**200 mm**			**300 mm**		
Defect density		9.5	5.0	1.0	9.5	5.0	1.0	9.5	5.0	1.0
Number of regular dies per wafers		26,489	26,489	26,489	10,6781	10,6781	10,6781	24,0876	24,0876	24,0876
Die yield of original die		0.9727	0.9855	0.9971	0.9727	0.9855	0.9971	0.9727	0.9855	0.9971
Number of regular working dies per wafer		25,767	26,106	26,412	10,3870	10,5237	10,6470	23,4309	23,7391	24,0174
Number of regular fault tolerant dies per wafer	One OIC	25,768	25,768	25,768	103,884	103,884	103,884	234,347	234,347	234,347
	Two OICs	25,084	25,084	25,084	101,139	101,139	101,139	228,163	228,163	228,163
	Four OICs	23,820	23,820	23,820	96,061	96,061	96,061	216,723	216,723	216,723
Die yield of fault tolerant die	One OIC	0.9719	0.9851	0.9970	0.9719	0.9851	0.9970	0.9719	0.9851	0.9970
	Two OICs	0.9712	0.9847	0.9969	0.9712	0.9847	0.9969	0.9712	0.9847	0.9969
	Four OICs	0.9697	0.9839	0.9967	0.9697	0.9839	0.9967	0.9697	0.9839	0.9967
Number of regular working fault dies per wafer	One OIC	25,046	25,385	25,691	10,0974	10,2340	103,573	227,784	230,864	233,645
	Two OICs	24,363	24,701	25,007	98,231	99,595	100,828	221,603	224,681	227,461
	Four OICs	23,100	23,437	23,743	93,157	94,518	95,750	210,170	213,243	216,021

Table 7.
Die yield for fault tolerant die consisting of one MIPS core with one/two/four OICs.

Wafer diameter		**100 mm**			**200 mm**			**300 mm**		
Defect density		9.5	5.0	1.0	9.5	5.0	1.0	9.5	5.0	1.0
Number of regular dies per wafers		13,159	13,159	13,159	53,220	53,220	53,220	120,182	120,182	120,182
Die yield for original die		0.9463	0.9713	0.9942	0.9463	0.9713	0.9942	0.9463	0.9713	0.9942
Number of regular working dies per wafer		12,454	12,782	13,082	50,368	51,694	52,911	113,740	116,736	119,484
Number of regular fault tolerant dies per wafer	One OIC	12,977	12,977	12,977	52,487	52,487	52,487	118,529	118,529	118,529
	Two OICs	12,494	12,494	12,494	50,544	50,544	50,544	114,150	114,150	114,150
	Four OICs	12,459	12,459	12,459	50,403	50,403	50,403	113,833	113,833	113,833
Die yield for fault tolerant die	One OIC	0.9456	0.9709	0.9941	0.9456	0.9709	0.9941	0.9456	0.9709	0.9941
	Two OICs	0.9449	0.9705	0.9940	0.9449	0.9705	0.9940	0.9449	0.9705	0.9940
	Four OICs	0.9428	0.9693	0.9937	0.9428	0.9693	0.9937	0.9428	0.9693	0.9937
Number of regular working fault tolerant dies per wafer	One OIC	12,272	12,600	12,900	49,636	50,962	52,177	112,091	115,085	117,830
	Two OICs	12,095	12,423	12,723	48,924	50,249	51,464	110,486	113,478	116,222
	Four OICs	11,592	11,919	12,219	46,900	48,222	49,435	105,923	108,908	111,650

Table 8.
Die yield for fault tolerant die consisting of two MIPS core with one/two/four OICs.

The die yield of the fault tolerant dies each consisting of two MIPS cores with < one/two/four> OICS with defect density 1.0 is <0.9941, 0.9940, 0.9937> respectively as shown in **Table 8**. The die yield of the original die at defect densities 1.0, 5.0, and 9.5 is 0.9942, 0.9713, and 0.9463 slightly higher than yield of fault tolerant dies. The average of the differences between yield of original die and fault tolerant dies is 0.00026. The average of the differences between yield of original die and fault tolerant dies increases by 0.0009 and 0.0018 for defect density 5.0 and 9.5 respectively.

The die yields of the original die at defect densities 1.0, 5.0, and 9.5 are 0.9884, 0.9436, and 0.8963 respectively. From **Table 9**, the die yield of the fault tolerant dies each consisting of four MIPS cores with <one/two/four> OICS with defect density 1.0 are <0.9883, 0.9882, 0.9880> respectively. It is observed that average of the differences between yield of the original die and fault tolerant dies at varying defect densities is similar with other alternatives discussed above.

From **Table 10**, the die yield of the fault tolerant dies each consisting of eight MIPS cores with <one/two/four/six> OICS with defect density 1.0 is <0.9769, 0.9767, 0.9765, 0.9764> respectively. The die yield of the original die at defect densities 1.0, 5.0, and 9.5 is 0.9769, 0.8912, and 0.8057 respectively. The average of the differences between the original die and fault tolerant dies with defect density of 9.5 is 0.0031, is the highest among the averages. From this data, it is inferred that larger chips with increasing redundancy widens gap between the yield of the original dies and fault

Wafer diameter		100 mm			200 mm			300 mm		
Defect density		9.5	5.0	1.0	9.5	5.0	1.0	9.5	5.0	1.0
Number of regular dies per wafers		6519	6519	6519	26,489	26,489	26,489	59,910	59,910	59,910
Die yield for original die		0.8963	0.9436	0.9984	0.8963	0.9436	0.9984	0.8963	0.9436	0.9984
Number of regular working dies per wafer		5843	6152	6444	23,744	24,997	26,182	53,700	56,536	59,216
Number of working fault tolerant dies per wafer	One OIC	6474	6474	6474	26,305	26,305	26,305	59,495	59,495	59,495
	Two OICs	6428	6428	6428	26,124	26,124	26,124	59,085	59,085	59,085
	Four OICs	6340	6340	6340	25,768	25,768	25,768	58,282	58,282	58,282
Die yield for fault tolerant die	One OIC	0.8956	0.9433	0.9883	0.8956	0.9433	0.9883	0.8956	0.9433	0.9883
	Two OICs	0.8949	0.9429	0.9882	0.8949	0.9429	0.9882	0.8949	0.9429	0.9882
	Four OICs	0.8929	0.9417	0.9880	0.8929	0.9417	0.9880	0.8929	0.9417	0.9880
Number of regular working fault tolerant dies per wafer	One OIC	5798	6106	6398	23,561	24,814	25,998	53,288	56,122	58,800
	Two OICs	5753	6062	6353	23,381	24,633	25,817	52,881	55,713	58,391
	Four OICs	5623	5930	6221	22,855	24,104	25,286	51,694	54,520	57,195

Table 9.
Die yield for fault tolerant die consisting of four MIPS core with one/two/four OICs.

Wafer diameter		100 mm			200 mm			300 mm		
Defect density		9.5	5.0	1.0	9.5	5.0	1.0	9.5	5.0	1.0
Number of regular dies per wafers		3217	3217	3217	13,159	13,159	13,159	29,827	29,827	29,827
Die yield for original die		0.8057	0.8912	0.9770	0.8057	0.8912	0.9770	0.8057	0.8912	0.9770
Number of regular working dies per wafer		2592	2867	3143	10,603	11,728	12,856	24,034	26,584	29,141
Number of regular fault tolerant dies per wafer	One OIC	3205	3205	3205	13,113	13,113	13,113	29,723	29,723	29,723
	Two OICs	3194	3194	3194	13,068	13,068	13,068	29,620	29,620	29,620
	Four OICs	3172	3172	3172	12,977	12,977	12,977	29,415	29,415	29,415
	Six OICs	3150	3150	3150	12,888	12,888	12,888	29,214	29,214	29,214
Die yield for fault tolerant die	One OIC	0.8051	0.8909	0.9769	0.8051	0.8909	0.9769	0.8051	0.8909	0.9769
	Two OICs	0.8040	0.8902	0.9767	0.8040	0.8902	0.9767	0.8040	0.8902	0.9767
	Four OICs	0.8028	0.8895	0.9765	0.8028	0.8895	0.9765	0.8028	0.8895	0.9765
	Six OICs	0.8016	0.8888	0.9764	0.8016	0.8888	0.9764	0.8016	0.8888	0.9764
Number of regular working fault tolerant dies per wafer	One OIC	2581	2856	3131	10,559	11,683	12,810	23,933	26,481	29,037
	Two OICs	2559	2833	3109	10,470	11,592	12,719	23,732	26,277	28,831
	Four OICs	2537	2811	3087	10,382	11,503	12,629	23,535	26,075	28,628
	Six OICs	2516	2790	3065	10,296	11,415	12,541	23,340	25,877	28,428

Table 10.
Die yield for fault tolerant die consisting of eight MIPS core with one/two/four/six OICs.

tolerant dies. Thus, a trade-off exists between the die yield and fault tolerance provided by the design alternatives (discussed above) having redundancy ranging between 2% and 11%.

7. Reliability analysis of 32-bit OIC

In order to assess the endurance for the four modes of OIC, reliability is evaluated and compared. The reliability, denoted by $R(\mathrm{t})$, is defined as the probability of its survival at least until time t, which is estimated using Weibull distribution and can be determined in the following manner:

$$R(t) = P(T > t) = e^{-\lambda t^{\beta}} \tag{6}$$

where β is the shape parameter, T denotes the lifetime and λ denotes the failure rate of a component. Defect induced faults occur in the early stage of the lifetime, but the wear-out induced faults increase in the tail end of the lifetime. $\beta < 1$, is used to model infant mortality and it is a period of growing reliability and decreasing failure rate. When $\beta = 1$, the $R(\mathrm{t})$ of Weibull distribution and exponential distribution are identical. $\beta > 1$, is used to model wear out and the end of useful life where failure rate is increasing. The initial failure rate is computed using the failure rate formula:

$$\lambda = (C_1 \pi_{\mathrm{T}} \pi_{\mathrm{V}} + C_2 \pi_{\mathrm{E}}) \pi_{\mathrm{Q}} \pi_{\mathrm{L}} \tag{7}$$

here, C_1, C_2 are the complexity factors, $\pi_{\mathrm{T}}, \pi_{\mathrm{V}}, \pi_{\mathrm{E}}, \pi_{\mathrm{Q}}, \pi_{\mathrm{L}}$ are temperature, voltage stress, environment, quality and learning factors respectively. Failure rate λ is assumed as a function of the number of logical elements in the micro-architectural components.

The reliabilities of the four modes of OIC given in the Eqs. (8)–(11) are expressed in terms of $R_{\text{select logic}}(t), R_{\text{sub}}(t), R_{\text{sub-sc}}(t), R_{\text{comp}}(t) R_{\text{voter}}$ which denote the reliabilities of select logic, subtractor, SCS, comparator and voters logic respectively.

TMR + SCS mode reliability is expressed as:

$$R_{\text{TMT+SCS}}(t) = R_{\text{sub-sc}}(t) R_{\text{select logic}}(t) R_{\text{comp}}(t) R_{\text{voter}}(t) \sum_{i=2}^{4} \binom{4}{i} (R_{\text{sub}}(t))^i (1 - R_{\text{sub}}(t))^{4-i} \tag{8}$$

TMR mode reliability is expressed as:

$$R_{\text{TMR}}(t) = R_{\text{select logic}}(t) R_{\text{comp}}(t) R_{\text{voter}}(t) \sum_{i=2}^{3} \binom{3}{i} (R_{\text{sub}}(t))^i (1 - R_{\text{sub}}(t))^{3-i} \tag{9}$$

DMR mode reliability is expressed as:

$$R_{\text{DMR}}(t) = R_{\text{select logic}}(t) R_{\text{comp}}(t) \sum_{i=1}^{2} \binom{2}{i} (R_{\text{sub}}(t))^i (1 - R_{\text{sub}}(t))^{2-i} \tag{10}$$

Baseline mode reliability is expressed as:

$$R_{\text{baseline}}(t) = R_{\text{select logic}}(t) R_{\text{sub-sc}}(t) \quad (11)$$

The reliabilities are plotted for TMR + SCS, TMR, DMR and Baseline modes in **Figure 12** for β = 0.9 and 1.0, (which denote defect induced fault phase) and in **Figure 13** for β = 1.1 and 1.2 (which denote wear out induced fault phase). λ is a function of number of logical elements as given in the **Table 3**.

In all these cases, TMR + SCS mode is observed to have a better failure tolerance when compared to all other modes. For β = 1.2, the reliabilities of TMR mode and DMR mode are less than that of TMR + SCS mode during the interval 3×10^4 to 15×10^4 hours, as illustrated in **Figure 13**. The levels of reliability of TMR modes decline far below DMR, and baseline modes in wear out induced fault phase due to the

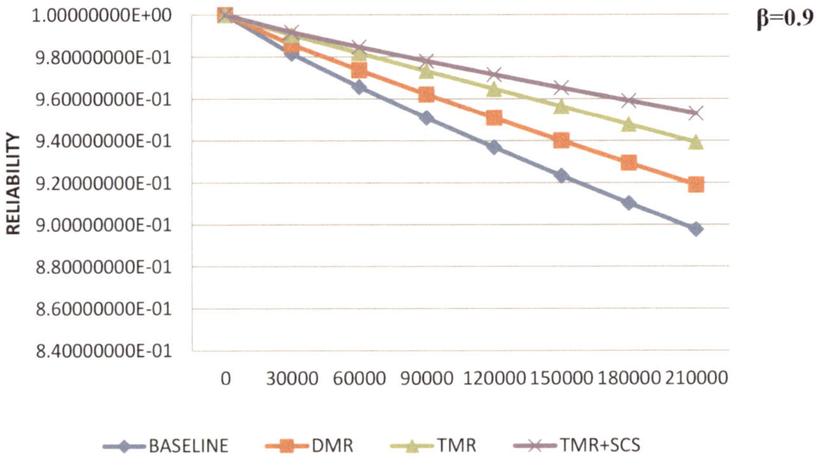

(a)

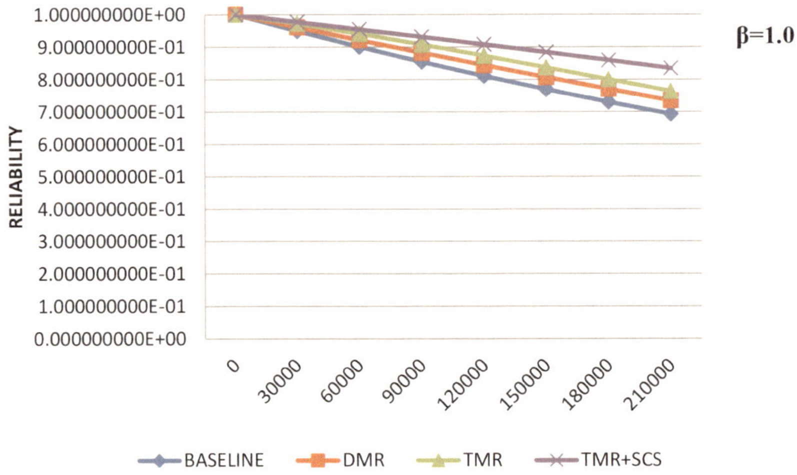

(b)

Figure 12.
Reliability vs. time for (a) β = 0.9 and (b) β = 1.0.

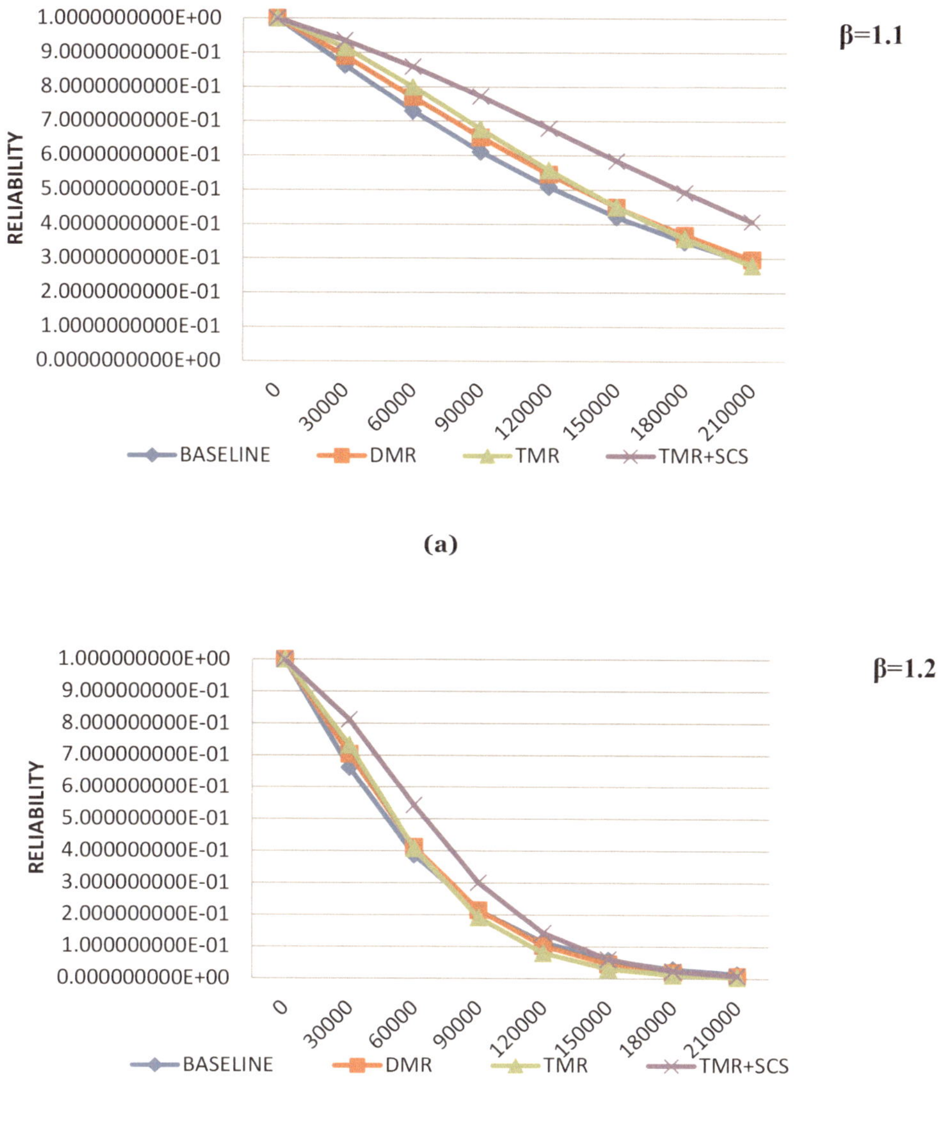

(b)

Figure 13.
Reliability vs. time for (a) $\beta = 1.1$ *and (b)* $\beta = 1.2$

fact that a single component reliability is below 0.5 and that the redundancy does not have any merit in the TMR mode. In **Table 11**, reliability of subtractor goes below 0.5 at t = 180,000 h or 20.5 years and reliability gap between TMR and DMR widens endorses the above argument.

7.1 Comparative analysis: OIC and URISC/URISC++

In this section, reliability of OIC is compared with that of URISC++. The reliability function of Weibull distribution with λ as a function of number of logical elements is used to estimate the reliability of URISC/URISC++. The number of logical elements in

t (h)	*R* (subtractor)	*R* (comparator)	*R* (TMR)	*R* (DMR)
120,000 (13.7 years)	0.6256	0.6948	0.16931	0.2112
150,000 (17.12 years)	0.5417	0.6226	0.09060	0.1272
180,000 (20.5 years)	0.4663	0.5545	0.04633	0.07706

Table 11.
Reliabilities of components in OIC for β = 1.2.

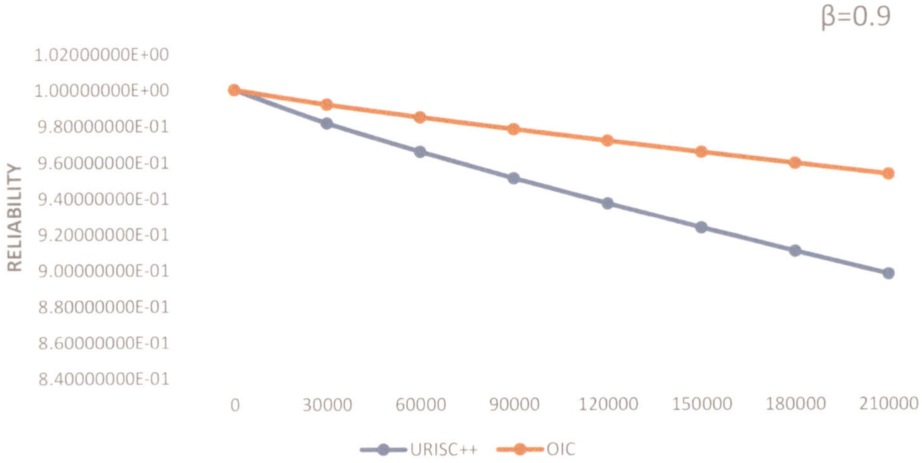

Figure 14.
β = 0.9 *reliability vs. time (hours).*

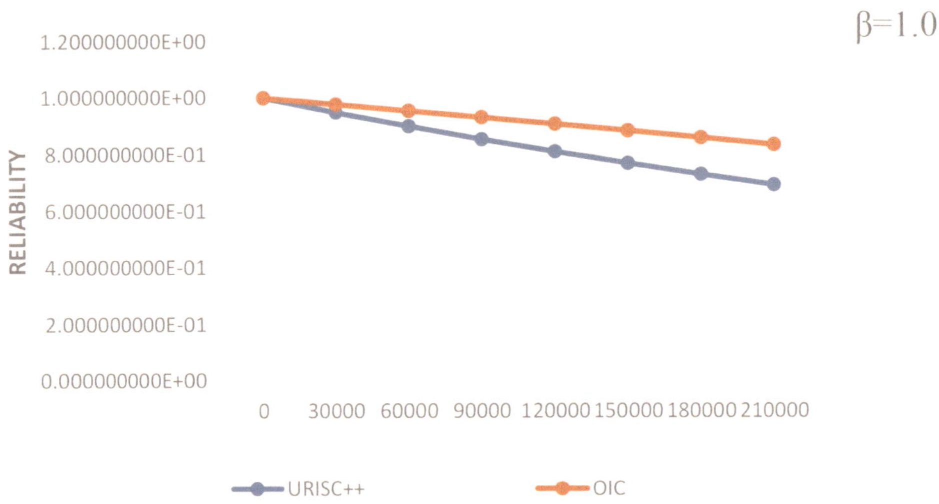

Figure 15.
β = 1.0 *reliability vs. time (hours).*

OIC and URISC++ are given in **Table 4**. In the defect induced fault phase (β = 0.9 and β = 1.0), a drastic fall in the URISC++ reliability is observed as shown in **Figures 14** and **15**. OIC continues to maintain a reliability of 0.96, unlike URISC++ with endurance reaching 0.87 after 210,000 hours. In the wear-out induced fault phase, the

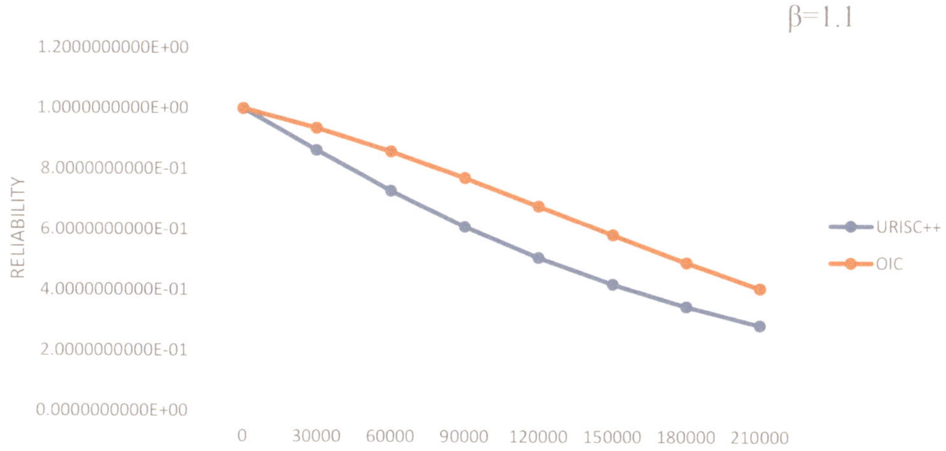

Figure 16.
β = 1.1 reliability vs. time (hours).

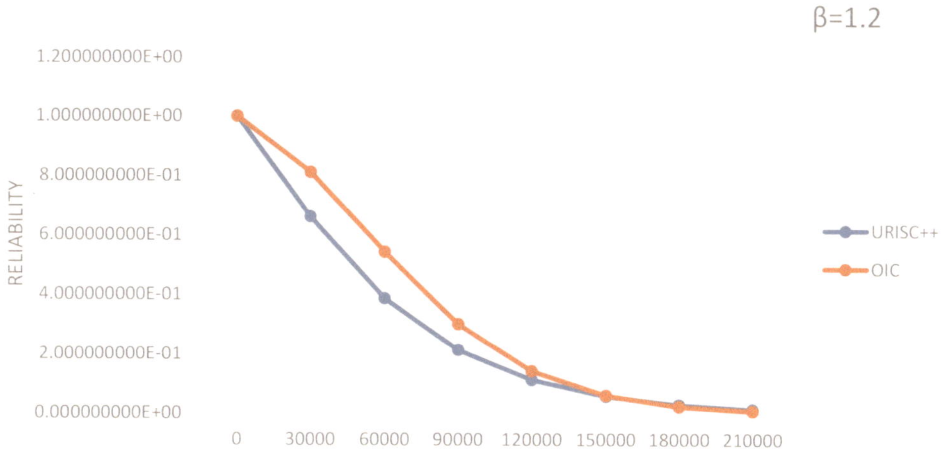

Figure 17.
β = 1.2 Reliability vs. time (hours).

reliability gap widens between 32-bit OIC and URISC++ when β = 1.1 (**Figure 16**) after 60,000 hours or 6.84 years. The reliability levels of OIC fall below that of URISC++ because single component reliability reduces below 0.5 after 23.4 years as shown in **Figure 17** and the redundancy in the OIC does not have any merit thereafter.

8. Conclusion

1. Power, area and total power for OIC and for its contender URISC++ are evaluated. OIC consumes less power and area compared to its contender. The registers count in OICs is significantly less compared to URISC++. It is observed that two large register files in URISC++ consume more power, unlike OIC which does not maintain register files.

2. The performance overheads at instruction level and application level are evaluated. In terms of performance overhead, based on the analysis in the Section 5, performance loss is incurred in compute intensive and memory intensive micro-benchmarks mainly due to MUL and DIV instructions in the programs. But the performance loss will not be high in programs with right mix of arithmetic instructions.

3. In 1:1 configuration of multi-core system with OICs i.e., one conventional core with one OIC, all the emulation request from the conventional core is handled by OIC. In 2:1 configuration (two cores and one OIC), simultaneous failures in two conventional cores results in higher performance loss for the application executing in the system. This performance loss can be reduced by augmenting the multi-core configuration with an additional OIC. That is, 1:1 model proves to be a viable solution with minimal performance loss. This is validated by the simulation results presented in this chapter. On 1:1 and 1: N basis i.e., one MIPS core with one or more OICs can scale to 100 MIPS core with 100 or more OICs. Hence, MCS-OIC model is a scalable design alternative.

4. As expected, it is observed from the reliability analysis of OIC that an increase in the number of subtractors results in higher reliability. Alternatively, it can be understood that replication of functional units improves reliability of the OIC significantly. Hence, TMR + SCS mode has higher reliability compared to the other modes.

5. The yield of the fault tolerant die is slightly lesser than the original die for all the design alternatives of MCS-OIC. It is inferred that larger chips with increasing redundancy widens gap between the yield of the original dies and fault tolerant dies. Thus, a trade-off exists between the die yield and fault tolerance provided by the design alternatives (discussed above) having redundancy ranging between 2% and 11%.

6. Reliability of OIC and URISC++ are evaluated and compared. Evaluation results indicate that OIC is more reliable than URISC++ both in the defect induced phase and the wear out induced phase. It can be understood that the level of redundancy is significantly less in URISC++ compared to OIC.

Author details

Shashikiran Venkatesha[1]* and Ranjani Parthasarathi[2]

1 Vellore Institute of Technology, Vellore, Tamil Nadu, India

2 Department of Information Science and Technology, College of Engineering Guindy, Anna University, Chennai, Tamil Nadu, India

*Address all correspondence to: shashikiran.annauniv@gmail.com

References

[1] Borkar S. Designing reliable systems from unreliable components: The challenges of transistor variability and degradation. IEEE Micro. 2005;**25**(6): 10-16

[2] Shivakumar P, Kistler M, Keckler SW, Burger D, Alvisi L. Modeling the effect of technology trends on the soft error rate of combinational logic. In: Proceedings of International Conference on Dependable Systems and Networks. IEEE Explorer. 2002. pp. 389-398. DOI: 10.1109/DSN.2002.1028924

[3] Feng S, Gupta S, Ansari A, Mahlke S. Shoestring: Probabilistic soft error reliability on the cheap'. ACM SIGARCH Computer Architecture News. 2010; **38**(1):385-396

[4] Li T, Ambrose JA, Ragel R, Parameswaran S. Processor design for soft errors: Challenges and state of the art. ACM Computing Surveys. 2016; **49**(3):1-44

[5] Mittal S. A survey of techniques for managing and leveraging caches in GPUs. Journal of Circuits, Systems, and Computers. 2014;**23**(08):1430002

[6] Rusu S, Muljono H, Ayers D, Tam S, Chen W, Martin A, et al. 5.4 Ivytown: A 22 nm 15-core enterprise Xeon® processor family. In: 2014 IEEE International Solid-State Circuits Conference Digest of Technical Papers (ISSCC). IEEE Explorer; 2014. pp. 102-103. DOI: 10.1109/ISSCC.2014.6757356

[7] Zyuban V, Taylor SA, Christensen B, Hall AR, Gonzalez CJ, Friedrich J, et al. IBM POWER7+ design for higher frequency at fixed power. IBM Journal of Research and Development. 2013;**57**(6): 1-1

[8] Postman J, Chiang P. A survey addressing on-chip interconnect: Energy and reliability considerations. International Scholarly Research Notices. 2012;**2012**:1-9. Article ID: 916259. DOI: 10.5402/2012/916259

[9] Nassif SR, Mehta N, Cao Y. A resilience roadmap. In: 2010 Design, Automation & Test in Europe Conference & Exhibition (DATE 2010). IEEE Explorer; 2010. pp. 1011-1016. DOI: 10.1109/DATE.2010.5456958

[10] Karnik T, Tschanz J, Borkar N, Howard J, Vangal S, De V, et al. Resiliency for many-core system on a chip. In: 2014 19th Asia and South Pacific Design Automation Conference (ASP-DAC). IEEE Explorer; 2014. pp. 388-389. DOI: 10.1109/ASPDAC.2014.6742921

[11] Gaisler J. A portable and fault-tolerant microprocessor based on the SPARC v8 architecture. In: Proceedings International Conference on Dependable Systems and Networks. IEEE Explorer; 2002. pp. 409-415. DOI: 10.1109/DSN.2002.1028926

[12] Lin S, Kim YB, Lombardi F. Design and performance evaluation of radiation hardened latches for nanoscale CMOS. IEEE Transactions on Very Large-scale Integration Systems. 2010;**19**(7): 1315-1319

[13] Slayman CW. Cache and memory error detection, correction, and reduction techniques for terrestrial servers and workstations. IEEE Transactions on Device and Materials Reliability. 2005;**5**(3):397-404

[14] Pomeranz I, Vijaykumar TN. FaultHound: Value-locality-based soft-fault tolerance. In: Proceedings of the 42nd Annual International Symposium

on Computer Architecture. ACM Digital Library; 2015. pp. 668-681. DOI: 10.1145/ 2749469.2750372

[15] Meaney PJ, Swaney SB, Sanda PN, Spainhower L. IBM z990 soft error detection and recovery. IEEE Transactions on Device and Materials Reliability. 2005;**5**(3):419-427

[16] Stackhouse B, Bhimji S, Bostak C, Bradley D, Cherkauer B, Desai J, et al. A 65 nm 2-billion transistor quad-core Itanium processor. IEEE Journal of Solid-State Circuits. 2008;**44**(1):18-31

[17] Venkatesha S, Parthasarathi R. 32-Bit one instruction core: A low-cost, reliable, and fault-tolerant core for multicore systems. Journal of Testing and Evaluation. 2019;**47**(6):3941-3962. DOI: 10.1520/JTE20180492. ISSN 0090-3973

[18] Hamming RW. Error detecting and error correcting codes'. The Bell System Technical Journal. 1950;**29**(2): 147-160

[19] Rajendiran A, Ananthanarayanan S, Patel HD, Tripunitara MV, Garg S. Reliable computing with ultra-reduced instruction set co-processors. In: DAC Design Automation Conference 2012. ACM Digital Library; 2012. pp. 697-702. DOI: 10.1145/2228360.2228485

[20] Ananthanarayan S, Garg S, Patel HD. Low-cost permanent fault detection using ultra-reduced instruction set co-processors. In: 2013 Design, Automation & Test in Europe Conference & Exhibition (DATE). IEEE Explorer; 2013. pp. 933-938. DOI: 10.7873/ DATE.2013.196

[21] Sundaramoorthy K, Purser Z, Rotenberg E. Slipstream processors: Improving both performance and fault tolerance'. ACM SIGPLAN Notices. 2000;**35**(11):257-268

[22] LaFrieda C, Ipek E, Martinez JF, Manohar R. Utilizing dynamically coupled cores to form a resilient chip multiprocessor. In: 37th Annual IEEE/ IFIP International Conference on Dependable Systems and Networks (DSN'07). IEEE Explorer; 2007. pp. 317-326. DOI: 10.1109/DSN.2007.100

[23] Aggarwal N, Ranganathan P, Jouppi NP, Smith JE. Configurable isolation: Building high availability systems with commodity multi-core processors. ACM SIGARCH Computer Architecture News. 2007;**35**(2):470-481

[24] Smolens JC, Gold BT, Falsafi B, Hoe JC. Reunion: Complexity-effective multicore redundancy. In: 2006 39th Annual IEEE/ACM International Symposium on Microarchitecture (MICRO'06). IEEE Explorer; 2006. pp. 223-234. DOI: 10.1109/MICRO.2006.42

Chapter 10

Computer Vision-Based Techniques for Quality Inspection of Concrete Building Structures

Siwei Chang and Ming-Fung Francis Siu

Abstract

Quality performance of building construction is frequently assessed throughout the construction life cycle. In Hong Kong, quality management system must be established before commencing new building works. Regular building inspections are conducted in accordance with the code of practice of new building works. Quality managers are deployed in construction sites to inspect and record any building defects. The concrete cracks must be identified, which is usually followed by proposed rectifications, in order to protect the public and occupants from dangers. This chapter is structured as follows: Background information of concrete cracks is firstly given. Traditional technique of conducting regular manual inspection is introduced, in accordance with Hong Kong's code of practice "Building Performance Assessment Scoring System (PASS)". Then, an advanced technique of conducting crack inspection intelligently based on computer vision is introduced. The procedures of defining, training, and benchmarking the architecture of convolutional neural network models are presented. The calculation steps are detailed and illustrated using a simple text-book example. An experiment case study is used to compare the time, cost of inspecting concrete cracks using both manual and advanced technique. The study concludes with a presentation of the future vision of robot-human collaboration for inspecting concrete cracks in building construction.

Keywords: building quality control, concrete crack, quality inspection, computer vision, artificial intelligence

1. Introduction

Throughout the entire construction life cycle, quality assessment plays an important role in ensuring the safety, economy, and long-term viability of construction activities. Construction products that have been completely inspected and certificated by quality inspectors are more inclined to be chosen by developers and buyers. Typically, the structural work is considered as an essential aspect for quality assessment because structural problems directly influence the construction stability and integrity. Among the construction structural forms, concrete structures are adopted as the most common and basic construction structure. Therefore, exploring advanced technologies that enable effective concrete defect inspection can be deemed a worthwhile endeavor.

Normally, the types of concrete defects include blistering, delamination, dusting, etc. Among them, concrete cracks, usually caused by deformation, shrinkage, swelling, or hydraulic, appear most frequently in concrete components. Concrete cracking is considered the first sign of deterioration. As reported by the BRE Group [1], cracks up to 5 mm in width simply need to be re-decorated because they only affect the appearance of the concrete. However, cracks with a width of 5–25 mm have the possibility to trigger structural damage to concrete structures [2]. A 40-year-old oceanfront condo building collapsed on June 27, 2021, in Florida because of the neglect of cracks. Experienced engineers noticed the cracked or crumbling concrete, the interior cracks, and the cracks at the corners of windows and doors are the significant and earliest signs of this tragedy. Therefore, in order to prevent potential failures that may pose a loss to society, crack problems should be thoroughly examined and resolved.

In general, construction works are divided into two categories: new building works and existing building works. The new works refer to a building that will be constructed from scratch. The existing building works mean that a building has existed for many years and residents are living inside. In Hong Kong, quality assurance and control should be conducted by full-time quality managers on-site for both new and existing buildings. Normally, the quality managers visually inspect implied build quality and by appointing a score to the building's quality in accordance to the Building Performance Assessment Scoring System (PASS) for new buildings, the Mandatory Building Inspection Scheme (MBIS), and the Mandatory Window Inspection Scheme (MWIS) for existing buildings. Meanwhile, to ensure a continuous and in-depth inspection, Non-destructive (NDT) methods e.g., eddy current testing, ultrasonic testing are also commonly applied in the quality inspection process.

Quality managers are commonly obliged to work 8 hours per day. Their salary ranges from HKD 30,000 to HKD 50,000 per month. In PASS, more than 300 quality assessment items are related to cracking-related problems. Cracks in all building components, including floors, internal and external walls, ceilings, and others are required to be strictly inspected during both structural and architecture engineering stages. Therefore, both manual and NDT inspections are considered time-consuming, costly, and dangerous, especially for large-scale and high-rise structures. To tackle this issue, computer-vision technique is increasingly introduced for automated crack inspection. For example, various convolutional neural network (CNN) architectures have been developed and implemented to increase the efficiency of manual crack inspection [3, 4].

Considering the aforementioned context, computer-vision-based automated crack inspection techniques were introduced by the authors in 2022. To achieve this, the theoretical background of CNN networks is firstly explained in the context of convolution, pooling, fully-connected, and benchmarking processes. AlexNet and VGG16 models were then implemented and tested to detail and illustrate the calculation steps. Meanwhile, a practical case study is used to compare the difference between manual and computer-vision-based crack inspection. The future directions of combining robotics and computer-vision for automated crack inspection are discussed. This study gives a comprehensive overview and solid foundation for a computer-vision-based automated crack inspection technique that contributes to high efficiency, cost-effectiveness, and low-risk quality assessment of buildings.

2. Computer vision-based automated concrete crack inspection

The term *computer vision* is defined as an interdisciplinary field that enables computers to recognize and interpret environments from digital images or videos [5]. Computer vision techniques are rapidly being used to detect, locate, and quantify concrete defects to reduce the limitations of manual visual inspection. By automatically processing images and videos, computer vision-based defect detection technologies enable efficient, accurate, and low-cost concrete quality inspection. Various techniques in the computer vision field, such as semantic segmentation and object detection, have been developed and applied to date [6]. Among them, image classification is considered the most basic computer vision technique and has been introduced most frequently to predict and target concrete defects.

The motivation of image classification is to identify the categories of input images. Different from human recognition, an image is first presented as a three-dimensional array of numbers to a computer. The value of each number ranges from 0 (black) to 255 (white). An example is shown in **Figure 1**. The crack image is 256 pixels wide, 256 pixels tall, and has three color channels RGB (Red, Green, and Blue). Therefore, this image generates $256 \times 256 \times 3 = 196{,}608$ input numbers.

The input array is then computed using computer vision algorithms to transform the numbers to a specific label that belongs to an assigned set of categories. One of the computer vision algorithms is CNN, which has become dominant in image classification tasks [7]. CNN is a form of a deep learning model for computing grid-shaped data. The central idea of CNN is to identify the image classification by capturing its features using filters. The features are then output to a specific classification by a trained weight and biases matrix.

There are three main modules included in a CNN model: convolution, pooling, and fully connected layer. The convolution and pooling layers are used to extract image features. The fully connected layer is used to determine the weight and biases matrix and to map the extracted features into specific labels.

Convolution layer is the first processing block in CNN. During the convolution process, a set of convolution filters is used to compute the input array $A = (a_{ij})_{m \times n}$, $m, n \in \left(width_{image}, height_{image}\right)$. After computing, a new image $A^* = (a^*_{ij})_{n \times n}$, is

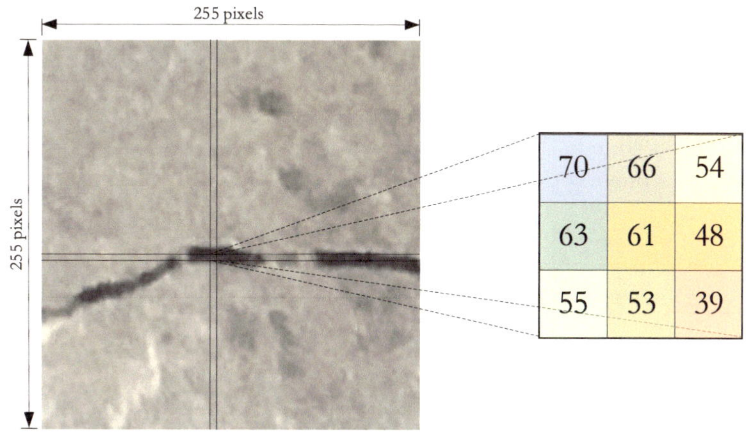

Figure 1.
An example of the input number array.

output and passed to the next processing layers. The size of the output image can be calculated with Eq. (1). The values of output image pixels can be calculated with Eq. (2). The output images are known as convolution feature map.

$$n = ((m - f + 2p)/s) + 1 \tag{1}$$

Here: n refers to the size of output image, m refers to the size of input image, f refers to the size of convolution filter, p refers to the number of pooling layer, s refers to the stride of convolution filter.

$$\mathrm{A}_o^* = f\left(\sum_k W_o \times \mathrm{A}_o + b_o\right) \tag{2}$$

Here: A_o^* refers to the pixels of output image, f refers to an applied non-linear function, W_o refers to the values of convolution filter matrix, k refers to the number of convolution filters. A_o refers to the pixels of input image, and b_o is an arbitrary real number.

An example of a convolution process is shown in **Figure 2**. In this example, both the width and height of the input image is 5. The pixels of the image are shown in **Figure 2**. The convolution filter is in a shape of 3×3. In this example, only one filter is used. The initial value of the convolution filter is set randomly. The filter matrix is adjusted and optimized in the following backpropagation process. In this example, the non-linear function, padding layer is not used, and the biases value b_o is set as 0. The stride of convolution filter is set as 1. The convolution filter moves from left to right, and from top to bottom. The size and value of the output feature map can be computed using Eqs. (1) and (2). The detailed calculation process of the example feature maps value and size is shown in **Table 1**. Seen from **Figure 2**, the value of size of input image, size of filter is 5, 3, respectively. Suppose the number of the pooling layer, the convolution stride is 0, 1, respectively.

A pooling layer is used to refine the feature maps. After pooling, the dimensions of the feature maps can be simplified. In doing so, the computation cost can be effectively decreased by reducing the number of learning parameters, whilst allowing only the essential information of feature maps to be presented. Usually, pooling layers follow behind convolution layers. Average pooling and maximum pooling are the main pooling operations. Similar to convolution layers, pooling filters are used to refine feature maps. For maximum pooling, the maximum value from the regions in feature map that is covered by pooling filters is extracted. For average pooling, the average value of the regions in feature maps covered by pooling filters is computed. The pooling filters slide in the feature map from top to bottom, and from left to right. The output of the pooling process is new feature maps that contain the most

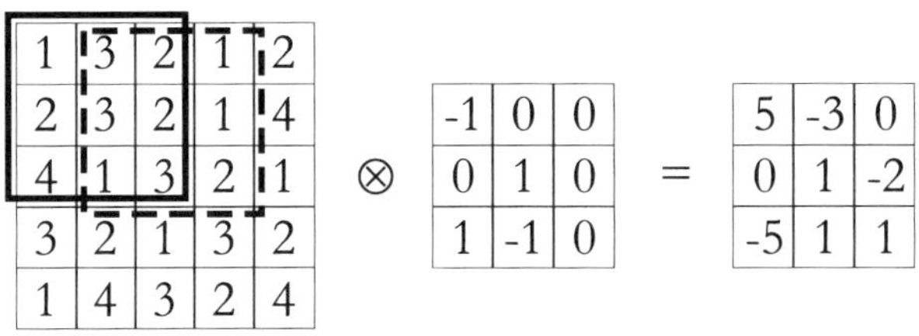

Figure 2.
An example of convolution process.

Variable	Equation	Calculation process
Size of feature map	$n = ((m - f + 2p)/s) + 1$	$((5–3 + 2 \times 0)/1) + 1 = 3$
Value of feature map	$A_o^* = f(\sum_k W_o \times A_o + b_o)$	$(-1) \times 1 + 0 \times 3 + 0 \times 2 + 0 \times 2 + 1 \times 3 + 0 \times 2 + 1 \times 4 + (-1) \times 1 + 0 \times 3 = 5$
		$(-1) \times 3 + 0 \times 2 + 0 \times 1 + 0 \times 3 + 1 \times 2 + 0 \times 1 + 1 \times 1 + (-1) \times 3 + 0 \times 2 = -3$
		$(-1) \times 2 + 0 \times 1 + 0 \times 2 + 0 \times 2 + 1 \times 1 + 0 \times 4 + 1 \times 3 + (-1) \times 2 + 0 \times 1 = 0$
		$(-1) \times 2 + 0 \times 3 + 0 \times 2 + 0 \times 4 + 1 \times 1 + 0 \times 3 + 1 \times 3 + (-1) \times 2 + 0 \times 1 = 0$
		$(-1) \times 3 + 0 \times 2 + 0 \times 1 + 0 \times 1 + 1 \times 3 + 0 \times 2 + 1 \times 2 + (-1) \times 1 + 0 \times 3 = 1$
		$(-1) \times 2 + 0 \times 1 + 0 \times 4 + 0 \times 3 + 1 \times 2 + 0 \times 1 + 1 \times 1 + (-1) \times 3 + 0 \times 2 = -2$
		$(-1) \times 4 + 0 \times 1 + 0 \times 3 + 0 \times 3 + 1 \times 2 + 0 \times 1 + 1 \times 1 + (-1) \times 4 + 0 \times 3 = -5$
		$(-1) \times 1 + 0 \times 3 + 0 \times 2 + 0 \times 2 + 1 \times 1 + 0 \times 3 + 1 \times 4 + (-1) \times 3 + 0 \times 2 = 1$
		$(-1) \times 3 + 0 \times 2 + 0 \times 1 + 0 \times 1 + 1 \times 3 + 0 \times 2 + 1 \times 3 + (-1) \times 2 + 0 \times 4 = 1$

Table 1.
Detailed calculation process of feature map value and size.

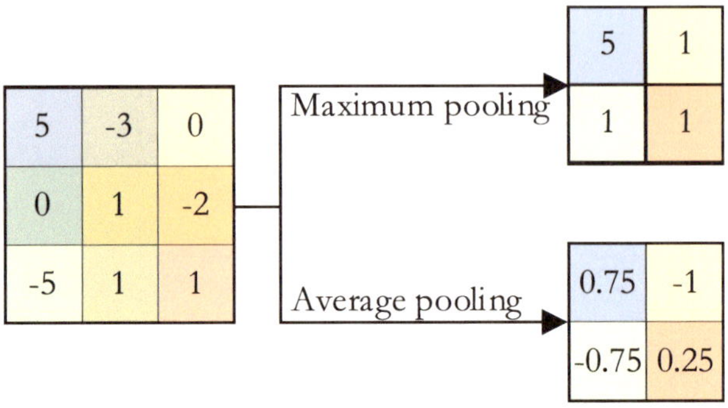

Figure 3.
An example of max pooling and average pooling.

prominent features or average features. An example of maximum pooling and average pooling is shown in **Figure 3**.

After extracting image features, the fully connected layers are applied to map these features with classification labels. The relationship between input feature maps and output classifications is calculated using an artificial neural network (ANN). The ANN is structured into input layers, hidden layers, and output layers. A group of neurons is included in the three layers. The neurons connect to one another in a processed weight matrix. The weights present the importance of input feature maps to classification labels. Therefore, the relationships between inputs and outputs can be obtained by calculating a weight matrix that connects image feature neurons and classification neurons.

To achieve this, the cube-shaped feature maps are first flattened into one-dimension vectors. The values of transformed vectors represent the values of input neurons. Then Eq. (3) is applied to calculate the value of new neurons that connect with input neurons. The initial weights and biases values are chosen at random.

$$y_j(x) = f\left(\sum_{i=1}^{n} w_j x_i + b\right) \tag{3}$$

Here: y_j refers to the weights of output neurons, w_j refers to the weights that connect different neurons, x_i refers to the values of input neurons, b refers to the biases.

A Back-Propagation algorithm (BP) is commonly used to train and modify weights and biases. BP updates weights and biases by computing the gradient of loss function. In doing this, the optimal weights and biases matrix that enable the minimum loss between model outputs and actual value are identified. For now, various loss functions are developed and applied. For example, the mean square error (MSE), shown in Eq. (4), is one of the most frequently used loss functions to calculate loss value. Stochastic gradient descent (SGD) is then processed to determine updated weights and biases using the gradient of loss function, shown as Eq. (5).

$$Loss = \frac{1}{n}\sum_{i=1}^{n} (y_{\mathrm{i}} - \hat{y}_{\mathrm{i}})^2 \tag{4}$$

Here: *Loss* refers to the loss value of output neuron and actual value, n refers to the number of neurons that connect to one specific output neuron, y refers to the actual value, $\hat{y}$ refers to the value of one output neuron.

$$w' = w - \eta \frac{\partial L}{\partial w} \quad (5)$$
$$b' = b - \eta \frac{\partial L}{\partial b}$$

Here: w', b' refers to updated weights and biases, η, η refers to former weights and biases, η refers to the learning rate, $\frac{\partial L}{\partial w}$, $\frac{\partial L}{\partial b}$ refers to the partial score of the loss function for weights and biases, respectively.

An example of feature map updating using BP is explained. **Figure 4** depicts an example of a fully connected process. The initial weights and biases in this process are determined randomly. Suppose the value of w_{11}, w_{12}, w_{21}, w_{22}, w_5, w_6 is 0.1, 0.2, 0.3, 0.4, 0.5, 0.6, respectively. The value of x_1, x_2, actual output value is 5, 1, 0.24. The detailed calculation of the updated weights, biases, feature map is shown in **Table 2**.

In conclusion, during the convolution and pooling processes in CNN, the features of the input image are extracted first. The pooled feature maps are then flattened and considered as input neurons in fully connected process. After several training periods, the appropriate weights and biases can be determined using BP. The classifications of input images can be predicted automatically and reliably using the optimal weights and biases.

A confusion matrix is a table structure that permits the viewing of CNN performance [8]. Each row of the matrix records the number of images from actual classes, while each column records the number of images from predicted classes. There are four type indicators in the matrix: (1) True positive (TP) represents the images that are predicted correctly as the actual class; (2) False positive (FP) represents the images that are wrongly predicted; (3) True negative (TN) represents the images that are correctly predicted as another actual class; (4) False negative (FN) represents the images that are wrongly predicted as another actual class. TP, FP, TN, FN can be expressed in a 2×2 confusion matrix, shown in **Figure 5**.

Based on TP, FP, FN, and TN, four typical CNN performance evaluation indexes: accuracy, precision, recall, and F1-score can be calculated using Eqs. (6)–(9). For the crack inspection problem, accuracy shows how many images can be predicted correctly. The percentage of actual cracked photos to all predicted cracked images is shown by precision. CNNs with a high precision score indicate a better inspection ability of cracked images. Recall shows the ratio of predicted cracked images to all actual cracked images. CNNs with a high recall score indicate a better distinguishing capacity between cracked and uncracked images. F1-score shows the comprehensive

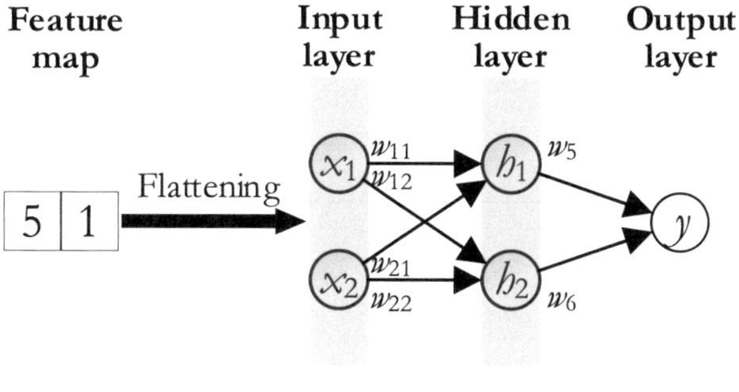

Figure 4.
An example of a fully connected process.

Variable	Equation	Calculation process
h_1	$h_1 = w_{11} \times 5 + w_{21} \times 1$	5 × 0.1 + 1 × 0.3 = 0.8
h_2	$h_2 = w_{12} \times 5 + w_{22} \times 1$	5 × 0.2 + 1 × 0.4 = 1.4
y	$y = w_5 \times h_1 + w_6 \times h_2$	0.8 × 0.5 + 1.4 × 0.6 = 1.24
Loss	$\text{Loss} = \frac{1}{2} \times \left(y_{\text{actual}} - y_{output}\right)^2$	1/2 × (0.24–1.24)^2 = 0.5
w_5’	$\frac{\partial L}{\partial w_5} = \frac{\partial L}{\partial y} \times \frac{\partial y}{\partial w_5} = 2 \times \frac{1}{2} \times \left(y_{\text{actual}} - y_{output}\right) \times h_1 \times (-1)$	2 × 1/2 × (0.24–1.24) × 0.8 × (−1) = 0.8
	$w_5' = w_5 - \eta\frac{\partial L}{\partial w5}$	0.5–0.1 × 0.8 = 0.42
w_6’	$\frac{\partial L}{\partial w_6} = \frac{\partial L}{\partial y} \times \frac{\partial y}{\partial w_6} = 2 \times \frac{1}{2} \times \left(y_{\text{actual}} - y_{output}\right) \times h_2 \times (-1)$	2 × 1/2 × (0.24–1.24) × 1.8 × (−1) = 1.8
	$w_6' = w_6 - \eta\frac{\partial L}{\partial w6}$	0.6–0.1 × 1.8 = 0.42
w_{11}’	$\frac{\partial L}{\partial w_{11}} = \frac{\partial L}{\partial y} \times \frac{\partial y}{\partial h_1} \times \frac{\partial h_1}{\partial w_{11}} = 2 \times \frac{1}{2} \times \left(y_{\text{actual}} - y_{output}\right) \times (-1) \times w_5 \times x_1$	2 × 1/2 × (0.24–1.24) × (−1) × 0.5 × 5 = 2.5
	$w_{11}' = w_{11} - \eta\frac{\partial L}{\partial w11}$	0.1–0.1 × 2.5 = −0.15
w_{12}’	$\frac{\partial L}{\partial w_{12}} = \frac{\partial L}{\partial y} \times \frac{\partial y}{\partial h_2} \times \frac{\partial h_2}{\partial w_{12}} = 2 \times \frac{1}{2} \times \left(y_{\text{actual}} - y_{output}\right) \times (-1) \times w_6 \times x_1$	2 × 1/2 × (0.24–1.24) × (−1) × 0.6 × 5 = 3
	$w_{12}' = w_{12} - \eta\frac{\partial L}{\partial w12}$	0.2–0.1 × 3 = −0.1
w_{21}’	$\frac{\partial L}{\partial w_{21}} = \frac{\partial L}{\partial y} \times \frac{\partial y}{\partial h_1} \times \frac{\partial h_1}{\partial w_{21}} = 2 \times \frac{1}{2} \times \left(y_{\text{actual}} - y_{output}\right) \times (-1) \times w_5 \times x_2$	2 × 1/2 × (0.24–1.24) × (−1) × 0.5 × 1 = 0.5
	$w_{21}' = w_{21} - \eta\frac{\partial L}{\partial w21}$	0.3–0.1 × 0.5 = 0.25
w_{22}’	$\frac{\partial L}{\partial w_{22}} = \frac{\partial L}{\partial y} \times \frac{\partial y}{\partial h_2} \times \frac{\partial h_2}{\partial w_{22}} = 2 \times \frac{1}{2} \times \left(y_{\text{actual}} - y_{output}\right) \times (-1) \times w_6 \times x_2$	2 × 1/2 × (0.24–1.24) × (−1) × 0.6 × 1 = 0.6
	$w_{22}' = w_{22} - \eta\frac{\partial L}{\partial w22}$	0.4–0.1 × 0.6 = 0.34
Updated feature map	$h_1' = y_{output} \times w_5'$	1.24 × 0.42 = 0.5208
	$h_2' = y_{output} \times w_6'$	1.24 × 0.42 = 0.5208
	$x_1' = h_1' \times w_{11}' + h_2' \times w_{12}'$	0.5208 × (−0.15) + 0.5208 × (−0.1) = 0.1302
	$x_2' = h_1' \times w_{21}' + h_2' \times w_{22}'$	0.5208 × 0.5+ 0.5208 × 0.6 = 0.57288

Table 2.
Detailed calculation process of feature map updating using BP.

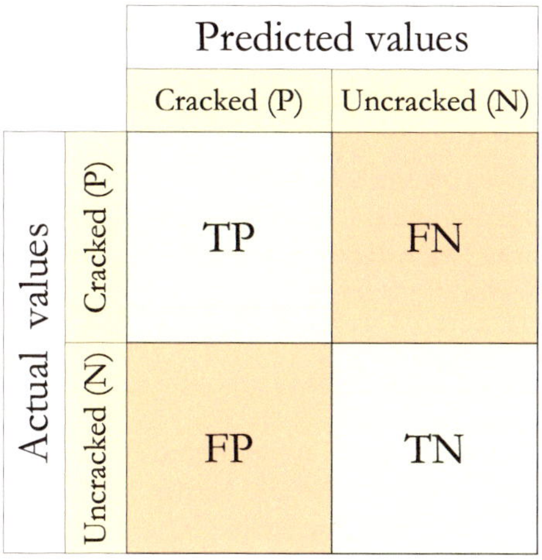

Figure 5.
An example of a fully connected process.

performance of precision and recall. A CNN with a high F1-score indicates stronger robustness.

$$Accuracy = \frac{TP + TN}{TP + TN + FP + FN} \times 100 \tag{6}$$

$$Precision = \frac{TP}{TP + FP} \times 100 \tag{7}$$

$$Recall = \frac{TP}{TP + FN} \times 100 \tag{8}$$

$$F1 - score = 2 \times \frac{Precision \times Recall}{Precision + Recall} \times 100 \tag{9}$$

For example, the prepared dataset contains 10,000 photos, with 32,000 and 7000 cracked surface images and uncracked surface images, respectively. After CNN processing, 2700 images are correctly predicted as cracked surfaces, 300 images out of the 3000 real cracked surfaces are wrongly predicted as uncracked surfaces. 6500 images are correctly predicted as uncracked surfaces, and 500 images out of the 7000 uncracked surfaces are wrongly predicted as cracked surfaces. Then, based on above-mentioned concepts, the values of TP, FN, FP, TN is 2700, 300, 500, 6500, respectively. **Table 3** shows the details of the accuracy, precision, recall, and F1 score calculations.

Variable	Equation	Calculation process
Accuracy	$\frac{TP+TN}{TP+TN+FP+FN} \times 100$	(2700 + 6500) /(2700 + 300 + 500 + 6500) × 100 = 92%
Precision	$\frac{TP}{TP+FP} \times 100$	2700/(2700 + 500) × 100 = 84.375%
Recall	$\frac{TP}{TP+FN} \times 100$	2700/(2700 + 300) × 100 = 90%
F1-score	$2 \times \frac{Precision \times Recall}{Precision+Recall} \times 100$	2 × ((0.84375 × 0.9)/(0.84375 + 0.9)) × 100 = 87.23%

Table 3.
Detailed calculation process of accuracy, precision, recall, and F1 score.

3. Example of concrete crack inspection using CNN

3.1 Textbook example of crack inspection using CNN

This chapter provides an example of how convolution, pooling, fully connected, and benchmarking can be demonstrated in real-world concrete crack inspection using CNN. The above-mentioned calculation was carried out using the Python programming language and the Pytorch package.

3.1.1 Dataset

In this example, the input images were gathered from Kaggle, the world's most well-known data science community. Kaggle allows access to thousands of public datasets covering a wide range of topics, including medical, agriculture, and construction [9]. By searching "concrete crack" in Kaggle datasets module, 12 datasets were found. The "SDNET2018" dataset was chosen from among them since it comprises sufficient and clean concrete surface images with and without cracks [10]. In "SDNET2018", 56,096 images were captured in the Utah State University Campus using a 16-megapixel Nikon digital camera, including 54 bridge decks, 72 walls, and 104 pavements. In this example, only images of walls and pavements were used to demonstrate the comparison analysis between manual inspection and CNN-based automatic inspection. Therefore, 42,472 images were used as training and testing dataset. Among them, 6459 cracked concrete surfaces are considered as positive class. The captured cracks are as narrow as 0.06 mm and as wide as 25 mm, while 36,013 uncracked concrete surfaces are considered as negative class. Images in this dataset contain a range of impediments, such as shadows, surface roughness, scaling, edges, and holes. The diverse photographing backgrounds contribute to ensuring the robustness of the designed CNN architecture. At a ratio of 80/20, the cracked and uncracked concrete photos were randomly separated into training and testing datasets. The input images' pixels were standardized to $227 \times 227 \times 3$ for AlexNet, and $224 \times 224 \times 3$ for VGG16. **Table 4** shows the details of the input images. **Figure 6** shows the examples of the input images.

3.1.2 CNN architecture

In this section, two pre-trained CNN networks, AlexNet and VGG16, were introduced to illustrate CNN computation process. AlexNet was designed as an eight-layer architecture. VGG16 has a depth that is two twice that of AlexNet. According to [11, 12], the depth of CNN network has a significant impact on model performance.

	Total dataset	Training dataset	Testing dataset
Total images	42,472	33,978	8494
Cracked images	6227	4986	1238
Non-cracked images	36,245	28,992	7256
Image pixels		AlexNet: $227 \times 227 \times 3$ VGG16: $224 \times 224 \times 3$	

Table 4.
Details of prepared dataset.

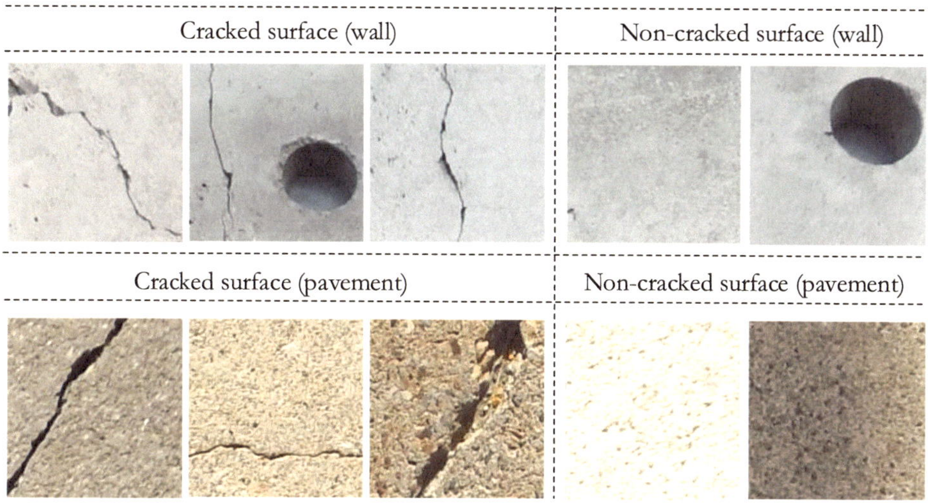

Figure 6.
Examples of cracked and non-cracked surface.

Therefore, by training and testing the prepared dataset with AlexNet and VGG16, the comparison of network depth to prediction performance and computation cost can be further highlighted.

1. AlexNet architecture

The AlexNet architecture, developed by Alex Krizhevsky, Ilya Sutskever, and Geoffrey E. Hinton in 2012, is considered one of the most influential CNN architectures [13]. AlexNet consists of five convolution layers and three fully-connected layers. The max-pooling layers follow the first, second, and fifth convolution layers. AlexNet was designed to predict 1000 object classifications. 1.2 million images with a pixel size of 2,242,243 were used as input images. As a result, 60 million parameters and 650,000 neurons are included in the computation process. The details of the AlexNet architecture are shown in **Figure 7**.

In the first convolution stage, 96 convolution filters with size of 11×11 were applied; they move with a stride with four pixels. The size of pooling filters is 3×3. The pooling filters move with a stride of two. It is worth noticing that the error rate can be reduced by applying overlapping pooling technique (the size of pooling filters is smaller than its stride). In the second convolution stage, the size of convolution filters becomes smaller from 11×11 to 5×5 while its number becomes larger from 96 to 256. The convolution filters in the third and the fourth convolution stage keep minimizing, from 5×5 to 3×3, while its number keeps increasing from 256 to 384. In the last convolution stage, the size of convolution filters remains same as 3×3, and its number turns back to 256. The size and stride of pooling filters also remain the same in the second and fifth convolution stage. Finally, 4096 neurons are included for both first and second fully-connected layers. The final fully-connected layer contains 1000 neurons to output the probabilities of 1000 classifications. The 1000 neurons are activated by softmax function.

The outputs of each convolution and fully-connected layer are activated by a non-linear function, namely the Rectified Linear Units (ReLU) [14]. It is proved in

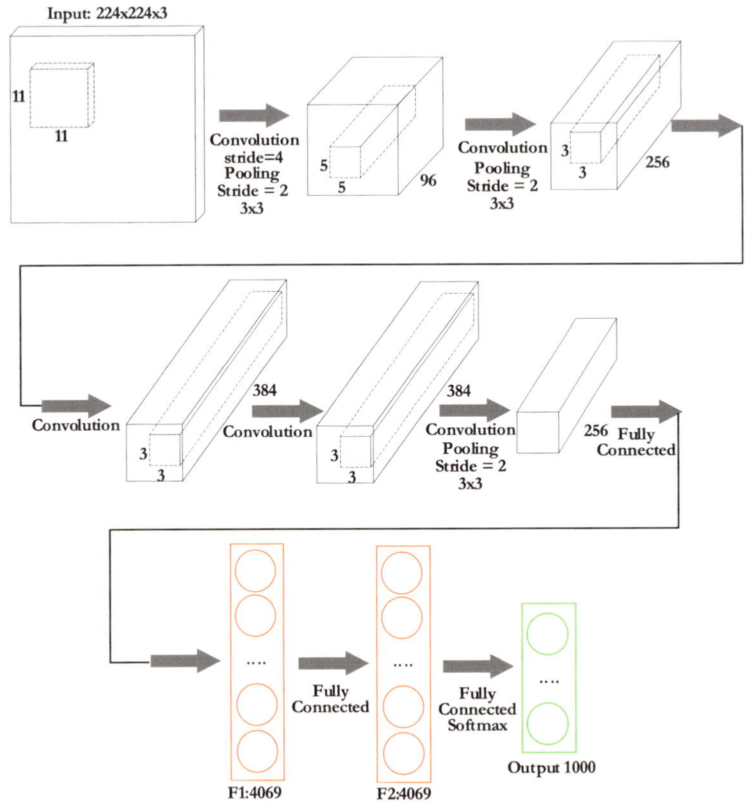

Figure 7.
Details of AlexNet architecture.

AlexNet that using ReLU instead of other activation functions effectively solves the overfitting problem and improves computation efficiency. Especially for the larger architectures trained on larger datasets. The local response normalization [15] technique (LRN) is also applied following ReLUs to reduce the error rate. Moreover, to avoid overfitting, drop-out techniques [16] are also applied in the first two fully-connected layers. The dropout criteria was set at 0.5.

AlexNet was computed using SGD. The batch size, momentum [17], and weight decay [18] were set as 128, 0.9, and 0.0005, respectively. The learning rate was set as 0.00001. AlexNet was computed for roughly 90 periods in NVIDIA GTX 580 3GB GPUs. As a result, the error rate of AlexNet on test set of top-1 and top-5 achieved 37.5% and 17.0%, which was 10% lower than the out-performed CNN architecture at that time.

2. VGG16 architecture

VGG16, designed by Karen Simonyan and Andrew Zisserman in 2015, was developed to investigate the influence of convolution network depth on prediction accuracy in larger datasets [19]. Therefore, VGG16 was designed as a deep architecture with 16 weight layers, including 13 convolution layers and three fully-connected layers. Convolution layers in VGG16 are presented as five convolution blocks. The details of the VGG16 architecture are shown in **Figure 8**.

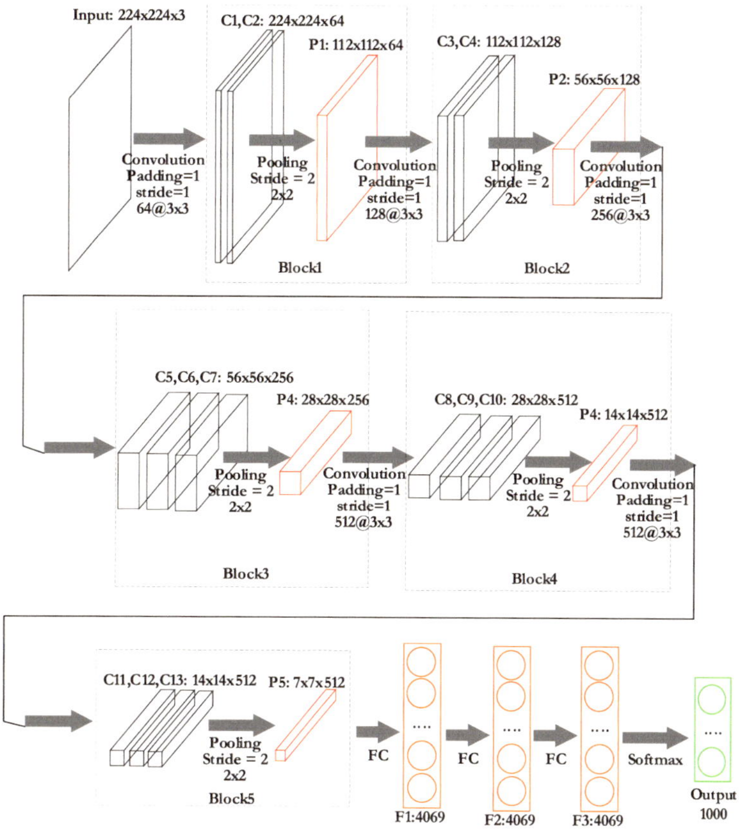

Figure 8.
Details of VGG16 architecture.

As seen from **Figure 8**, there are two convolution layers in the first two convolution blocks, respectively, and three convolution layers in the following three convolution blocks, respectively. The size of all convolution filters is uniformly 3×3. All the convolution filters move with a stride of one. The number of convolution filters increases gradually from 64 to 128, 256, and 512 in the five convolution blocks. To preserve information about image boundaries as completely as possible, spatial padding is applied [20]. As with AlexNet, ReLU is applied as a non-linearity function for convolution and fully-connected outputs to avoid overfitting problems. However, unlike in AlexNet, LRN is not used in VGG16 because the authors stated that LRN has no influence on model performance and increases memory consumption and computation time.

Five max-pooling layers follow the last convolution layer in each block. The max-pooling filters are uniformed with a size of 2×2, and a stride of two. As with AlexNet, the first two fully-connected layers have 4096 neurons and 1000 output neurons. The output neurons are activated by softmax. To avoid overfitting problems, drop-out technique is also applied in the first two fully-connected layers. The dropout ratio is set at 0.5. It can be concluded that the most important novelty of VGG16 compared with AlexNet are: (1) the designed deep architecture; (2) the uniformed and small size convolution filters.

In the training process, the training batch size, momentum, weight decay, and learning rate were set as 256, 0.9, and 0.0005, 0.0001, respectively. As a result, the top-1 and top-5 errors of VGG16 achieved 24.4% and 7.2%, which is 13% and 9.8% lower than AlexNet. The result proved that the deep architecture and small convolution filters have positive influences on CNN performance.

3.1.3 Training and benchmarking

Finally, the prepared dataset mentioned in Section 3.3.1 was used to train and test AlexNet and VGG16, respectively. The training and testing process was conducted in Kaggle kernels [21]. Kaggle kernel, provided by Kaggle community, is a virtual environment equipped with NVIDIA Tesla K80, a dual GPU design, and 24GB of GDDR5 memory. This high computing performance enables 5–10 times faster training and testing processes than CPU-only devices. Both AlexNet and VGG16 were trained using SGD. Batch size was and learning rate set as 64, 0.0001, respectively. To avoid overfitting problem, dropout was applied at the fully-connected stage, dropout probability was set as 0.5.

Python was used to program the computing process. Pytorch library was imported. The whole computation time of AlexNet was roughly 2 h, and 4 h for VGG16. The model's performance in the training and testing datasets is shown in **Figures 9** and **10**,

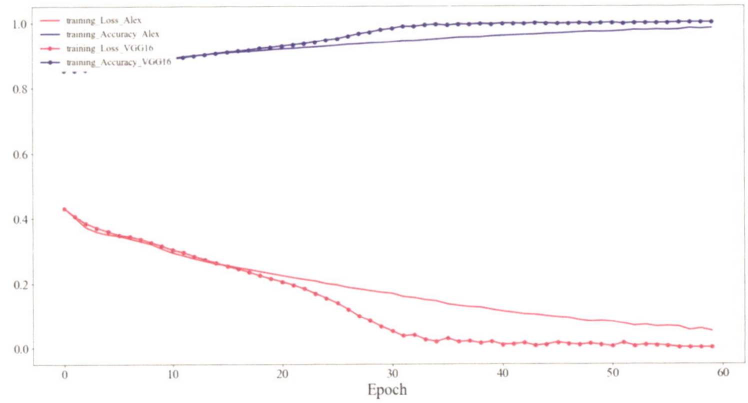

Figure 9.
Training loss and accuracy of AlexNet and VGG16.

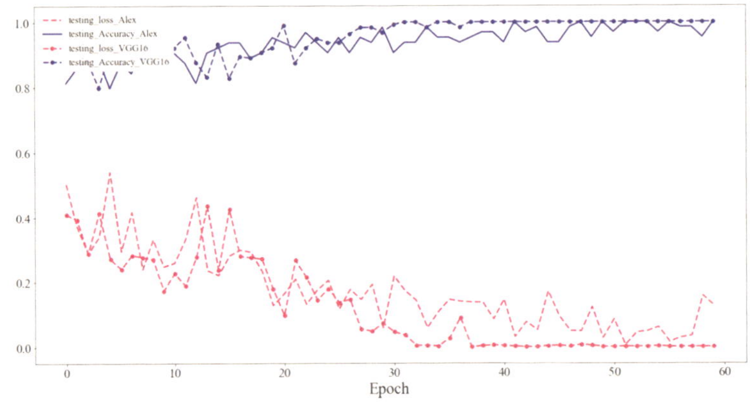

Figure 10.
Testing loss and accuracy of AlexNet and VGG16.

respectively. The training and testing loss and accuracy values are represented on the vertical axis, while the processing epochs are represented on the horizontal axis. Since the loss and accuracy variation remained consistent after the 60th epoch overtraining the model could lead to an overfitting problem [22]. The training epoch was set to 60 epochs.

As shown in **Figure 9**, both AlexNet and VGG16 converged successfully. The training loss for AlexNet reduced steadily from 0.43 to 0.05 in the 58th epoch and then remained constant in subsequent epochs. Similarly, at the 58th epoch, AlexNet's training accuracy increased from 0.85 to 0.98. At the 35th epoch, the training loss for VGG16 dropped from 0.42 to 0.01 and subsequently stayed steady at approximately 0.008–0.01 in following epochs. At the 34th epoch, the training accuracy of VGG16 increased from 0.85 to 0.99 and then remained at 0.99. The results revealed that VGG16 performed better during the training procedure. VGG16's convergence speed is roughly two times that of AlexNet. VGG16's minimum training loss is 0.04 lower than AlexNet's, while its maximum accuracy is 0.01 times higher. It is observed that deeper CNN designs assist in the faster processing of larger datasets, which contributes to producing more trustworthy weights and biased matrices. These results are in accordance with those proposed by [23].

Figure 10 shows the loss and accuracy variations of AlexNet and VGG16 in the testing dataset. The testing loss and accuracy consist of the fluctuation tendency of training loss and accuracy. It indicated that neither AlexNet nor VGG16 had overfitting or underfitting problems. VGG16 also out-performed AlexNet in the testing process. AlexNet and VGG16 have minimum testing losses of 0.01 and 0.00003, respectively. AlexNet's maximum accuracy was 0.98, and VGG16's was 0.99. In the testing dataset, VGG16 converges at the 34th epoch, which is nearly 2 times faster than AlexNet.

The confusion matrix of AlexNet and VGG16 is shown in **Table 5**. It can be shown that the accuracy scores of AlexNet and VGG16 are nearly identical, indicating that AlexNet and VGG16 have similar prediction abilities for cracked and uncracked concrete surfaces. VGG16 has a precision and recall of 96.5% and 89.6%, respectively, which is nearly 1% and 5% greater than AlexNet. The results show that VGG16 outperforms AlexNet for predicted positive variables (cracked surfaces). Meanwhile, more cracked images from actual datasets can be correctly identified by applying VGG16. AlexNet and VGG16 have F1-scores of 89.6% and 92.9%, respectively, indicating that the VGG16 model is more robust.

	AlexNet	**VGG16**
TP	5242	5579
FN	985	648
TN	36,007	36,040
FP	238	205
Accuracy	0.971204558	0.97991618
Precision	0.956569343	0.9645574
Recall	0.84181789	0.895937048
F1-score	0.895532587	0.928981767

Table 5.
Confusion matrix of AlexNet and VGG16.

In conclusion, VGG16 demonstrates better performance. Since it is important to avoid ignoring any cracked surfaces, the model with the highest recall and F1-score is more worthwhile. Meanwhile, AlexNet is also a preferable option when the number of cracked and uncracked images is balanced because it shows a similar accuracy score as VGG16 and has a lower computation cost.

3.2 Comparison of CNN and manual inspection

During on-site construction quality management process, quality control managers (QCM) or registered inspectors (RI) are responsible for personally inspecting and reporting quality problems with forms, reports, and photocopies. According to the Mandatory Building Inspection Scheme (MBIS) and related contract regulations, QCMs and RIs are obliged to examine cracks and other defects in building components visually or with non-destructive equipment [24]. For example (1) cracks on the structural components, e.g., structural beam, column, (2) cracks on the external finishes, e.g., tiling, rendering, and cladding, (3) cracks on the fins, grilles, windows, curtain walls.

When using computer-vision-based inspection techniques, differently, there is no necessity for QCMs and RIs to conduct the aforementioned inspection tasks on-site. Instead, their primary responsibilities may switch to (1) taking photos or videos of building components, and (2) inputting the images and videos into pre-trained CNN models. To highlight the differences between manual and computer-vision-based crack inspection, an experiment was set up to calculate and compare inspection time and cost.

The layout of the experiment is shown in **Figure 11**. Suppose this experiment case is a 15 m × 15 m × 2 m residential building that is located in San Bernardino. The inspection items include cracks on slab, internal walls, and external walls. According to Dohm, John Carl [25], the total manual inspection time for 1600–2600ft^2 home in San Bernardino is around 13.65 h, including inspection items of building slab, shear walls, etc. The manual inspection service cost is around $85.9 per hour.

Referring to the computer-vision-based inspection process described above, the total inspection time includes the time of taking images or videos and CNN processing. Assume that the input videos are obtained with handheld camera devices while QCMs or RIs are by means of walking. Then, the time of taking videos can be considered as the time of walking.

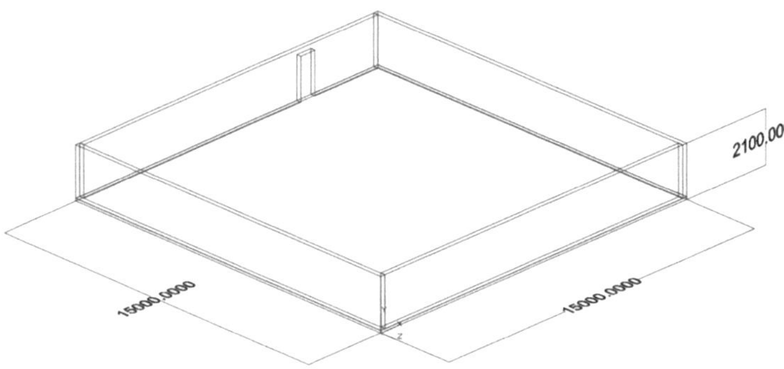

Figure 11.
Layout of the experiment case.

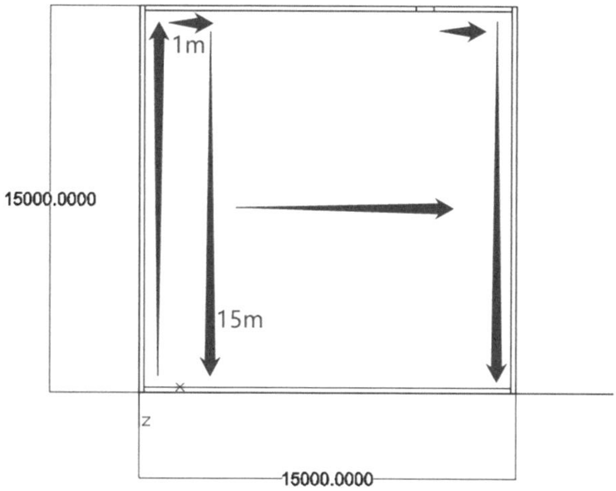

Figure 12.
Walking path of the inspectors.

Manual inspection		Computer-vision based inspection		
Time	13.65 × 3600 = 49,140 s	Time	Taking video	(1/0.1 × 15) × 15) +1/0.1 × 14 = 2390 s
			CNN processing	(2390 × 24)/100 = 573.6 s
Cost	(85.9/3600) × 49,140 = $1172.5	Cost		(573.6 + 2390) × (85.9/3600) = $70.7

Table 6.
Time and cost of manual and computer-vision based crack inspection.

Normally, the average walking speed between the age of 20–49 is around 1.42 m/s [26]. Considering the time delays of taking videos, the walking speed can be considered as 0.1 m/s. Suppose the walking path follows an S-curve, shown in **Figure 12**.

According to [27], the universally accepted frame rate is 24 FPS per second. Suppose the inspector begins to record video while taking the first step. Then the time of captured video equals the time of walking. The number of the input images that converted from the captured video can be calculated as 2390 s × 24FPS = 57,360. According to the testing time of the textbook examples mentioned in Section 3.1, and the study outcomes of [28], the time of CNN processing is around 100 images per second. Then, the time of CNN processing can be calculated as 57,360/100 = 573.6 s. Therefore, the cost of computer-vision based crack inspection can be calculated as (2390 s + 573.6 s) × (85.9/3600) = $70.7.

Table 6 summarizes the calculation process of time and cost of manual and computer-vision-based crack inspection. It can be seen that using CNN-based technique can effectively reduce inspection time and cost. The inspection time decreases from 13.65 to 0.8 h in total, the inspection cost decreases from $1172.5 to $70.7.

4. Conclusion

To facilitate automatic building quality inspection and management, this study introduced a computer-vision-based automated concrete crack inspection technique. In order to demonstrate the computing and benchmarking process, the mathematical understanding of one of the most essential computer vision algorithms, convolution neural network, was first detailed.

The theoretical foundation was then explained using a textbook example. In this case, the input dataset "SDNET2018" was obtained from the Kaggle community. A digital camera was used to acquire the 56,096 photos from the Utah State University campus. To train the input images, the two most basic CNN architectures, AlexNet and VGG16, were chosen. The Pytorch library was used to carry out the training process in the Kaggle kernel. The model's performance was evaluated using a confusion matrix. The results revealed that the prediction accuracy of AlexNet and VGG16 is nearly identical. However, VGG16's precision and recall are higher than AlexNet's, indicating that VGG16 has a stronger capacity to identify cracked surfaces. VGG16's F1 score is also greater than AlexNet's, signifying that VGG16 is more robust. VGG16 is deemed to have a better significance since it has higher precision, recall, and F1-score, which is crucial when distinguishing cracked and uncracked surfaces. When the ratio of cracked and uncracked images is almost the same, however, AlexNet is a feasible alternative because of its high accuracy score and low computation cost. It's worth noting that, when compared to shallow CNN architectures, deeper and broader CNN architectures outperform shallow CNN architectures for larger datasets.

Next, an experimental case was designed to compare manual and computer-vision-based crack inspection in terms of time and cost. The results showed that the efficiency and cost-effectiveness can be effectively improved when adopting computer-vision-based techniques. The inspection time and cost or the designed case can nearly decrease from 13.65 to 0.8 h, and from $1172.5 to $70.7, respectively.

The findings help to demonstrate the computer-vision-based quality inspection technique in both theory and practice. Although the recently developed computer-vision-based technology improves the efficiency, cost-effectiveness, and safety of human quality inspection, it still relies primarily on the collected image quality. Some concrete surface images are difficult to capture in real-life situations, including among others high-rise buildings, component corners, and buildings in extremely harsh environments. To address this issue, robotics techniques are growing rapidly as a means of upgrading computer-vision-based quality inspection [29]. Previous research has begun to use mobile robots, such as UAVs in order to gather surface images [30–32]. Some studies have focused on exploring robotic inspection systems to raise the automatic level of quality inspection [33, 34]. Therefore, merging robotics and computer vision approaches may be considered as a worthwhile future research direction to improve the efficiency and accuracy of manual quality control and management.

Acknowledgements

The authors highly appreciate the full support funding of the full-time PhD research studentship under the auspice of the Department of Building and Real Estate, The Hong Kong Polytechnic University, Hong Kong. The authors would like to express their deepest gratitude to Prof. Heng Li for his guidance. Finally, the authors would like to acknowledge the research team members (Mr. King Chi Lo, Mr. Qi Kai) and anyone who provided help and comments to improve the content of this article.

Conflict of interest

All authors declare that they have no conflicts of interest.

List of abbreviations

PASS	Building Performance Assessment Scoring System
MBIS	Mandatory Building Inspection Scheme
MWIS	Mandatory Window Inspection Scheme
NDT	Non-Destructive
CNN	Convolutional Neural Network
RGB	Red, Green, and Blue
ANN	Artificial Neural Network
BP	Back-Propagation
MSE	Mean Square Error
SGD	Stochastic Gradient Descent
ReLU	Rectified Linear Units
LRN	Local Response Normalization
QCM	Quality Control Managers
RI	Registered Inspectors

Author details

Siwei Chang and Ming-Fung Francis Siu*
The Hong Kong Polytechnic University, Hung Hom, Hong Kong

*Address all correspondence to: francis.siu@polyu.edu.hk

References

[1] Driscoll R. Assessment of damage in low-rise buildings, with particular reference to progressive foundation movement. In: Digest. London: H.M.S. O.; 1981. p. 251

[2] Chitte CJ, Sonawane YN. Study on causes and prevention of cracks in building. International Journal for Research in Applied Sciences and Engineering Technology. 2018;**6**(3): 453-461. DOI: 10.22214/ijraset.2018.3073

[3] Kim B, Cho S. Image-based concrete crack assessment using mask and region-based convolutional neural network. Structural Control and Health Monitoring. 2019;**26**:e231. DOI: 10.1002/stc.2381

[4] Rao A, Nguyen T, Palaniswami M, Ngo T. Vision-based automated crack detection using convolutional neural networks for condition assessment of infrastructure. Structural Health Monitoring. 2020;**20**:1475921720 96544. DOI: 10.1177/14759217209 65445

[5] Vandoni HT, Carlo E. Computer Vision: Evolution and promise. 19th CERN School of Computing. Geneva: CERN; 1996. pp. 21-25. DOI: 10.5170/CERN-1996-008.21 ISBN 978-9290 830955

[6] Feng X, Jiang Y, Yang X, Du M, Li X. Computer vision algorithms and hardware implementations: A survey. Integration. 2019;**69**:309-320. DOI: 10.1016/j.vlsi.2019.07.005

[7] Yamashita R, Nishio M, Do RKG, Togashi K. Convolutional neural networks: an overview and application in radiology. Insights into Imaging. 2018; **9**(4):611-629. DOI: 10.1007/s13244-018-0639-9

[8] Stehman SV. Selecting and interpreting measures of thematic classification accuracy. Remote Sensing of Environment. 1997;**62**(1): 77-89. DOI: 10.1016/S0034-4257(97) 00083-7

[9] Banachewicz K, Massaron L. Data Analysis and Machine Learning with Kaggle: How to Compete on Kaggle and Build a Successful Career in Data Science. Birmingham, United Kingdom: Publisher Packt Publishing Limited. 2021. Available from: https://www.bookdepository.com/Data-Analysis-Machine-Learning-with-Kaggle-Konrad-Banachewicz/9781801817479

[10] Dorafshan S, Robert JT, Marc M. SDNET2018: An annotated image dataset for non-contact concrete crack detection using deep convolutional neural networks. Data in Brief. 2018;**21**: 1664-1668. DOI: 10.1016/j.dib.2018. 11.015

[11] OrShea A, Lightbody G, Boylan G, Temko A. Investigating the impact of CNN depth on neonatal seizure detection performance. In: 2018 40th Annual International Conference of the IEEE Engineering in Medicine and Biology Society (EMBC); 18–21 July 2018; USA, New York: IEEE; 2018. pp. 5862-5865

[12] Pasupa K, Sunhem W. A comparison between shallow and deep architecture classifiers on small dataset. In: 2016 8th International Conference on Information Technology and Electrical Engineering (ICITEE); 5–6 October 2016; Indonesia, New York: IEEE; 2016. pp. 1-6

[13] Krizhevsky A, Ilya S, Hinton GE. Imagenet classification with deep convolutional neural networks.

Advances in Neural Information Processing Systems. 2018;**25**:1097-1105. DOI: 10.1145/3065386

[14] Nair V, Hinton GE. Rectified linear units improve restricted boltzmann machines. In: Proceedings 27th International conference on machine learning; 21–24 June. Israel: International Machine Learning Society; 2010. pp. 417-425

[15] Kim GB, Jung KH, Lee Y, Kim HJ, Kim N, Jun S, et al. Comparison of shallow and deep learning methods on classifying the regional pattern of diffuse lung disease. Journal of Digital Imaging. 2018;**31**(4):415-424. DOI: 10.1007/s10278-017-0028-9

[16] Srivastava N, Hinton G, Krizhevsky A, Sutskever I, Salakhutdinov R. Dropout: A simple way to prevent neural networks from overfitting. The Journal of Machine Learning Research. 2014;**15**(1):1929-1958. DOI: 10.5555/2627435.2670313

[17] Smith LN A Disciplined Approach to Neural Network Hyper-Parameters: Part 1: Learning Rate, Batch Size, Momentum, and Weight Decay. arXiv preprint arXiv:1803.09820. 2018

[18] Gnecco G, Sanguineti M. The weight-decay technique in learning from data: An optimization point of view. Computational Management Science. 2008;**6**(1):53-79. DOI: 10.1007/s10287-008-0072-5

[19] Simonyan K & Zisserman A. Very Deep Convolutional Networks for Large-Scale Image Recognition. arXiv preprint arXiv; 1409.1556. 2014

[20] He K, Zhang X, Ren S, Sun J. Spatial pyramid pooling in deep convolutional networks for visual recognition. IEEE Transactions on Pattern Analysis and Machine Intelligence. 2015;**37**(9): 1904-1916. DOI: 10.1109/TPAMI.2015.2389824

[21] Banachewicz K. Data Analysis and Machine Learning with Kaggle. Birmingham: Packet Publishing Limited; 2021

[22] Al Haque AF, Rahman MR, Al Marouf A, Khan MAAA. Computer vision system for Bangladeshi local mango breed detection using convolutional neural network (CNN) models. In: 4th International Conference on Electrical Information and Communication Technology (EICT); 20–22 December 2019; Bangladesh, New York: IEEE; 2019. pp. 1-6

[23] Zheng Z, Yang Y, Niu X, Dai H, Zhou Y. Wide and deep convolutional neural networks for electricity-theft detection to secure smart grids. IEEE Transactions on Industrial Informatics. 2018;**14**(4):1606-1615. DOI: 10.1109/tii.2017.2785963c

[24] Buildings Department. Code of Practice for Mandatory Building Inspection Scheme and Mandatory Window Inspection Scheme. Hong Kong: Hong Kong Government. 2012. Available from: https://www.bd.gov.hk/doc/en/resources/codes-and-references/code-and-design-manuals/CoP_MBIS_MWISe.pdf, https://www.bd.gov.hk/en/safety-inspection/mbis/index.html

[25] Dohm JC. Building Inspection Fee Analysis. Theses Digitization Project. San Bernardino, California, United States: California State University. 2007. Available from: https://scholarworks.lib.csusb.edu/etd-project/3249

[26] Mohler BJ, Thompson WB, Creem-Regehr SH, Pick HL, Warren WH. Visual flow influences gait transition speed and

preferred walking speed. Experimental Brain Research. 2008;**181**(2):221-228. DOI: 10.1007/s00221-007-0917-0

[27] Apple Inc. Final Cut Pro User Guide. Apple, One Apple Park Way, Cupertino, CA 95014, United States: Apple Inc. 2021. Available from: https://support.apple.com/en-hk/guide/final-cut-pro/ver917522c9/mac

[28] Ma D, Fang H, Wang N, Xue B, Dong J, Wang F. A real-time crack detection algorithm for pavement based on CNN with multiple feature layers. Road Materials and Pavement Design. 2021:1-17. DOI: 10.1080/14680629.2021.1925578

[29] Chang S, Siu MFF, Li H, Luo X. Evolution pathways of robotic technologies and applications in construction. Advanced Engineering Informatics. 2022;**51**:101529

[30] Seo J, Duque L, Wacker J. Drone-enabled bridge inspection methodology and application. Automation in Construction. 2018;**94**:112-126. DOI: 10.1016/j.autcon.2018.06.006

[31] Humpe A. Bridge inspection with an off-the-shelf 360° camera drone. Drones. 2020;**4**(4):67. DOI: 10.3390/drones4040067

[32] Liu Y, Nie X, Fan J, Liu X. Image-based crack assessment of bridge piers using unmanned aerial vehicles and three-dimensional scene reconstruction. Computer-aided Civil and Infrastructure Engineering. 2020;**35**(5):511-529. DOI: 10.1111/mice.12501

[33] La H, Gucunski N, Dana K, Kee S. Development of an autonomous bridge deck inspection robotic system. Journal of Field Robotics. 2017;**34**(8):1489-1504. DOI: 10.1002/rob.21725

[34] Montero R, Menendez E, Victores JG, Balaguer C. Intelligent robotic system for autonomous crack detection and characterization in concrete tunnels. In: 2017 IEEE International Conference on Autonomous Robot Systems and Competitions (ICARSC); 26–28 April 2017; Portugal, New York: IEEE; 2017. pp. 316-321

Chapter 11

Development and Usage of Electronic Teaching Technologies for the Economic Training of Students in a Technical University

Valeryi Semenov

Abstract

In this chapter, the experience of the Department of Economic Theory in the development and use of electronic technologies in teaching economic theory for students of technical directions is described. The necessity of electronic testing in the context of the concept of practice-oriented teaching has been substantiated. The stages of development and structure of electronic testing are presented. The process of forming the base of test tasks is described. The structure of the software is stated. The experience of approbation and application of testing technology is presented. The influence of electronic testing technology on teaching methods is shown. The issues of electronic support of business games are considered. Electronic technologies are considered as a necessary and essential element in the organization and implementation of business games developed at the department. An assessment of the impact of electronic testing and electronic support of business games on the quality of the educational process is given.

Keywords: economic theory, students of technical specialties, practice-oriented teaching, business games, electronic support, electronic testing technology, quality control of the educational process

1. Introduction

The development of the system of economic education for students in technical areas involves the following in particular: taking into account the requirements of employers, compliance of domestic standards with foreign ones, creativity in teaching, and the use of electronic learning technologies.

The subject "Economic theory" is included in the block of humanitarian and socioeconomic disciplines that provide students of technical stream with the necessary competencies. There are two main problems in the teaching of economic theory in a technical university:

- Uninterested perception of the subject as "not basic" and not relevant to the disciplines in the specialty

- Small amount of auditor hours

To solve these problems, the concept of practice-oriented teaching has been implemented at the Department of Economic Theory in recent years [1].

A necessary element in the implementation of the concept of practice-oriented learning is the use of electronic learning technologies.

2. The structure of a practice-oriented approach to teaching economic theory

It consists of the fact that along with the consideration of questions of a theoretical nature, it is obligatory to consider concrete and real data on lectures and seminars on all the topics under study. At the same time, the main methodological principle is the maximum possible usage of examples corresponding to the streams of student training.

A practically oriented approach in teaching the subject "Economic theory" is implemented in the following streams:

1. Development of presentations of lectures containing real data. This may include—but be not limited to—the following practical applications, addressing the following topics:

   - Automotive market

   - Oil and gas market

   - Information engineering and technology market

The choice of the automotive market as a topic for practical application is justified by acknowledging the following circumstances: widespread use, a lot of information, the ability to historically analyze the evolution of the market and the structure of competition, tracking the effects of mass production, the availability of data on prices, production volumes, and technologies.

In addition, this topic correlates with the use of the business game "Formation of the automotive market" in practical classes.

The subject of the oil and gas market is related to the peculiarities of the Russian economy and is characterized by the possibility of obtaining various data for analysis, thus making it possible to analyze the activities of monopolies and oligopolies.

The choice of the information technology and technology market as a topic for practical application is due to its relevance, the widespread use of digital technologies, and their development. This topic corresponds to the areas of training of one of the faculties of the university and correlates with the business game "Digital Economy" developed within the department.

2. Change the methodology for conducting practical exercises.

As far as the change in the methodology of teaching is concerned, at first, the technology of preparing and presenting reports by students is introduced. The structure of the report should contain a brief summary of the theoretical content of the key aspect of the topic (5–6 min), as well as bringing statistical and other data on the topic (4–5 min). The recommended subject of reports on all topics coincides with the directions chosen by the department.

In addition to this topic, students are offered other areas in accordance with their specialty.

3. Development and application of business games and specific situations for analysis in the educational process.

The department has developed a number of business games and specific situations for analysis [2].

Therefore, when studying the topic of supply and demand, the business game "Demand and Supply in the Automotive Market" is used. In a playful way, students analyze the formation of the automotive market in the United States, supply and demand factors, market structure, and the strategy of competing firms.

When reviewing the topic of the production factors market, the business game "Real estate: rent or buy" is used. The game deals with supply and demand in the labor market of engineers, wage dynamics, and housing market data. Students analyze the possibility of buying or renting an apartment depending on the level and dynamics of their future income, other life criteria, and factors of the real estate market.

The study of the topic "Cost theory of the firm" is also conducted using a business game. By imitating small businesses, students are divided into subgroups, creating small enterprises in the field of catering. In doing so, they analyze the main costs, their structure, and dynamics.

Studying the topics "Fiscal policy of the state" and "Monetary policy of the state," the business game "State regulation of the automotive market" is used. The actions of the "AvtoVAZ Bank" as well as the state policy on supporting the car industry and AvtoVAZ's efforts to attract investments and implement its production program are considered.

When studying the topic "Economic functions of the state," the business game "Digital Economy" is used. During the business game, four teams of students interact, representing the positions of the state, business, consumers, and experts, assessing the socioeconomic consequences of implementing programs for the development of the digital economy.

4. Attraction of students of technical faculties for participation in conferences with reports on economic theory of practical orientation.

In the context of the practical orientation of teaching, students of technical faculties are involved in terms of participating in conferences. They present reports on economic topics with the involvement of specific materials. Some of the students present their own reports, while others act in collaboration with the teachers of the department.

The experience of students participating in conferences held at the university showed that students readily present reports on specific topics.

It is worth noting at this point that at the Russian scientific and practical conference of students, that is, postgraduate students of the "Modern problems of management" (2019), 34 reports of students were presented by the department (which accounted for more than 40% of all the reports presented by four departments of the Faculty of Economics and Management).

In particular, students presented reports on the following topics:

- Demand and supply in the Russian electronics market
- Competition in the electronics market in Russia
- Domestic market of laser equipment and laser technology
- On the state of the instrument making and measuring equipment market
- Russian market of medical technologies and medical equipment
- On the development of the market for ultrasonic non-destructive testing
- Effect of scale in the development of the automotive industry
- On factors affecting the oil market
- Demand and supply in the world oil market
- Structure of the information technology market in Russia
- Change in supply and demand in the information technology market

5. Development of technologies for electronic testing and support of business games.

Electronic teaching technologies are considered in the context of a practice-oriented approach. The technology of electronic testing allows one to save time of class hours, increasing the possibilities for meaningful studying. It furthermore improves the quality of testing in assessing students' mastery of the main theoretical material.

Indeed, it may be stated that the electronic support of business games is an essential element of their conduct.

6. Development of online courses and use of other forms of distance learning.

3. Structure of the score-rating system for assessing students' knowledge

At the Department of Economic Theory, a point-rating system for assessing students' knowledge has been used for a long time [3].

At present, this system has the following structure:

Basic controls.

1. Test papers:

- Two for "Microeconomics" and two for "Macroeconomics."
- The maximum number of points for each control work is 10.

- Final examination—the maximum number of points is 15.
- The relevant control works are carried out in the form of test tasks and include tests and tasks.

2. Report

- This is rated from 0 to 10 points.
- Each student has the right to make only one report during the semester.
- Preparation of one report by a group of two students is allowed.

3. Variably considered forms of educational work.

- Solving problems and test tasks at the seminar. Points allocated range from 0 to 3 points.
- Performances: evaluated from 0 to 4 points.
- Answers to questions for discussion and control questions, participation in the discussion of the report, and the discussion: evaluated for each type of activity and making-up between 0 to 3 points.

The approximate structure of time spent in a practical lesson (one topic is considered for 40–45 min):

- report—no more than 10 min.
- discussion of the report: 5 min.
- speeches on the topic of the report or the topic of the seminar: 5 min.
- discussion of the issues of the topic and answers to control questions: 10 min.
- problem solving: 10 min.
- Total: 40 min.

The final rating is formed as follows:
The mark "satisfactory" is set if the student scores 55–69 points, "good" is awarded between 70 and 84 points, and "excellent" from 85 points upwards.
If a student does not score 55 points, he/she receives an unsatisfactory grade and must pass the exam with the performance of test tasks.

4. Testing stages of development and technology structure

1. Urgency of development

An urgency of development is related, first of all, with the reduction of time of class hours. The working program on economic theory provides the study of 12 topics.

The lectures and seminars are allocated for 18 h. Practically, for each topic, both the lecture and the practical lesson have one academic hour (45 min). Such a structure of classes assumes a significant change and improvement in the methods of lecturing and conducting practical classes.

Before the introduction of the electronic testing system, knowledge control was carried out only during practical classes, and this was done by using paper carriers for test tasks.

Its main purpose was to eliminate these problems that the development of a system of electronic tests was primarily aimed at.

The following shortcomings of this system were revealed:

- too much time-related expenditures,
- limited variants,
- cribbing,
- replicating the right answers,
- difficulty in controlling the independent work of students due to lack of time in practical classes.

2. Formation of the base of test tasks.

As the experience of other developers of similar methods and our estimates has shown, in order to avoid repetition of the questions, it was necessary to form at least 40 test tasks for each topic. Taking into account that all the topics of the course were subject to testing, the total number of generated test tasks was more than 500. As such, the author of this research work used previously proven test tasks, newly developed, as well as tasks taken from other sources, in particular from websites attnica.ru, i-exam.ru, and fepo.ru, which was a time-consuming and laborious work. The formation of the base of test tasks was done by a group of three people.

3. Software.

To develop computer tests and questions, students used the database management system Question Mark Perception, which allows them to reproduce memories, as well as generate the results of various forms of reports for data analysis.

The Database Management System Question Mark Perception supports different types of questions:

- open-ended questions included in the tested text of the answer;
- question with one choice—the subject must choose one correct answer from several;
- multiple choice questions—the subject must choose at least the correct answer from the proposed ones;
- a question with a Likert scale—the subject chooses one of several values based on the scale;

- filling in the void—the subject must enter the missing word in a paragraph of the text;
- selection of words from the drop-down list—the subject must choose the correct answer from the drop-down list;
- a question about the details of the text;
- a question about entering a number—it involves entering the answer in a countable form;
- a question about the transfer of objects—the subject must reassign a lot of markers on the image;
- a question with a graphic choice—the subject must reassign the marker on the image
- matrix question—it is a table in each series of which the subject must choose one answer (column);
- ordering question—the subject must put the options in the correct order.

When setting rules for checking answers, many options are also possible, for example, correct answer, wrong answer, partially correct answer, etc. For each individual answer option, one needs to create one's own rule. It is also possible to set up a feedback, for example, a message with comments on the answer.

The evaluation of each question is also configured by the author, and the points awarded for the answer depend on the specified conditions.

Test events can be taken on any device that has access to the Internet. This allows testing not only in the classroom, but also to give homework in the form of independent work.

For testing within the disciplines of the Department of Economic Theory, questions with a single choice and a question for entering a number and choosing a word from a drop-down list were mainly used. Evaluation is carried out according to the system wherein the correct answer is linked to a simple question—one point, the correct answer to the problem—two to four points (depending on the level of complexity).

4. Approbation of technology.

The technology of electronic testing was proved during practical classes in economic theory at the Faculty of Economics and Management, the Humanities Faculty, and a number of groups of other technical faculties.

Testing was carried out for each topic of the course of economic theory in computer labs during the practical classes immediately after passing the corresponding topic. For a number of groups, testing on the same topics was conducted on an extracurricular basis (online).

The analysis of the results of testing on the parameters noted above was carried out, and the answers of students in computer classes and at home were compared.

A number of test tasks have been adjusted.

There was a slight repetition of the questions and some unevenness in the complexity of some tests.

It turned out that the percentage of failure to perform tests in extracurricular time for various reasons (including technical ones) was about 7%. This made it possible to compile a real timetable for retesting on the examination week and to estimate the scope of retesting with a larger number of students.

The total number of students tested in each academic year was more than a thousand people. On average, about 1–2% of students tested outside the classroom had problems during testing for technical reasons. They were promptly retested.

At present, the system is fully debugged.

5. Application of testing technology and changing teaching methods.

The technology of electronic testing for students of technical faculties is introduced from the 2014/2015 academic year. In this regard, the system of monitoring was changed. Four tests are carried out in the test form (two under the section "Microeconomics" and two under the section "Macroeconomics"). Test works included test tasks and exercises on all topics of the course. For each section, one control work was done in a computer lab (or in the classroom—in written form), and the other—on an extracurricular time (online). A test schedule was prepared, students were given logins and passwords, and consultations were conducted on testing technology. As previously mentioned, the total number of students tested in each academic year was more than a thousand people. On average, about 1–2% of students tested on an extracurricular time had problems in testing for technical reasons. Retesting was reopened for them promptly. Currently, the system is completely debugged.

5. Electronic support of business games

A business game is conducted in the form of an analysis of a specific situation using the role method. The business game is directly related to the lecture's material and the topic of the practical class.

The goal of the business game is to assimilate learning issues by students, to teach the skills of applying knowledge in economic theory to the analysis of the real situation, and to develop the abilities of working in small groups.

The business game is based on the careful study of a specific situation by students, searching for and analyzing additional materials, organizing individual and group work, preparing a report in accordance with questions and tasks for the business game, presenting the report, and discussing it at the seminar.

Electronic technologies are an essential component in the preparation, organization, and implementation of business games developed at the department.

This is determined by the following specific circumstances:

- The need to present materials in the form of presentations and texts,

- The training schedule, according to which the practical classes are held in a week, and therefore, a high efficiency is required in the preparation for the conduct of a business game

The time for conducting a business game is one to two academic hours; therefore, strict requirements are imposed on the representativeness, volume, and quality of the information provided.

Introductory instructive explanations of the lecturer on the organization of the business game are accompanied by the issuance of a task, a scenario, an algorithm for conducting, and the necessary materials in electronic form. A schedule is established for presenting the reports of students to the lecturer. In the process of preparing for a business game, consultations and necessary adjustments of the submitted materials take place. When conducting a business game, it is necessary to introduce presentations of the final report and other materials.

6. Development of online courses and the use of distance technologies

The departments are focused on the development of online courses and other products using the university's technological and organizational resources.

The St. Petersburg Electrotechnical University "LETI" has developed a strategy in the field of distance learning and distance learning technologies.

The strategy is a formalized set of approaches consistent with the development priorities of the university, on the basis of which an action plan is implemented to saturate the educational process with information and communication technologies [4].

The technical infrastructure is:

- computer network, including wireless access equipment;
- computer equipment, telecommunication and communication devices, presentation, and video equipment;
- mobile devices for access to digital resources;
- systems for monitoring and managing access to resources, alarm, and video surveillance systems.

The information infrastructure consists of the following systems:

- intrauniversity SPOC3 platform on open edX;
- automated e-learning system (LMS);
- interactive media library;
- system for online conferences;
- electronic library platforms;
- the "Electronic Dean's Office" system for the implementation of educational programs of all forms of education.

The information infrastructure is implemented in the form of digital resources and services of the corporate information environment. A single attribute for accessing university resources is the personal IDs of students and staff.

The Center for New Educational Technologies and Distance Learning and the Department of Educational Programs are responsible for the development of the online education system at LETI [5].

During the implementation, the following organizational technology was used:

- At the beginning of the semester, consultative meetings were held with all the students (in groups and on streams) on the organization of the educational process. Students were provided with information about the procedure for registering for online courses, the training schedule, types of classes, reporting, and the assessment system.

- During the semester, several times a week, mailing lists were sent with information about the opening of new course materials and deadlines for completing tests and handing in practical assignments.

- On a weekly basis, the Dean's offices were provided with detailed information on the development of the online course by students. At the end of the semester, all the students passed the final certification for their courses in the format of offline proctoring.

The following models of embedding distance learning technologies are assumed [6]:

- full distance learning;

- express distance learning;

- full or partial reduction of lecture classes;

- partial replacement of classroom hours with in-depth study of the material;

- flipped learning;

- automated issuance of individual homework assignments and carrying out control activities in the form of computer tests;

- controlled independent work on discipline.

When developing online courses, there is a need for their expertise. The university has the following system of peer review, which includes the following two stages:

1. Preliminary technical expertise: It includes the availability and performance of the online course components.

2. Comprehensive expertise: The subjects of a comprehensive examination are:

    - assessment of the compliance of the course structure and its content with the goals and objectives in the development of the academic discipline for which the online course is being created;

    - assessment of the presentation of text and presentation materials, various audiovisual aspects of the online course;

- evaluation of the control materials used in the online course.

Before creating the online course "Economics," the following main issues were analyzed:

- the educational objectives of the course,
- the audience of the course,
- technical capabilities.

Depending on the objectives of the course, audience, and technical capabilities, the format was chosen and the components of the course were determined, i.e.:

- who will be involved in creating the course,
- what the length of the course and its volume will be,
- whether the course provides additional classroom lessons,
- whether a situational assessment with instant feedback is needed.

As a result, it was decided that the course would be recorded by an employee of the department who would work together with the author and producer of the course. In the case described herein, the producer was also a member of the department.

The online course is seen as an addition to the classroom.

The course contains 13 topics and corresponds to the program of the discipline "Economic theory." The course structure includes video content and testing based on the results of mastering. The online course developed at the department is an integral part of the educational process and can be used as an independent material at the same time.

It should be noted that the development of the course required a significant amount of time. In addition, there was a need to master new competencies in content design and lecture recording.

7. Conclusion and further work

Electronic teaching technologies are considered in the context of a practice-oriented approach.

The used score-rating system for assessing students' knowledge has become even more effective. Furthermore, the developed technology of electronic testing has allowed to improve essentially the technique of conducting practical classes.

The quality of testing has significantly improved. The time for discussing issues of the topic, presenting reports, and solving problems has increased. The control of independent work of students became better.

Experience has shown that the electronic support of business games is an indispensable element of their conduct. The developed technology of electronic support has allowed to considerably increase the effectiveness of the conducted business games.

As a result of the introduction of electronic testing and business games support technologies, the perception of subject has expressively been enhanced, and the quality and effectiveness of training has advanced.

A short online course developed at the Department of Economic Theory is included in the system of the educational process and is its important component.

According to the results of a sociological survey [7, 8], the main advantages of online learning as named by students were:

- the ability to independently organize the process of mastering the academic discipline;
- the rational use of time in the development of the course;
- the use of video materials in the educational process;
- the possibility of returning to lecture materials and restudying these.

The most important aspects of online learning for students were: the clarity and consistency of the presentation of the educational material, the usefulness of the course for the specialty, and the pleasure of attending the course.

Main directions of further work may include but not be limited to the following areas:

- inclusion in the online course of business games and case studies;
- creation of a number of webinars and forums on the problems of economic theory;
- organization of design work in groups using remote
- technologies;
- creating a massive online course in economics for engineering students.

Author details

Valeryi Semenov
Department of the Economic Theory, Saint Petersburg Electrotechnical University "LETI", Saint Petersburg, Russia

*Address all correspondence to: vps290446@mail.ru

References

[1] Semenov VP. Practical-oriented approach in the teaching of economic theory. Sovremennoe obrazovanie: soderzhanie, tekhnologii, kachestvo. Sbornik materialov XXI Mezhdunarodnoy nauchno-prakticheskoy konferentsii. [Modern education: content, technology, quality. Information package of the XXI International Scientific and Practical Conference.]. Vol. 1. Saint Petersburg Electrotechnical University; 2017. pp. 256-258. (In Russian)

[2] Semenov VP. Experience of business games in the teaching of economic theory for students of technical streams. Sovremennoe obrazovanie: soderzhanie, tekhnologii, kachestvo. Sbornik materialov XXI Mezhdunarodnoy nauchno-prakticheskoy konferentsii. [Modern education: content, technology, quality. Information package of the XXI International Scientific and Practical Conference.]. Vol. 1. Saint Petersburg Electrotechnical University; 2016. pp. 249-252. (In Russian)

[3] Semenov VP, Nikishin VM, Baranova LY. Experience in the development and use of technologies for teaching economic theory for students of technical areas // Modern education: content, technology, quality. Collection of materials of the XXI international scientific-practical conference. Vol. 2. St. Petersburg Electrotechnical University; 2015. pp. 108-110. (In Russian)

[4] Strategy for the development of e-learning and distance learning technologies in St. Petersburg Electrotechnical University "LETI" for 2018-2020. Available from: https://etu.ru/assets/files/university/normativnye-dokumenty/strategiya-razvitiya-eoidot.pdf [Accessed: 2021-12-15]

[5] Detailed typology of electronic educational resources. 2017. Available from: https://etu.ru/ru/on-line-learning/digest-e-learning/such-different-online-courses

[6] Regulations on the organization of the educational process at St. Petersburg Electrotechnical University "LETI" using distance learning and distance technologies. 2018. Available from: https://etu.ru/assets/files/ru/universitet/normativnye-dokumenty/Prikaz%20OD0328/001-poryadok-primeneniya-eo-i-dot.pdf [Accessed: 2022-01-12]

[7] Kustov TV, Timofeev AV, Room PV. Experience of using online courses in the implementation of the main educational programs of the Electrotechnical University. In: Ram VN, editor. Distance Educational Technologies. Proceedings of the IV All-Russian Scientific and Practical Conference (with the participation of RIAC), Dedicated to the 75th anniversary of the GPA. St. Petersburg: GPA; 2019. pp. 68-75. (In Russian)

[8] Strogetskaya EV, Pashkovsky EA, Kazarinova NV, Betiger IB, Timofeev AV. The experience of teaching students in the new digital paradigm of education. Discourse. 2019;**5**(6):91-107. (In Russian)

Chapter 12

Exploring the Effects of Learning Capability and Innovation on Quality Management-Organizational Performance Relationship

Mohsen Modarres

Abstract

Management scholars should further study the scientific area concerning the contingent effects of learning capability and organizational innovations on the relationship between quality management organizational performance. This chapter examines the interactive effects of quality management with organizational learning capability and innovations on organizational performance. Indeed, it may be argued that within quality management theory and methodology, the need to consider the contingency approach may result in an in-depth understanding of how the intersection of constituent elements associated with quality management influences organizational performance. Results revealed that the interaction of quality management and learning capability explained higher variance in organizational performance than the direct effect of quality management on performance. Similarly, interactions between quality management and innovations explained more significant variance in organizational performance than the direct effect of quality management on performance. Outcomes showed that quality management might not directly impact organizational performance. Findings underscore the importance of interactive effects of innovation and organizational learning capability with quality management in explaining the relationship between quality management and organizational performance.

Keywords: strategy, integrated quality management, contingency theory, innovation, learning capability

1. Introduction

Organizations competing in dynamic industries are required to be cognizant of challenges and complexity in maintaining a balance between initiating changes through innovations and maintaining stability in their existing processes. Unpredictability within dynamic competitive markets creates a paradox between replicating stable processes or re-allocating resources toward innovation [1]. Hence, organizational success tends to be contingent on organizational commitment and capability to continuously explore a new way of doing things and exploit existing competencies [2]. Markets in dynamic industries tend to exert more significant pressure on competing firms to sense and respond to cues in their environment by creating flexible and adaptable core capabilities. The recent trends toward the adoption and implementation of total quality management have been indicative of competitive challenges in dynamic industries. As competitive advantage tends to erode at an accelerated pace [1], organizations that are responsive to intra-organizational cues and shifts in elements within the immediate organizational environment may have a better chance of success and prosperity [3, 4]. A healthy competitive position in the marketplace requires managers to coordinate among various internal processes, such as continuous improvements, innovations, and efficiency, through enhanced organizational learning capability. Moreover, internal coordination among process improvement, innovation, and organizational learning may lead to equilibria between continuous changes in various constituents in quality management and maintaining stability in existing processes. Integrated total quality management strategies enable managers to explore and implement a novel way of doing things and maintain stable and standard processes by repetition and duplication of high-performing processes. A number of researchers have posited that the performance outcome of quality management strategies tends to be contingent on the managerial capability to coordinate among timely innovations, investment in human capital, enhanced learning capability, and knowledge collaboration among organizational members and subunits [5–8]. Moreover, integrated quality management enables organizations to exploit the existing core capabilities and channel organizational knowledge into individual and team cognitive energy to gain competitive advantage and enhance organizational performance e.g., [8, 9] and organizational excellence [10]. Moreover, integrated quality management provides a window of opportunity for managers to detect and adapt to the external environment contingencies in a timely fashion [11]. The inconsistency in the causal linkage between desired performance outcome [12, 13] and integrated quality management strategies and practices at the operational level remain inconsistent [14, 15]. Past studies have shown inconsistent results in the relationship between performance and integrated quality management. For example, research by Powell [16] and Westphal et al. [17] revealed no statistical significance between performance and total quality management. In contrast, few researchers have reported a direct and positive association [18, 19] or a mediated relationship between organizational performance and quality management. Previous researchers have parsed and identified various components of integrated quality management and investigated each component's relationship with performance.

In this body of work, the financial measure of organizational success [8], human resource capability [20], research and development were explored as firm-specific capability [9]. Furthermore, integrated total quality management draws upon firm-specific resources and capabilities and coordinates a strategic balance between exploring new ideas and exploiting existing firm-specific capabilities [9, 21]. Such capabilities developed within integrated quality management tend to be non-imitable and sources of competitive advantage and higher performance [22, 23]. The causal ambiguity in the relationship between quality management and performance led to failures in the implementation of quality management [16]. Furthermore, causal ambiguity in the quality management-performance relationship has refocused research studies on the interrelationship between constituent elements of quality management and organizational performance. For instance, research by Modarres and Pezeshk indicated that the relationship between total quality management and organizational performance is mediated by organizational learning and innovation performance. Similarly, Huang et al. [6] argued that individual interactions mediate the innovation performance in the quality management method and the degree of the team learning that may result from team member interactions.

Another body of research centered on the interrelationship between investment in human capital and success in the implementation outcome of quality management [7]. Other researchers have discussed that the quality management-performance relationship tends to be contingent on creating a culture of dyadic trust among organizational members and promoting knowledge sharing among the organizational members [6]. Both dyadic trust and knowledge sharing create an internal organizational environment that generates enhanced cognitive learning. Furthermore, knowledge sharing allows accumulated knowledge by members of the organization to become the basis for diverse ideas and explorations of novel routines. Within this body of research, the relationship between quality management and performance tends to be contingent on a culture of employee empowerment within organizations [24]. Such a culture promotes an environment of learning and interaction, mutual trust, and information sharing among organizational members that may lead to the introduction of new products and services and the implementation of new codes in the organization.

Parsing quality management into its constituent parts and their synthetic roles within quality management have partially contributed to our understanding of the performance-quality management relationship. However, previous researchers have provided little information about the interactive effects of quality management with organizational learning and innovations to explain performance variations within corporations. This chapter derives from contingency theory to examine the contingency theory, neglected in recent quality management studies, to examine the interactions between quality management and two important variables, organizational learning, and innovations in explaining variations in organizational performance.

The proposed model (**Figure 1**) and hypotheses tested both direct and interaction effects between quality management, organizational learning, and innovation on various organizational performance levels. In contrast to parsing the constituent parts and their synthetic roles within quality management, the present research proposes that the interactions between quality management and learning capability and innovation tend to positively impact organizational performance. The present research views quality management as an integrated, gestalt, and adaptive method capable of continuously learning [25] and innovating novel routines and new core competencies. Furthermore, present research argues that integrated quality management allows for incremental modifications and radical reengineering of existing operations and enables managers to be flexible and enable the transformation and enhancement of internal capabilities.

2. Interaction effects of quality management with learning capability

Integrated quality management practices promote cross-functional communication and frequent exchanges of complex information among individuals and teams.

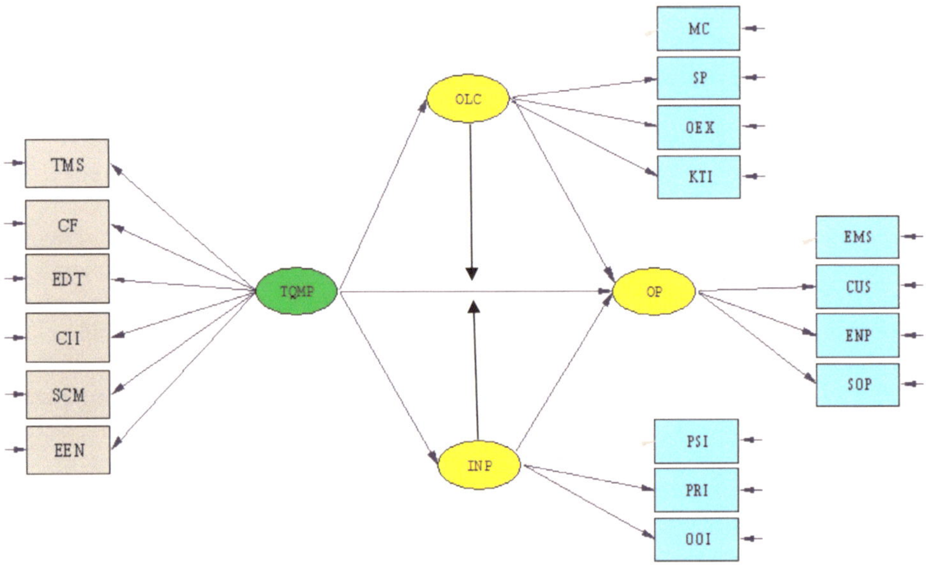

Figure 1.
Macro model.

Interaction between quality management and learning capability across subunits is likely to result in a novel way of doing things. Such knowledge creation commits top executives to allocate resources to employees' education, expression of new ideas, and team learning.

Furthermore, the managerial challenge in establishing a stable and reliable process tends to be contingent on creating an organizational culture. Such culture focuses on creating new knowledge and continuous organizational learning and the existing experience curve accumulated through information flow across subunits [6]. Such a seamless flow of information across subunits allows organizational members and managers to explore novel routines and exploit existing knowledge. Integrated quality management enables top managers to invest in continuous education and learning through employees' interactions. Over time, the accumulated education and learning become the basis for organizational learning capability [25, 26] and the flexibility to explore new routines and continuous process improvement [27]. According to Jerez-Gomez et al. [28], the interactions between top management commitment to employees' education and employee involvement in strategic directions of the organization enhance learning as one of the organization's core competencies. Moreover, higher levels of learning and education tend to lead to better implementation of quality management, greater innovation, higher quality of products and services, and higher organizational performance [26].

Moreover, high levels of learning capability within quality management enhance organizational awareness and ability to absorb new knowledge and transform the collective organizational know-how into new products and competitive advantage [9]. In contrast, low adaptive learning and low organizational performance tend to be attributed to parochial organizational practices and the inability to absorb new knowledge [29]. Similarly, the interaction between quality management and organizational innovations is likely to allow exploration for the opportunity to develop new products and services. Innovation tends to be among the success factors that contribute to high corporate performance [9, 22]. Previous researchers have argued that a positive association between innovation and organizational performance tends to be contingent on the flexible structural design that facilitates subunits innovations and interconnectedness, decentralized decision-making, and accumulated organizational learning [13, 30–32]. According to Singh and Smith [33], quality management practices promote an organic environment within organizations that is conducive to innovation and high levels of learning. Such organic structural design promotes employee interactions and cross-functional links and interactions. Furthermore, the organic structural design creates greater flexibility [34], that facilitates the speed and extent of innovations, and timely adaptation to changes in the firm's industry environment.

Moreover, quality management practices that promote the timely introduction of products and services to the marketplace can lead to competitive advantage and high organizational performance [8]. Similarly, entrepreneurial mindset within organizations tends to be a key factor in technological and product innovations. Furthermore, entrepreneurial mindset enables managers to respond to environmental changes by reallocating valued resources within the organization toward new products and services and enhancing corporate performance [22, 30, 35, 36]. Finally, quality management creates a culture of collaborations and exchanges of new ideas as employees interact within each function and cross-functionally. Researchers must identify the interrelationship among quality management, learning capability, and innovations to realize a deeper understanding of how employee interaction may lead to higher organizational learning capability and innovations. Furthermore, research studies should explore the interactive effects of quality management, learning capability, and innovation on organizational performance. Given the above, this study hypothesizes the main and intersection effects between integrated quality management, organizational learning, and innovations in the following manner:

H1: There will be a positive and significant relationship between quality management, organizational and organizational performance.

H1a: There will be a positive relationship between quality management, organizational learning.

H1b: There will be a positive relationship between quality management, organizational, and innovation.

H2: There will be a positive relationship between organizational learning and organizational performance.

H3: There will be a positive relationship between innovation and organizational performance.

H4: The interactions between quality management and organizational learning positively influence the relationship between quality management and organizational performance.

H5: The interactions between quality management and innovation positively influence the relationship between quality management and organizational performance.

3. Methodology

3.1 Sample and data

Data. The data used in this study were collected by the survey method. The survey was carried out during the year 2015 and provided information on Iran's food business environment, quality management, organizational learning, innovation performance, and organizational performance. Top executives and senior managers represent the most appropriate sources of information for this study. The population of top executives and managers was determined to be 400. A questionnaire and cover letter were mailed to the managing director or chief executive officer of each company from the Food Industry in Iran. A total of 37% of the 400 mailed surveys was completed and returned, a sample of 148. All 148 completed surveys were used in this investigation. Given the population of $N = 400$, the Cochran sample size formula indicated a sample of $n = 148$ allows the study to draw correct inferences from the population.

3.2 Measurement of variables

A survey method was used for all the variables in the present study. Respondents were asked to indicate their levels of agreement with descriptive statements using a 5-point Likert scale (range, 1 = strongly disagree to 5 = strongly agree).

Quality management. To measure the effectiveness of integrated quality management, following the study by Vanichchinchai and Igel [37] and Coyle-Shapiro [38], the present research employed the following seven variables:

- top management support
- employee involvement
- continuous improvement
- customer focus
- education and training
- supply management

Organizational learning capability. Based on the study by [28] learning capability was operationalized as top executive commitment, system perspectives, organizational experimentations, and knowledge transfer initiatives.

Organizational Innovations. Exploring new ways of things in the organization requires managerial decisions on innovations and reallocation of valued resources toward new processes, products, and services [9, 35]. Following the study by Parjogo et al., innovation performance was operationalized as product/service innovation, performance innovation, and overall organizational innovations.

Organizational Performance. Organizational performance can be defined as the desired outcome within organizations. Performance is multi-dimensional and may be measured as such. Following the study by Santos and Brito [39], the present research operationalized performance as employee satisfaction, response to environmental changes, sustainability, customer satisfaction, and projected revenue from new products/services.

3.3 Procedures and design

Congruent with the previous research in contingency theory [8], the present research considers quality management as an integrated organizational strategy. As such, the study used structural equation modeling to explore the independent and interaction effects of integrated quality management, innovations, and organizational learning on organizational performance. For parsimony, and to reduce the number of relationships, a hierarchical component model was created. Model I (**Table 1**, **Figure 1**) shows the results of the structural equation modeling analysis of the high component model, and standardized regression weights showing integrated quality management association with organizational learning capability, products and services innovations, and organizational performance. The hierarchical analysis of Model I also shows the relationship between each of the four constructs in this study with their sub-constructs.

4. Analysis

4.1 Main constructs, sub-constructs, and variables

Integrated quality management. According to the results shown in Model 1 (shown in **Table 1**), integrated quality management is positively and significantly associated with continuous education and employee training ($B = 0.90$), and continuous long-term employee support after the implementation of quality management ($B = 0.72$). Furthermore, continuous improvement programs (B = 0.61), managing the supplier relationship ($B = 0.55$), customer relations and satisfaction ($B = 0.45$), and employee involvement in the decision-making processes ($B = 0.41$) were positively and significantly associated with quality management.

Organizational learning capability. As shown in Model I (shown in **Table 1**), the organizational learning capability construct has a positive and significant relationship with long-term management commitment to involve employees in decision-making processes ($B = 0.84$). Results also indicated that within and inter-subunit knowledge transfer significantly influenced and enhanced organizational learning capability ($B = 0.83$). Furthermore, organizational learning capability is positively and significantly associated with the exploration of new ideas, and exploitation of the existing process to enhance further process improvements ($B = 0.66$). Results also indicated that subunits independently set divisional strategies and goals, and were responsible for co-align their strategies and goals with overall organizational strategies, goals, and mission ($B = 0.71$).

Innovation. Results (shown in **Table 1**) revealed that top managers encouraged and permitted exploration of new products and services ($B = 0.92$) and process innovation ($B = 0.78$). Furthermore, overall organizational innovations were significantly related to operational cost reductions and revenue generations ($B = 0.79$).

Organizational Performance. The results (shown in **Table 1**) indicated that the implementation of integrated quality management instituted organizational performance assessments were associated with continuous monitoring of the competitive dynamics in the marketplace ($B = 0.63$). Furthermore, results also indicated that the performance construct has a significant relationship with employee work satisfaction ($B = 0.75$), employee participation in decision-making processes ($B = 0.63$), customer expectations and satisfaction ($B = 0.59$), projected financial post-integrated quality management implementation ($B = 0.65$), and implementation of environmental sustainability programs ($B = 0.92$).

	Standardized regression weight	Standardized bias	t-value
Quality management → Organizational learning	0.95	0.08	13.41*
Quality management → Innovation performance	0.91	0.08	12.41*
Quality management → Organizational performance	0.43	0.08	1.13
Quality management → Education and training	0.90	0.08	14.20*
Quality management → Total management support	0.72	0.08	10.41*
Quality management → Continuous improvement	0.61	0.08	8.60*
Quality management → Supply chain mgt	0.55	0.08	7.67*
Quality management → Customer focus	0.45	0.08	6.29*
Quality management → Employee involvement	0.41	0.08	6.21*
Organizational learning → Management commitment	0.84	—	—
Organizational learning → System perspective	0.71	0.08	10.34*
Organizational learning → Organizational experiment	0.66	0.08	9.31*
Organizational learning → knowledge transfer	0.83	0.08	8.21*
Organizational learning → Organizational performance	0.58	0.08	6.89*
Innovation performance → Product/service	0.92	—	—
Innovation performance → Process innovation	0.78	0.08	12.24*
Innovation performance → Overall organizational innovation	0.79	0.08	12.57*
Innovation performance → Organizational performance	0.62	0.08	9.17*
Innovation performance → Product/service	0.92	—	—
Innovation performance → Process innovation	0.78	0.08	12.24*
Innovation performance → Overall organizational innovation	0.79	0.08	12.57*
Innovation performance → Organizational performance	0.62	0.08	9.17*
Organization Performance → Post TQM financial expectation	0.65	—	—
Organization Performance → Employee participation	0.63	—	—
Organization Performance → Customer satisfaction	0.59	0.08	7.79*
Organizational performance → Employee satisfaction	0.75	0.08	8.57*
Organizational performance → Sustainability	0.92	0.08	13.16*

Chi-square = 247.24; df = 114; GFI = 0.92; AGFI = 0.86; RMSEA = 0.08.
*p < .05.

Table 1.
Results of structural equation modeling-model I.

For the accuracy of the constructed model and to make sure the data is presenting accurate and reliable drawing from the population under study the Kolmogrov-Smrinov (KS) test was performed [40, 41]. **Table 2** shows that all four variables' data are normally distributed.

		TQM	OLC	INP	OP
		149	**149**	**149**	**149**
Normal Parametes[a]	Mean	3.62	3.37	3.34	3.38
	Std. Deviation	0.60	0.65	0.67	0.66
Most Extreme Difference	Absolute	0.059	0.059	0.076	0.057
	Positive	0.059	0.051	0.075	0.039
	Negative	−0.054	−0.059	−0.079	−0.057
Kolmogrov-Smrinov Z		0.751	0.721	0.926	0.691
Asymp. Sig (2-tailed)		0.687	0.676	0.358	0.726

[a]*Test distribution is normal*

Table 2.
One-sample Kolmogrov-Smrinov biased analysis.

5. Explanation of latent constructs

In this section of the chapter, complex hierarchical constructs, sub-constructs, and related subset variables are disentangled and discussed.

5.1 Integrated quality management

Table 3 presents the result of an orthogonal (VARIMAX) rotation of the factor matrix underlying the quality management items. Based on the six-independent factor solution suggested by the eigenvalue pattern (i.e., greater than 1.0), 25 items were identified so that each of which loaded at least cleanly on only one of the six factors. A cut-off of 0.50 was used for item-scale selection. These factors accounted for over 78% of the variance in the quality management scale items. Following an inspection of the factor loadings, the six factors were subsequently labeled:

- Total management support
- customer focus
- education and training
- continuous improvement and innovation
- supply chain management
- employee participation

Table 4 shows an examination of the Kaiser-Meyer Olkin measure of sampling adequacy suggested that the sample was factorable. The results reasonably describe each set of items as being indicative of an underlying factor for quality management.

(KMO = 0.833); χ^2 = 3485, df, 300, sig 0.000).

	Derived factors[c]					
Quality management[b]	EEN[b1]	TMS[b2]	SM[b3]	CII[b4]	CF[b5]	EDT[b6]
TMS1	0.219	**0.850**	0.201	0.077	0.057	0.188
TMS2	0.140	**0.794**	0.200	0.310	0.245	0.074
TMS3	0.217	**0.848**	0.194	0.115	0.113	0.176
TMS4	0.251	**0.822**	0.147	0.194	0.199	0.141
CF5	0.037	0.141	0.196	0.051	**0.859**	0.136
CF6	0.147	0.271	0.116	0.113	**0.821**	−0.002
CF7	0.187	0.072	0.069	0.099	**0.887**	0.141
EDT8	0.110	0.413	0.359	0.347	0.297	**0.568**
EDT9	0.057	0.131	0.222	0.287	0.065	**0.791**
EDT10	0.176	0.363	0.384	0.404	0.207	**0.526**
EDT11	0.231	0.334	0.321	0.169	0.165	**0.670**
CII12	0.109	0.108	0.209	**0.828**	0.206	0.188
CII13	0.002	0.193	0.156	**0.818**	0.15	0.158
CII14	0.063	0.207	0.227	**0.845**	0.096	0.177
SM15	0.021	0.210	**0.867**	0.144	0.111	0.072
SM16	0.010	0.303	**0.793**	0.204	0.004	0.146
SM17	0.016	0.031	**0.836**	0.148	0.159	0.130
SM18	0.084	0.165	**0.820**	0.151	0.153	0.267
EEN19	**0.674**	0.103	0.062	0.223	0.003	0.021
EEN20	**0.899**	0.199	0.035	0.020	0.135	0.009
EEN21	**0.908**	0.150	0.054	0.033	0.061	0.037
EEN22	**0.719**	0.015	0.050	0.042	0.051	0.349
EEN23	**0.785**	0.025	0.126	0.199	0.154	0.018
EEN24	**0.780**	0.109	0.038	0.093	0.029	0.197
EEN25	**0.806**	0.276	0.064	0.015	0.099	0.027
Eigenvalue	9.73	3.99	1.84	1.60	1.56	1.09
Variance explained	19.35	14.78	14.09	11.40	10.61	8.67

[a]A VARIMAX orthogonal rotation is performed on the initial factor matrix.
[b]Factors derived from quality management.
[c]Loadings above 0.50 are in boldface.

Factors	Cronbach's alphas	Scales included
[b1] Employee involvement		0
[b2] Total Management Support		0
[b3] Supply Management		0
[b4] Continuous improvements		0
[b5] Customer Focus		0
[b6] Education Training		0

Table 3.
Factor analysis of quality management scales.[a]

Kaiser-Meyer-Olkin Measure of Sampling Adequacy		0.833
Bartlett's Test of Sphericity	Approx. Chi-Square	3485
DF		300
Sig		0.000

Table 4.
KMO and Bartlett's test of quality management variable.

Results of second-order confirmatory factor analysis (**Table 3**) present the scale reliability on quality management dimensions that reached statistical significance. This indicates that criteria had a significant correlation with appropriate dimensions and scales had convergent validity [42].

Association of the latent constructs and quality management.

Findings (shown in **Table 5**) also indicated that integrated quality management is positively and significantly associated with human resource development through continuous education and training ($B = 0.94$). Findings also indicted executives' commitment to coordinate and support continuous improvements, post quality management implementation ($B = 0.73$), and employee involvements in implementation decision making ($B = 0.33$). Furthermore, findings indicated that top managers encouraged exploring new ideas and innovation ($B = 0.70$). Results revealed that managers were cognizant about immediate factors in the organization industry environment by managing supplier relationships ($B = 0.70$), focusing on customer relations ($B = 0.54$).

Analysis of subset variables and their relationship with quality management.

Education and training. Further analysis of the subset variables shows that long-term quality management training programs ($B = 0.96$) and employee know-how about the developments and changes in the industry ($B = 0.95$) as the most important variables. Work conditions and environment ($B = 0.76$) and customer relations ($B = 0.67$) were important factors in the implementation of quality management.

Top management support. As shown in **Table 5** top managers strategy, post quality management implementation focuses on continued investment in quality management programs ($B = 0.91$), coalignment of quality management strategies with changes in the industry ($B = 0.98$), and employee involvement in the implementation process ($B = 0.88$).

Continuous improvements. Top managers encouraged the employee for both input for new products and existing product improvements ($B = 0.94$), and research and development activities focusing on products and services improvements ($B = 0.75$). Furthermore, employees were encouraged to participate and suggest work environment improvements ($B = 0.88$).

Managing supplier relations. Results revealed top managers included supplier relations in their strategic plans for the long term ($B = 0.86$). Such a strategic plan was based on information sharing with the suppliers ($B = 0.88$), and assessment of the supply chain based on the long-term trend in the quality of the services and products the organization received ($B = 0.86$).

Customer focus. Top managers' decision-making process prioritized customer expectation ($B = 0.83$), contentedness with the quality of the product ($B = 0.84$), and importance to the organization ($B = 0.88$).

Employee involvement in decision-making processes. According to the results, employees were encouraged to form improvement circles and teams ($B = 0.96$) and

Items	First-order Standardized loading	t-value	Second-order Standardized loading	t-value
Total Quality Management-QM				
Education and Training				
1. Top managers' commitment to training employees in quality management	0.96	/[a]	0.94	
2. Top managers training in best conduct with employees and customers	0.67	10.36*		13.32*
3. Employees knowledge about food industry	0.95	15.66*		
4. Managers' commitment to providing employees essential needs at work	0.76	13.27*		
Top management support				
1. Top managers' commitment to post-implementation of quality management	0.86	/[a]	0.73	8.64*
2. Top managers' commitment to long-term investment in quality management	0.91	15.65*		
3. Top managers' support of employee involvement in quality management implementation	0.88	14.51*		
4. Top managers' strategic co-alignment of quality management with changes in market	0.98	16.49*		
Continuous improvement and innovation				
1. Employees are encouraged to make suggestions about work condition improvements	0.88	/[a]	0.70	8.23*
2. Employees are encouraged to research to improve products and services	0.75	11.21*		
3. Manager's consideration of suggestions for product/services improvement	0.94	15.44*		
Supply Management				
1. Coordination with the critical supplier through information sharing	0.88	/[a]	0.70	8.27*
2. Enhance the quality of suppliers post quality management implementation	0.86	13.84*		
3. Establish a win-win relation with suppliers	0.78	11.76*		
4. Strategic view on managing supply-chain	0.86	13.83*		
Customer Focus				
1. Center firm activities based on customer satisfaction	0.84	/[a]	.54	6.02*
2. Customer satisfaction and expectation as a top goal	0.83	11.52*		
3. Importance of customers in top managers' decisions	0.88	12.25*		
Employee Involvement				
1. Employee training and encouragement to participate in company programs	0.57	/[a]	0.33	3.50*
2. Creation of work improvement teams	0.96	7.09*		

Items	First-order	t-value	Second-order	t-value
	Standardized		Standardized	
	loading		loading	
3. Employees suggestions about improving supply-chain	0.96	7.99*		
4. Employees responsibility to inspect work outcome	0.66	6.47*		
5. Creation of quality circles to assist staff in problem-solving	0.70	6.71*		
6. Employee participation in management quality programs	0.75	7.00*		
7. Establishing a reward program for novel suggestions by employees	0.82	7.36*		

Chi-square = 670.02 (p < 0.001); df = 269; GFI = 0.93; AGFI = 0.88; RMSEA = 0.100.
/ᵃFixed parameter.
**p < 0.001.*

Table 5.
Results of the first-order and second-order confirmatory factor analysis of integrated quality management.

provide input about the supplier selection based on the quality of services and products (B = 0.96).

5.2 Organizational Learning capability

Results of an orthogonal (VARIMAX) rotation of the factor matrix (**Table 6**) indicate underlying organizational learning capability items. Based on the four-independent factor solution suggested by the eigenvalue pattern (i.e., greater than 1.0), 15 items were identified so that each of which loaded at least cleanly on only one of the four factors. A cut-off of 0.50 was used for item-scale selection. These factors accounted for over 75% of the variance in the organizational learning capability scale items. Following an inspection of the factor loadings, four factors were subsequently labeled "management commitment," "system perspectives," "organizational experiment," and "knowledge transfer initiative." After the initial component analysis number of items was reduced to 15 which explained the highest variation in organizational learning.

Table 7 shows the Kiser-Meyer-Olkin, and Bartlett test of sphericity utilized to measure four organizational learning dimensions, with each of the dimensions being measured by responses to several items. The results reasonably describe each set of items as being indicative of an underlying factor for learning capability (KMO > 0.818; χ^2 = 1843, df, 120, sig 0.000).

Results of second-order confirmatory factor analysis (**Table 6**) present the scale reliability on organizational learning dimensions that reached statistical significance. This indicates that criteria had a significant correlation with dimensions and scales had convergent validity [42].

Furthermore, results (shown in **Table 8**) indicated that organizational learning capability positive and significant relationship with management commitment to long-term investment in human resources development and organizational learning

	Derived Factors[c]			
Organizational Learning Capability[b]	**MC[b1]**	**SP[b2]**	**OEX[b3]**	**KTI[b4]**
MC1	**0.667**	0.286	0.335	0.132
MC2	**0.734**	0.174	0.317	0.152
MC3	**0.714**	0.351	0.135	0.222
MC4	**0.852**	0.059	0.154	0.098
MC5	**0.771**	0.295	0.269	0.080
SP6	0.237	**0.850**	0.162	0.035
SP7	0.255	**0.797**	0.247	0.168
SP8	0.199	**0.867**	0.226	0.199
OEX9	0.230	0.290	**0.845**	0.053
OEX10	0.186	0.374	**0.789**	0.087
OEX11	0.344	0.052	**0.800**	0.246
OEX12	0.502	0.086	**0.634**	0.211
KTI13	0.166	0.296	0.162	**0.765**
KTI14	0.296	0.164	0.010	**0.838**
KTI15	-0.079	-0.119	0.221	**0.726**
Eigenvalue	9.73	1.76	1.50	1.24
Variance explained	19.35	18.57	18.26	15.89

[a]A VARIMAX orthogonal rotation is performed on the initial factor matrix.
[b]Factors derived from organizational learning capability

Factors	*Cronbach's alphas*	*Scales included*
[b1]MC	*0*	
[b2]SP	*0*	
[b3]OEX	*0*	
[b4]KTI	*0*	

[c]Loadings above 0.50 are in boldface

Table 6.
Factor analysis of organizational learning Scales.[a]

Kaiser-Meyer-Olkin Measure of Sampling Adequacy		0.818
Bartlett's Test of Sphericity	Approx. Chi-Square	1843
Df		120
Sig		0.000

Table 7.
KMO and Bartlett's test of organizational learning variable.

(B = 0.88). Moreover, to enhance learning capability at all levels within organizations, top managers promoted a culture of information sharing and knowledge transfer at all levels (B = 0.63). Results showed that top managers encouraged individuals and teams to explore new ideas through open experimentation (B = 0.80). Findings also indicated that subunits were encouraged to adopt a system perspective notion, as it relates to the understanding of organizational goals and strategic orientation (B = 0.72).

Items	First-order Standardized loading	t-value	Second-order Standardized loading	t-value
Organizational Learning Capability				
Management commitment				
1. Employee participation in management decision making	0.81	/[a]	0.88	0.930*
2. Invest in employee learning	0.78	10.36*		
3. Embracing change to adapt to changing business environment	0.73	9.66*		
4. Employee learning as a key success factor in company	0.77	10.33*		
5. Rewarding novel ideas	0.86	11.89*		
Open experimentation				
1. Job expansion through creativity and experimentation	0.85	/[a]	0.80	9.0*
2. Adopting best practices in competitive field	.84	12.60*		
3. Considering expert views outside company to improve learning	0.85	12.64*		
4. Creating a culture of accepting ideas generated by employees	0.76	10.84*		
System perspective				
1. Employee knowledge about the strategic direction of company	0.83	/[a]	0.72	7.96*
2. Divisional participation in company goals	0.88	13.19*		
3. Communication among company divisions/ departments	0.94	14.30*		
Knowledge Transfer Initiative				
1. Discussion about shortcomings and mistakes at all levels	0.82	/[a]	0.63	6.66*
2. Discussions about ideas, programs, and activities among employees	0.83	10.43*		
3. Culture of teamwork	0.44	5.14*		
4. Maintenance of work process documentation	0.78	9.90*		

/[a] Fixed parameter
Chi-square = 235.64 ($p < 0.001$); df = 100; GFI = 0.95; AGFI = 0.86; RMSEA = 0.095
*$p < 0.001$.

Table 8.
Results of the first-order and second-order confirmatory factor analysis of organization learning.

Analysis of subset variables and their relationship with quality management.

Management commitment. Results shown in **Table 8** revealed that investment in human capital through learning programs ($B = 0.78$) will be considered a key success factor in the organization ($B = 0.77$). Furthermore, the analysis indicated that employees participating in the management decision-making process will be important ($B = 0.81$) and can contribute to decisions on how to adapt to changing industry environment ($B = 0.73$). Management also implemented a program to reward novel ideas by individuals and teams (B0.86).

Knowledge sharing and cross-functional transfer. Knowledge sharing within a subunit and among various subunits contributes to the generation of new ideas among employees ($B = 0.83$), proper documentation of work processes ($B = 0.78$), creates a culture of teamwork (B =0.44), also generates productive discussions about the subunits and top management shortcomings ($B = 0.82$).

System perspective. Establishing a system perspective requires a lateral and flexible organizational structure. Results of the data analysis showed that top executives implemented integrated quality management by designing a lateral organizational structure which enabled departments and divisions to participate in the strategic goals setting process of the organization.

($B = 0.88$). Furthermore, lateral structural design facilitated a more effective communicate cross-functionally to co-align division objects and goals ($B = 0.94$), and from conferences to educate employees about organizational strategic direction ($B = 0.83$).

Exploration and open experimentation. According to March [2], organizations engage in exploration to find a new way of doing things, creating products and services. The data analysis in the present research indicated that organizations pursued both internal strategy and external monitoring to explore and experiment with novel ideas. Data analysis showed that the organization created a culture of welcoming and accepting new ideas by employees ($B = 0.76$), also, employee job expansion employees were enabled to explore and experiment with new ideas ($B = 0.85$). Within business environment, results indicated that organizations monitored and adopted best practices ($B = 0.84$) and consulted with experts in the field outside the organization to improve learning capability ($B = 0.85$).

5.3 Organizational innovation

Table 9 presents the result of an orthogonal (VARIMAX) rotation of the factor matrix underlying organizational innovation items. Based on the three-independent factor solution suggested by the eigenvalue pattern (i.e., greater than 1.0), 17 items were identified so that each of which loaded at least cleanly on only one of three factors. A cut-off of 0.50 was used for item-scale selection. These factors accounted for over 74% of the variance in the organizational innovation scale items. Following the factor loadings, the three factors were subsequently labeled "product/services initiatives," "product innovation," and "overall organizational innovation."

Table 10 shows the Kiser-Meyer-Olkin and Bartlett test of sphericity. Results reasonably describe each set of items as being indicative of underlying factors for organizational innovation (KMO > 0.891; $\chi^2 = 2.418E3$, df, 136, Sig, 0.000). Furthermore, results are indicative of a relationship among the innovation components, "product innovation," "process innovation," and "organizational innovation."

Table 9 shows the results of second-order confirmatory factor analysis and the scale reliability on organizational innovation dimensions that reached statistical significance. This indicates that criteria had a significant correlation with dimensions and that the scales had convergent validity [42]. Results (Shown in **Table 8**) were also indicative of a significant and positive correlation between innovation and the introduction of new products and services ($B = 0.92$). Moreover, top managers allocated resources for

	Derived Factors[c]		
Organizational Innovation[b]	**PS[b1]**	**PR[b2]**	**OOI[b3]**
PS1	**0.760**	0.229	0.303
PS2	**0.795**	0.219	0.346
PS3	**0.818**	0.413	0.180
PS4	**0.550**	0.339	0.447
PS5	**0.781**	0.400	0.218
PS6	**0.659**	0.246	0.439
PR7	0.459	**0.718**	0.155
PR8	0.277	**0.730**	0.381
PR9	0.141	**0.805**	0.121
PR10	0.170	**0.757**	0.358
PR11	0.337	**0.738**	0.253
PR12	0.430	**0.757**	0.169
OOI13	0.398	0.169	**0.736**
OOI14	0.299	0.261	**0.784**
OOI15	0.415	0.298	**0.687**
OOI16	0.103	0.249	**0.855**
OOI17	0.241	0.162	**0.797**
Eigenvalue	9.84	1.66	1.18
Variance explained	25.57	25.09	24.01

[a]*A VARIMAX orthogonal rotation is performed on the initial factor matrix*
[b]*Factors derived from organizational innovation*

Factors	*Cronbach's alphas*	*Scales included*
[b1]*PS*	*0*	
[b2]*PR*	*0*	
[b3]*OOI*	*0*	

[c]*Loadings above 0.50 are in boldface*

Table 9.
Factor analysis of organizational innovation Scales[a].

Kaiser-Meyer-Olkin Measure of Sampling Adequacy		0.891
Bartlett's Test of Sphericity	Approx. Chi-Square	2.418E3
Df		136
Sig		0.000

Table 10.
KMO and Bartlett's test of innovation variable.

continuous process innovation (B = 0.78). Findings also revealed that top managers coordinated subunits efforts to enhance overall organizational innovations (B = 0.77).

Analysis of subset variables and their relationship with quality management.

Products and services innovation. Results for the subset variables of the innovation dimension (**Table 11**) reveal that executives place strategic importance on the first-

Items	First-order Standardized loading	t-value	Second-order Standardized loading	t-value
Overall Organizational Innovation				
Product and service innovation				
1. Higher rate of innovation in comparison to competitors	0.79	/[a]	0.92	9.88*
2. Higher production improvement in comparison to competitors	0.82	11.32*		
3. Faster acquisition of innovative ideas compare to competitors	0.94	13.55*		
4. Knowledge and skill improvement through R&D	0.72	9.56*		
5. Production of products that better fit customer needs	0.92	13.21*		
6. Introduction of new products to customers faster than competitors	0.95	10.10*		
Performance innovation				
1. Utilizing novel ideas to improve the product quality and speed of deliver	0.88	/[a]	0.78	10.25*
2. Utilizing quality resources in the production process	0.80	12.61*		
3. Flexibility in resources allocation	0.68	9.72*		
4. Cost reduction through efficient resource allocation	0.78	11.86*		
5. Adoption of human resources management	0.81	12.86*		
6. Flexibility in org-structure compare to competitors that allows innovation	0.89	15.42*		
Overall organization innovation				
1. Best use of organizational resources to implement quality management	0.77	/[a]	0.77	8.28*
2. Unit cost reduction after implementation of quality management	0.84	11.08*		
3. Financial improvement after quality management improvement	0.81	10.61*		
4. Increased employee productivity after quality management implementation	0.79	10.18*		

/[a]Fixed parameter.
Chi-square = 244.89 (p < 0.001); df = 116; GFI = 0.91; AGFI = 0.81; RMSEA = 0.086.
*p < 0.001.

Table 11.
Results of the first-order and second-order confirmatory factor analysis of organization innovation.

mover advantage and faster generation of new products and services compare to other rivals ($B = 0.94$). Furthermore, the first-mover advantage enabled the organization to present customers with products and services that best served their needs compared to other rivals in the marketplace ($B = 0.92$), at a higher rate of market presentation of innovative products compared to other rivals (B = 0.79). Results also indicated that as a first-mover strategy, top managers placed strategic emphasis on R&D and allocated greater resources toward research and development (B = 0.72). Congruent with results presented in the learning capability segment, flexible and lateral structural design and greater cross-functional communication and knowledge sharing, reduced process costs associated with the higher production improvements and efficiency, compared to other competitors (B = 0.82), and generating new products and services for customers (B = 0.75).

Innovation performance. Findings reveal that designing a lateral flexible organizational structure was highly correlated with innovations in the organization (B = 0.89). enabled subunits to transform the novel ideas into products and services and present them to the marketplace in a timely fashion (B = 0.88). Moreover, resources are to be allocated and reallocated cross-functionally (B = 0.68), with lower costs and more efficiency (B = 0.78). Findings also indicated that top managers focused on human resource development and management (B = 0.81) and acquire high-quality resources in the production processes (B = 0.80).

Organizational innovation. The results of the analysis of innovation showed that there are two important aspects of organizational innovation. The financial aspect indicated that innovation leads to a reduction in costs per unit (B = 0.84). Moreover, innovation enhances the employee productivity (B = 0.79), efficient resources allocation cross-functionally (B = 0.77), and prospects of healthier finances (B = 0.79).

5.4 Organizational performance

Table 12 presents the result of an orthogonal (VARIMAX) rotation of the factor matrix underlying organizational performance items. Based on the four-independent factor solution suggested by the eigenvalue pattern (i.e., greater than 1.0), 16 items were identified so that each of which loaded at least cleanly on only one of four factors. A cut-off of 0.50 was used for item-scale selection. These factors accounted for over 77% of the variance in the organizational performance scale items. Following an inspection of the factor loadings, the four factors were subsequently labeled "customer satisfaction," "employee satisfaction," "environmental performance," and "environmental sustainability."

Kaiser-Meyer-Olkin and Bartlett test of sphericity (shown in **Table 13**) was utilized to measure four organizational performance dimensions, with each of the dimensions being measured by responses to several items. Results (shown in Table E) reasonably describe each set of items as being indicative of an underlying factor for organizational performance (KMO > 0.862; χ^2 = 1.971E3, df, 120, Bartlett's Test of sphericity with significant of 0.000 (less than 0.05).

Table 12 shows the results of second-order confirmatory factor analysis and the scale reliability on organizational performance dimensions that reached statistical significance. This indicates that criteria had a significant correlation with dimensions and that the scale had convergent validity [42].

Results of path analysis indicated top echelon focus on reduced turnover rate by instituting a high remuneration policy and employee satisfaction (B = 0.75). Moreover, the data analysis indicated that customer contentment with products and services was high with little or no defect returns (B = 0.59). Findings also indicated that top managers monitored the industry environment and continuously selected best practices (B = 0.63). Furthermore, top managers were cognizant of the organization's

	Derived Factors[c]			
Organizational Performance[b]	**CUS**[b1]	**EMS**[b2]	**SOR**[b3]	**ENP**[b4]
EMS1	0.306	**0.731**	0.288	0.279
EMS2	0.211	**0.810**	0.297	0.210
EMS3	0.301	**0.844**	0.195	0.120
EMS4	0.400	**0.689**	0.089	0.319
CUS5	**0.668**	0.436	0.285	−0.016
CUS6	**0.819**	0.133	0.198	0.226
CUS7	**0.680**	0.090	0.542	0.095
CUS8	**0.797**	0.337	0.088	−0.038
CUS9	**0.876**	0.219	0.139	0.138
ENP10	0.163	0.157	0.202	**0.836**
ENP11	0.078	0.543	0.158	**0.680**
ENP12	0.024	0.184	0.306	**0.784**
SOR13	0.310	0.362	**0.724**	0.203
SOR14	0.218	0.014	**0.738**	0.307
SOR15	0.015	0.470	**0.689**	0.173
SOR16	0.270	0.267	**0.756**	0.196
Eigenvalue	8.20	1.90	1.30	1.05
Variance explained	22.62	22.06	18.32	14.36

[a]VARIMAX orthogonal rotation is performed on the initial factor matrix
[b]Factors derived from organizational performance

Factors	*Cronbach's alphas*	*Scales included*
[b1]CUS	*0*	
[b2]EMS	*0*	
[b3]SOR	*0*	
[b4]ENP		

[c]Loadings above 0.50 are in boldface

Table 12.
Factor analysis of organizational performance Scales[a].

Kaiser-Meyer-Olkin Measure of Sampling Adequacy		0.862
Bartlett's Test of Sphericity	Approx. Chi-Square	1.971E3
Df		120
Sig		0.000

Table 13.
KMO and Bartlett's test of organizational performance variable.

impact on the environment and negative externalities and pursued a sustainability strategy as a priority post integrated quality management implementation (B = 0.93).

Analysis of subset variables and their relationship with quality management.

Human resource management. Analysis of the subset variables (shown in **Table 14**) revealed that executives place strategic importance on employee retention (B = 0.88)

Items	First-order Standardized loading	t-value	Second-order Standardized loading	t-value
Organizational Performance				
Employee satisfaction				
1. Employee satisfaction	0.86	/[a]	0.75	10.88*
2. Ample remuneration for employees	0.87	14.12*		
3. Reducing turnover after quality management implementation	0.88	14.37*		
4. Reduction of absenteeism after quality management implementation	0.83	12.97*		
Customer satisfaction				
1. Customer satisfaction	0.86	/[a]	0.75	10.88*
2. Introduction of new product and services	0.80	10.91*		
3. Reduction of product defect returns after quality management implementation	0.75	10.03*		
4. Strategies to maintain customer base	0.82	11.41*		
5. Higher profitability after quality management implementation	0.88	11.74*		
6. Reducing customer complaints after quality management implementation	0.89	12.65*		
Sustainability and Environmental				
1. Consideration of environmental projects after implementation of quality management	0.74	/[a]	0.63	7.41*
2. Sustainability/Reducing production pollution	0.85	9.31*		
3. Reducing complains about environmental pollution	0.75	8.56*		
Social responsibility performance				
1. Sustainability	0.89	/[a]	0.93	10.04*

/[a]Fixed parameter.
Chi-square = 232.06 ($p < 0.001$); df = 100; GFI = 0.93; AGFI = 0.84; RMSEA = 0.094.a Fixed parameter.
$^{*}p < 0.001$.

Table 14.
Results of the first-order and second-order confirmatory factor analysis of organizational performance.

and reducing absenteeism ($B = 0.83$) by offering employees competitive remunerations ($B = 0.87$), and overall employee satisfaction of their jobs ($B = 0.86$).

Customer contentment. Results of data analysis indicated that investment in the introduction of new and high-quality products and services ($B = 0.80$) tend to reduce the rate of defected products ($B = 0.75$), consumer complaints ($B = 0.89$), maintain the market share ($B = 0.82$), and assure consumers are contented with the products and services ($B = 0.81$).

Monitoring environmental conditions and sustainability strategy. The analysis outcome also revealed that top executives were cognizant about the company's reputation by maintaining sustainability by considering environmental renewable energy projects ($B = 0.74$), reducing the negative externalities caused by production pollution ($B = 0.85$). Integrating the sustainability strategy with quality management enhanced the company's legitimacy and reputation for social responsibility by planning for environmentally friendly projects and sustainability ($B = 0.89$).

6. Results and discussion

Macro model. As shown in **Table 1** (**Figure 1**), the standard regression weight for the overall model indicated a positive and significant relationship between main variables, quality management, organizational learning, and innovations. According to the results, organizational integrated quality management is positively and significantly associated with organizational learning capability ($B = 0.95, p < 0.05$). Similarly, results showed a positive and significant relationship between innovation performance and integrated quality management.

($B = 0.91, p < 0.05$). Results indicated that when parsing the main effects of learning capability and innovation performance, the association between quality management and organizational performance remains positive but statistically non-significant ($B = 0.43$, n.s.) and does not explain significant variance ($R^2 = 0.18$) in organizational performance. A detailed analysis revealed that organizational learning capability is positively and significantly associated with organizational performance ($B = 0.58, p < 0.05$). Furthermore, innovation performance, according to findings, is also positively and significantly associated with organizational performance ($B = 0.62$, $p < 0.05$). Findings are congruent with hypotheses H1a and H1b. Findings, however, being partially congruent with hypothesis a, H1.

H1: There will be a positive and significant relationship between quality management, organizational and organizational performance.

H1a: There will be a positive relationship between quality management, organizational learning.

H1b: There will be a positive relationship between quality management, organizational, and innovation.

A detailed analysis revealed that organizational learning capability is positively and significantly associated with organizational performance ($B = 0.58, p < 0.05$). Furthermore, innovation performance, according to findings, is also positively and significantly associated with organizational performance ($B = 0.62, p < 0.05$).

Findings also supported hypotheses H2 and H3.

H2: There will be a positive relationship between organizational learning and organizational performance.

H3: There will be a positive relationship between innovation and organizational performance.

6.1 Interaction effects

As managers attempt to identify factors that influence organizations' performance, this research argued that it is important to gain a deeper understanding as to how interaction effects of quality management, learning capability, and innovations matter in influencing organizational performance. The hypothesis H4 specified that organizational performance would be affected by an interactive effect of quality management and organizational learning capability. The hypothesis H5 specified that organizational performance would be affected by an interactive effect of quality management and innovation. To test these hypotheses, I employed structural equation modeling analysis to reduce the number of variables and to capture the interrelations of measured variables and latent constructs, as suggested by Tarka [43]. Results indicated that compared to the effects of quality management and organizational performance (B = 0.43, n.s., R^2 = 0.18), the multiplicative interaction term for quality management and organizational learning capability increased explanatory variance in organizational performance (R^2 = 0.34, $p < 0.05$), significantly (0.95 × 0.58) = (0.55, $p < 0.05$). Similarly, the multiplicative term between quality management and increased the variance (R^2 = 0.38, $p < 0.05$) significantly (0.91 × 0.62) = (0.56, $P < 0.05$). Results of the analysis were congruent with H4 and H5.

H4: The interactions between quality management and organizational learning positively influence the relationship between quality management and organizational performance.

H5: The interactions between quality management and innovation positively influence the relationship between quality management and organizational performance.

7. Discussion

There are several important theoretical and practical implications that emerge from this research. Findings underscore the importance of the interaction of quality management elements. Over the past decade, researchers have systematically underplayed the interaction effects of quality management elements. The present research showed that the dominant impact on organizational performance, beyond external resource considerations, is the intersection of forces associated with quality management, organizational learning capability, and innovations within these organizations. It was argued earlier that within quality management theory and methodology, the need to consider the contingency approach might result in an in-depth understanding of the strategic allocation of resources and managing and coordinating among the interrelated constituent elements within quality management. Results suggested that organizational performance is positively influenced by the interaction of quality management and innovation and learning capability at organizational levels. It is also clear that there are distinct differences between parsed and integrated constituents within quality management with respect to explaining variations in organizational performance. This finding is of some theoretical significance.

As a strategy, quality management appears to have coordination challenges associated with learning capability and application of such learning to innovations of new products and services. This study found that organizational performance is significantly impacted by the interaction between quality management and learning capability. Similarly, findings indicated that interaction between innovation and quality management positively and significantly influences organizational performance. The strength of these findings, particularly in light of incorporating external environmental factors such as sustainability considerations, points to the potential importance of revitalizing the contingency theory perspective pertaining to integrated quality management. Such a revival would not necessarily imply that researchers "pit" internal elements influencing performance against external forces. Instead, more direct integration of contingency variables within quality management is suggested to better balance internal and external perspectives on organizational performance.

Nevertheless, any resurrection of this perspective within quality management theory and methodology may require changes in how contingency theory may be employed (e.g., Pfeffer 1997). This study did not limit its focus to examining the main effects of organizational learning capability, innovations, and quality management on performance. As I argued in theoretical development, one cannot easily specify the nature of these main effects. Instead, what may be as, if not more, important to consider is the interaction of these variables as previous organizational researchers have argued that internal and external characteristics of organizations and their members may cluster together in predictable patterns to explain a variety of micro to macro-level organizational processes and relationships [44]. Congruent with Meyer et al.'s findings on organizational learning capability showed top managerial commitment to implement a complex set of policies on the development of human resources. Such policies included learning based on system perspectives, learning associated with experimentation and exploration of a novel way of doing things, and knowledge transfer at various levels of individuals, teams, and organizational subunits. Furthermore, findings revealed managerial efforts to coordinate and co-align subunits' strategies with the organization at the macro level. Similarly, findings on innovations showed managerial commitment to implementing flexible resource allocation strategies for subunits to explore novel processes and ideas. Findings were congruent with the notion that integrating interrelated constituents of quality management at the micro and macro level require greater structural flexibility and high levels of coordination among organizational activities. While the explicit consideration of interactive variables in quality management theory adds complexity to the understanding and application of contingency theory, this type of complexity is what managers must face. Rarely, is there the luxury of focusing exclusively on one aspect of quality management, as has been the themes of previous research, in isolation from others? For the contingency theory to develop as a theoretical perspective and be relevant to the practical concerns of managers and executives, researchers may need to provide further attention to how constructs in quality management and their subset variables interact to influence organizational performance over time.

Employing contingency theory to conduct future research in the quality management field will also require making more direct connections between the results of studies and the organizational design concerns of managers. One important vehicle for doing this is by considering how quality management research findings can be connected to process considerations at various levels of organization. It is often organizational processes that are of most direct concern to managers adopting quality management practice. Perhaps, the most direct implication relates to the enhanced importance of managing and integrating complex processes within and between each constituent of quality management.

Therefore, it is critical for an organization adopting quality management to develop an organizational capability or competence for managing internal complex and interrelated process models. Without this capability, managerial policies and efforts can become misguided and create greater conflict, thereby undermining the effectiveness of coordination efforts among complex processes to achieve timely policy and strategy adjustments. Successful corporations such as Boeing and car manufacturers recognized the need to employ quality management and, as the corporation evolves, developed organizational capabilities to manage complex processes.

Future researchers may wish to create a matrix that examines the contingent effects of long-term variations in learning capability on innovations and assess variations in long-term innovations on organizational performance.

Author details

Mohsen Modarres
Management and Technology Consulting, Kirkland, WA, USA

*Address all correspondence to: modarresfm@gmail.com

References

[1] Raynor EM. The Strategy Paradox. London: Currency Doubleday; 2007

[2] March J. Exploration and exploitation in organization learning. Organization Science. 1991;**2**(1):71-87

[3] Ireland DR, Hoskisson RE, Hitt MA, Loomis CJ. Can anyone run Citigroup? Fortune. 2008:80-91

[4] Bartlett CA, Ghoshal A, Birkinshaw J. Transnational Management, Texts, Cases and Readings in Cross-Boarder Management. San Francisco: Irwin-McGraw-Hill; 2004

[5] Douglas TJ, Judge JQ. Total quality management and competitive advantage: The role of structural control and exploration. Academy of Management Journal. 2001;**44**(1):158-169

[6] Hung RYY, Lien BYH, Yang B, Wu CM, Kuo YM. Impact of TQM and organizational learning on innovation performance in the high-tech industry. International Business Review. 2011;**20**(2):213-225

[7] Luthans F. Organizational Behavior. 7th ed. New York, NY: Mcgraw Hill; 1995

[8] Joiner TA. Total quality management and performance: The role of organization support and co-worker support. International Journal of Quality and Reliability Management. 2006;**21**(6):617-627

[9] Hana U. Competitive advantage achievement through innovation and knowledge. Journal of Competitiveness. 2013;**5**(11):82-96

[10] Senge PM. The Fifth Discipline: The art & Practice of the Learning Organization. New York: Currency Doubleday; 1990

[11] Corredor P, Goñi S. TQM and performance: Is the relationship so obvious? Journal of Business Research. 2011;**64**:830-838

[12] Porter ME. What is strategy? Harvard Business Review. 1996;**74**(6):61-78

[13] Wheelen T, Hunger D, Hoffman A, Bamford C. Strategic Management and Business Policy: Globalization, Innovation and Sustainability. 14th ed. USA: Pearson; 2015

[14] Gomez-Gras JM, Verdu-Jover AJ. TQM, structural and strategic flexibility and performance: An empirical research study. Total Quality Management and Business Excellence. 2005;**16**(7):841-860

[15] Hasan M, Kerr RM. The relationship between TQM practices and organizational performance in service organizations. The TQM Magazine. 2003;**15**(4):286-291

[16] Powell T. Total quality management as competitive advantage: A review and empirical study. Strategic Management Journal. 1995;**16**(1):15-37

[17] Westphal JD, Gulati R, Shortell SM. The Institutionalization of Total Quality Management: The Emergence of Normative TQM Adoption and the Consequences for Organizational Legitimacy and Performance. August: Academy of Management Proceedings; 1996. pp. 249-253

[18] Demirbag M, Tatoglu E, Tekinkus M, Zaim S. An analysis of the relationship between TQM

implementation and organizational performance: Evidence from Turkish SMEs. Journal of Manufacturing Technology Management. 2006;**17**(6): 829-847

[19] Salaheldin I. Problems. success factors and benefits of QCs implementation: a case of QASCO. The TQM Journal. 2009;**21**:87-100

[20] Jaafreh AB, Al-Abedallat AZ. The effect of quality management practices on organizational performance in Jordan: an empirical study. International Journal of Financial Research. 2012;**4**(1):93

[21] Winter S. Knowledge and competence as strategic assets. In: Teece D, editor. The Competitive Challenge: Strategies for Industrial Innovation and Renewal. New York: Harper and Row; 1987. pp. 159-184

[22] Schumpeter JA. The Theory of Economic Development. Cambridge, MA: Harvard University Press; 1934

[23] Penrose E. The Theory of the Growth of the Firm. New York: Wiley; 1959

[24] Conner K, Prahalad C. A resource-based theory of the firm: Knowledge versus opportunism. Organization Science. 1996;**7**:477-501

[25] Cyert R, March J. A Behavioral Theory of the Firm. Englewood Cliffs: Prentice-Hall; 1963

[26] Kim DH. Link between individual and organizational learning. In: Klien D, editor. Strategic Management of Intellectual Capital. 1998. pp. 41-62

[27] Modarres M, Beheshtian M. Enterprise information systems: Integrating decision support systems, executive information systems, and simulation technology. . Vol. 31. International Journal of Technology Management. 2005;**31**(1–2):116-128

[28] Jerez-Gomez P, Cespedes-Lorente J, Valle-Cabrera R. Organizational learning capability: A proposal of measurement. Journal of Business Research. 2005;**58**: 715-725

[29] Jimenes D, Sans-Valle R. Innovation. organizational learning and performance, Journal of Business Research. 2011;**64**:408-417

[30] Atalay M, Anafarta N, Sarvan F. The relationship between innovation and firm performance: An empirical evidence from Turkish automotive supplier industry. Proceedings-Social and Behavioral Sciences. 2013;**75**:226-235

[31] Günday G, Ulusoy G, Kılıç K, Alpkan L. Effects of innovation types on firm performance. International Journal of Production Economics. 2011;**133**(2): 662-676

[32] Hoang D, Igel B, Laosirihongthong T. The impact of total quality management on innovation: Findings from a developing country. International Journal of Quality and Reliability Management. 2006;**23**(9):1092-1117

[33] Singh PJ, Smith AJ. Relationship between TQM and innovation: An empirical study. Journal of Manufacturing Technology Management. 2004;**15**(5):394-401

[34] Holbeche, L. 2017. Designing More Agile Organizational Structure and Ways of Working. CIPD Annual Conference and exhibition.

[35] Luecke R, Katz R. Managing Creativity and Innovation. Boston, MA: Harvard Business SchoolPress; 2003

[36] Tushman ML, Anderson P. Technological discontinuities and organizational environments. Administrative Science Quarterly. 1986; **31**(3):439-465

[37] Vanichchinchai A, Igel B. The impact quality management on supply chain management and firm's supply performance. International Journal of Production Research. 2011;**49**(11): 3405-3424

[38] Coyle-Shapiro J. A psychological contract perspective on organizational citizenship behavior. Journal of Organizational Behavior. 2002;**23**: 927-946

[39] Santos JB, Briot LAL. Toward a subjective measurement model for firm performance. Brazilian Administration Review (BAR). 2012;**9**(6):95-117

[40] Fields A. Discovering Statistics Using SPSS. 3rd ed. London: Sage Publications; 2009

[41] Royston P. Constructing time-specific reference ranges. Statistics in Medicine. 1991;**10**(5):675-690

[42] Anderson JC, Gerbing DW. Structural equation modeling in practice: A review and recommended two-step approach. Psychological Bulletin. 1988; **103**(3):411-423

[43] Tarka P. An overview of structural equation modeling: Its beginnings, historical development, usefulness and controversies in the social sciences. Quality & Quantity. 2018;**52**(1):313-354. DOI: 10.1007/s11135-017-0469-8

[44] Modarres M. Interactive Effects of Organizational Size and Structural Differentiation on Administrative Reorganization. Research in Progress. 2019

Chapter 13

Artificial Intelligence Deployment to Secure IoT in Industrial Environment

Shadha ALAmri, Fatima ALAbri and Tripti Sharma

Abstract

Performance enhancement and cost-effectiveness are the critical factors for most industries. There is a variation in the performance and cost matrices based on the industrial sectors; however, cybersecurity is required to be maintained since most of the 4th industrial revolution (4IR) are based on technology. Internet of Things, IoT, technology is one of the 4IR pillars that support enhancing performance and cost. Like most Internet-based technologies, IoT has some security challenges mostly related to access control and exposed services. Artificial intelligence (AI) is a promising approach that can enhance cybersecurity. This chapter explores industrial IoT (IIoT) from the business view and the security requirements. It also provides a critical analysis of the security challenges faced by IoT systems. Finally, it presents a comparative study of the advisable AI categories to be used in mitigating IoT security challenges.

Keywords: artificial intelligence, Internet of Things, cybersecurity, industry, industrial IoT (IIoT), 4th industrial revolution

1. Introduction

The 4th Industrial revolution (4IR) is the current era where industry is driven by technology. It encourages the co-operation between scientific knowledge and experience with business mindset and requirements. The key technologies that allow 4IR to be sustained are additive manufacturing techniques, Autonomous and collaborative robotics, Industrial Internet of Things (IIoT), Big data analytics, Cloud Manufacturing techniques [1]. The current scenarios show the benefits of IIoT in improving QoS industries, starting from predictive maintenance, reaching remote controlling of assets, and deploying Digital Twin concept that allows virtualizing the operations environment and permits the owner to be proactive when any anomalies are detected [2]. Even though IIoT adds value to the traditional industry, there should be a balance between the operational benefits and the security level.

Aims and objectives

- To study and compare the existing IoT architectures

- To explore industrial IoT (IIoT) from the business point of view
- To analyze various IIoT threats and security challenges and existing mitigation techniques
- To perform a comparative study of the different AI categories and their applicability in IIoT security
- To recommend the most convenient AI techniques for mitigation of IIoT security challenges

This chapter is designed to be used as a reference to study the effectiveness of Artificial intelligence (AI) and to enhance the security techniques for mitigating the threats faced by IIoT deployment. Section 2.1 discusses IoT architecture and Section 2.2 demonstrates the IoT security challenges. Section 2.3 describes the main AI categories and their subcategories. It also points out the appropriate and relevant situation to employ AI categories based on the available data and the type of intelligence needed. Section 3 explores IIoT details, its significant business model, and the added values. Section 4 focuses on IoT security in terms of threat model, threats classification, and common IoT security mitigations. This chapter ends with a comparative study of AI categories used to mitigate IIoT security challenges in Section 5.

2. Background

2.1 IoT architecture

Internet of Things (IoT) is a service-oriented paradigm that is built on the involvement of several technologies. Therefore, its architecture consists of layers starting from sensors and reaching to constructive data displayed on the system-analyzer screen.

In References [3–5], the main IoT architecture consists of devices that have sensors and edge computing which has embedded devices, fog computing such as gateway and servers, cloudlets such as base stations, and the last component being cloud computing, which can be any cloud platform. **Table 1** shows some IoT architectures with variations on the number of layers based on five different references. In general, there are three main layers that are devices, network, and cloud computing. However, the device layer can be divided into two sub-layers based on the type and functionality: The first sub-layer comprises of end-user devices that contain sensors; and the second sub-layer are devices that support machine-to-machine communication such

Ref	Number of layers	Layers title
[3]	5 layers	Devices, edge Computing, fog computing, cloudlets, cloud computing
[4]	3 layers	IoT layer, fog layer, cloud layer
[5]	3 layers	EoT(ecosystem of Things) layer, edge layer, cloud layer
[6]	4 layers	Fog network consist of (IoT layer, Mist, cloudlet/edge layer,) cloud
[7]	4 layers	Sensors and systems layer, far edge layer, near-edge layer, cloud layer

Table 1.
IoT ecosystem architecture comparison.

as an Arduino platform. The network layer can be divided as well into two sub-layers based on the communication characteristics such as the speed and bandwidth: fog computing and cloudlet. The third layer is the cloud computing layer. **Figure 1** illustrates the authors' insights into IoT architecture after studying the literature. Layer one consists of IoT devices, layer two covers all networking related technologies and devices, and the third layer consists of cloud computing and related data analytics technologies.

The IoT layers are connected through networking media using wireless or wired connections. However, wireless technology evolvement is critical to extend IoT deployment as the complexity of energy impact and processing capacity are getting worse at the sensor's layer [6]. The emerging of 5G in wireless communication adds an advantage to IoT architecture since it improves the performance by allowing the transformation of more data in less time, which technically reduces service-latency and enhances real-time access to data [6, 8].

2.2 IoT security challenges

The growing use of Internet of Things (IoT) technology in the industrial sector has posed new issues for the device and data security. Based on different world statistics, the number of devices connected to IoT networks is rapidly increasing. This expansion leads to experience different levels of vulnerabilities, which may—in turn—cause an increase in security threats and challenges. Security may be regarded as a big threat that leads to limitations of the IoT systems deployment. As a result thereof, it is the Authors' view that effective security practices may become more vital in the IoT industry.

The National Institute of Standards and Technology (NIST) designed programs to boost cybersecurity involvement in IoT [9]. This initiative promotes the development and implementation of cybersecurity standards, guidelines, and tools for IoT products, connected devices, and their deployment environment.

- Security Challenges: Some common challenges posed by the security requirements for the IoT systems are given as follows:
    - Because IoT involves various and diverse technologies, determining and understanding security needs is more complicated.

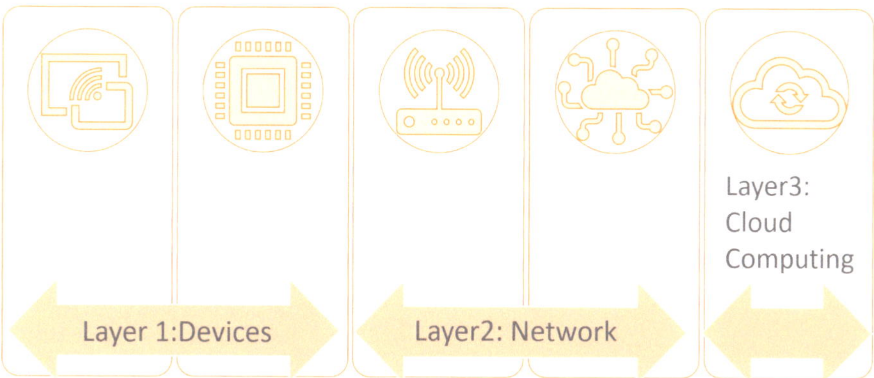

Figure 1.
IoT architecture.

- IoT networks typically consist of resource-constrained devices. Therefore, these devices became the weakest link for cyberattacks.
- The Internet of Things (IoT) may include mobile devices that demand adaptability, posing security vulnerabilities.
- IoT also generates a vast amount of data, which is referred to as Big data. The latter has its own set of security and management concerns.
- Security requirements: Because of the varied nature of IoT Applications, security requirements may also differ. Based on the scenarios from a specific industry and the infrastructure to which IoT is being applied to, the requirements and consequent security measures may be changed or adjusted. Nevertheless, the common security requirements [10–13] of IoT systems can be summarized as given in **Table 2**.

Satisfying all the above-mentioned requirements is a huge challenge because of the limitations and constraints associated with the IoT devices in terms of capability and capacity to deploy the conventional security solutions.

2.3 Artificial intelligence categories

When it comes to artificial intelligence (AI), there are several philosophical groundworks that have been done. As per Russel [14], there are two types of AI: weak AI where machine can act intelligently and strong AI where machine can really think. However, when hybrid mechanisms are used, the deployment of AI system features is enhanced.

Artificial intelligence (AI) can be divided into two main categories as per the mechanisms that are used to reach intelligence through data processing [14–16]. The

Security requirements	Example of mitigation techniques	Requirement description
Confidentiality	Encryption	Only authorized entities should be able to read it to ensure data protection.
Integrity	Hash generation	The data should be checked to ensure that it has not been tampered with.
Authentication, Authorization, Access control (AAA)	Implement policies, Security credentials, firewall, and authentication servers. Digital signature, etc.	• Identification of devices and users. • Special rights or privileges for authorized users; • Access to resources and data should be restricted.
Availability	Fault tolerance mechanism, clustering and high availability architecture, etc.	The ability to be accessed and used by an authorized entity on demand
Non-repudiation	Digital signature	Securing information transmission by supplying confirmation of delivery and identification to both sender and receiver so that neither can later deny processing it. It ensures data origin and integrity.

Table 2.
IoT security attributes, techniques and requirements.

first category is knowledge-based in which the main component is the existence of inference engine, and it is known as expert system (ES). The second category is machine learning (ML) where different algorithms are used to allow the machine to learn from the dataset. **Table 3** illustrates the main AI categories. The core element is knowledge engineering in order to build either the dataset for ML or the fact database for ES. The data preparation phase needs to make use of other technology such as data mining and Big data techniques. The ML sub-categories are supervised learning, reinforcement learning, and un-supervised learning. The ES types of systems are rule-based, Fuzzy-logic, and frame-based.

- Machine Learning types (ML): The intelligence behind ML is the ability to learn. ML involves adaptive mechanisms; therefore, it is considered as the basis of adaptive systems. In this context, the ML detects and extrapolates patterns by adapting to new circumstances. This learning process can be based on experience or examples or analogy. Therefore, ML has three sub-categories as follows:

    - **Supervised learning**: is learning from examples. This type is the easiest ML type in terms of mathematical complexity. The machine learns from *a behavior (labels)*.

    - **Reinforcement learning**: defined as learning from the environment based on experience. This type is based on an agent that can learn from *reward signal*. The machine learns from its mistake.

    - **un-supervised**: referred to as learning based on analogy and to find a pattern from a dataset. This type is used when there are no examples to learn from and no reward signal to get feedback.

Figure 2 shows examples of mechanisms for each ML sub-category.

- Expert System (ES): The Expert System, ES is dealing with uncertain knowledge and reasoning. Rule-based ES consist of five basic components that are shown in **Figure 3**: the knowledge base, the database, the inference engine, the explanation facility, and the user. ES intelligence resembles the way the expert human apply their knowledge and intelligence to solve the problem in a narrow domain. ES processes knowledge in the form of rules and uses symbolic reasoning to solve the problem. The main difference between ES and conventional programs (CP) is that the CP processes data using algorithms on well-defined operations to solve a problem in a general domain. Examples of ES are as follows:

Expert system (ES)	**Machine learning (ML)**
Rule-based	Supervised learning
Fuzzy logic	Reinforcement learning
Frame-based	Un-supervised Learning

Table 3.
Artificial intelligence (AI) main categories.

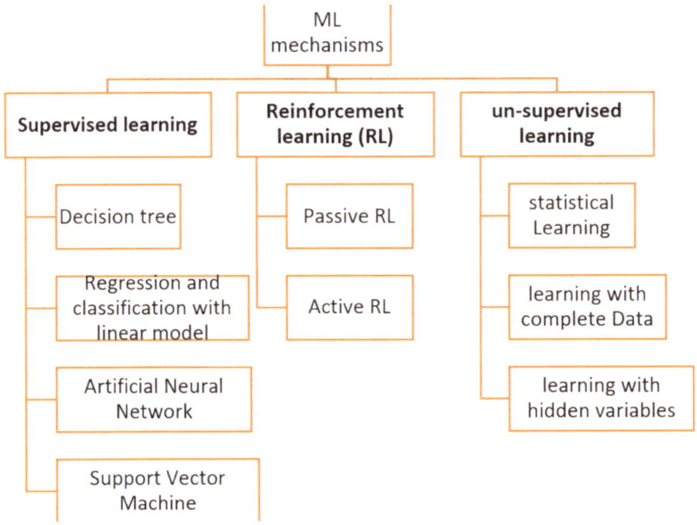

Figure 2.
Examples of ML sub-categories mechanisms.

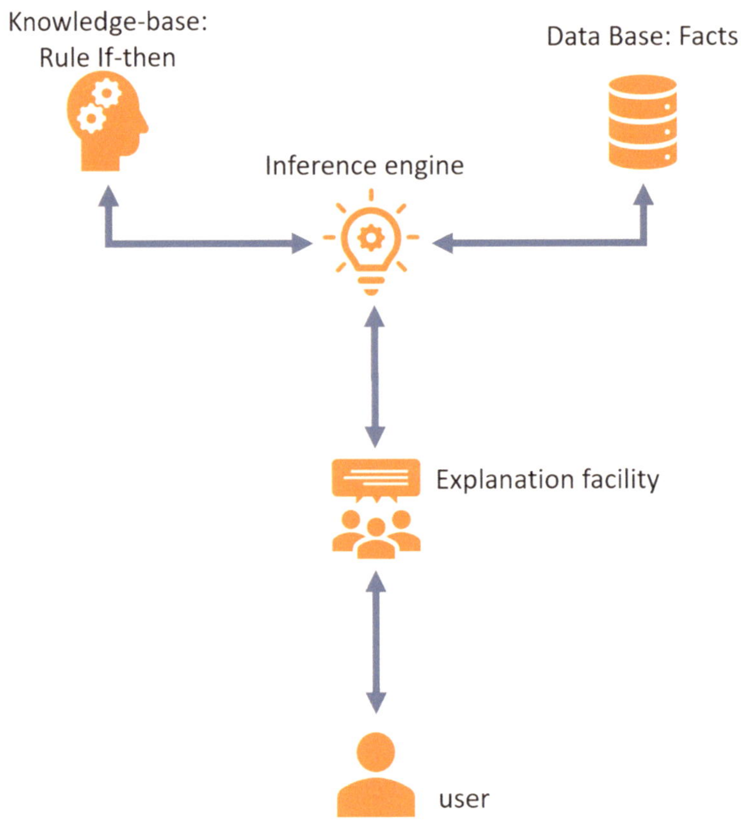

Figure 3.
Expert system (ES) rule based adapted from *[15]*.

- **Rule-based**: is based on logical rules. Its disadvantage comes from ineffective search strategy and inability to learn.

- **Fuzzy logic**: is centered on logic that describes fuzziness. It models the common sense of a human. Tuning is the most ponderous stage of building a fuzzy system.

- **Frame-based**: is constructed on structuring knowledge based on object attributes. It usually uses pattern matching but it has a limitation to make decisions about the hierarchical structuring during knowledge engineering.

3. Artificial intelligence in industrial IoT

3.1 The significance of AI in IIOT

Artificial intelligence (AI) deployment in Industrial IoT (IIoT) systems is very convenient due to the huge data generated by the IoT system. AI approaches are used to infer knowledge and support data analytics. The main areas requiring exploration and proposing solutions for intelligent IIoT systems are threat hunting and intelligence, blockchain, edge computing such as cloud computing, privacy preservation [17].

The generated big data from IIoT are due to real-time computation and the risk increases when the communicated data are critical and sensitive; therefore, AI can support the need of big data analysis with low latency [2]. Designing security and privacy solutions require to identify business processes and operations. However, this task is complex in the regular industrial system, and it comes more sophisticated in IIoT [18]. AI technology deployment has several implementations including computing paradigm and security; however, inter-operability issues are regarded as a critical challenge [3].

The Internet of Things (IoT) has grown from a concept used in research laboratories and technology companies to a reality in everyday lives. IoT has become embedded in the operations of some companies, enterprises, and governments [19]. Emerging IoT applications are spread out in all domains, and it has affected a variety of industries. **Figure 4** illustrates the examples of IoT technology applications, which include Smart Homes, Smart Health, Intelligent Transportation, Smart Cities, Smart Agriculture, and Factory Automation [3].

Indeed, the very same report by McKinsey & Company mentioned above [19] identifies the top five sectors where IoT adds the most economic value: factories that include all standardized production environments followed by human health, work sites, cities, and retail environment. Indeed, it has been estimated in this report that IoT could add a value of $5.5 trillion to $12.6 trillion by 2030, where the most value can be created in B2B type of applications.

3.2 The IoT business model

The term business model describes how an organization creates, delivers, and captures value [20]. The adoption of IoT technologies in an organization will most certainly affect the business relationships and the business model for that organization. In this section, the common business models used will be discussed.

One of the early initiatives to develop an IoT business model was published in 2015 [21]. The research focused on identifying the relevant building blocks that can fit in

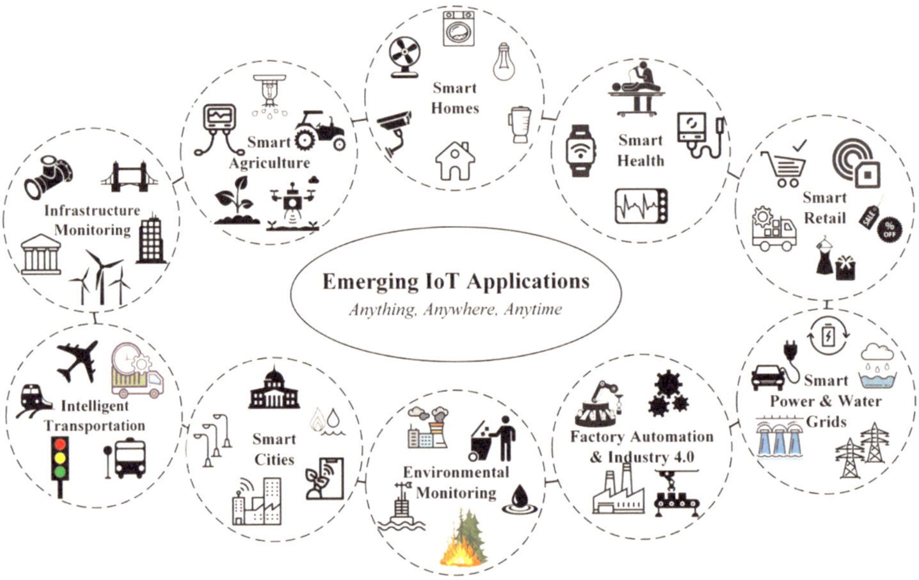

Figure 4.
Example of industry utilizing IoT technology [3].

IoT business models, as well as the types and importance of the building blocks. This framework identified value proposition as the most important building block for IoT business models. The entities "customer relationships" and "key partnerships" followed suit in terms of importance.

Another conceptual IoT Business Model is the AIC (Aspiration, Implementation and Contribution) model presented in [22], which focuses on context-specific implementation of IoT. This model consists of three interconnected phases: Aspiration, Implementation, and Contribution. The first phase "Aspiration" focuses on defining and predicting the value creation through adoption of IoT. The second phase Implementation includes strategy development in which an organization should investigate how IoT will improve the business by gaining competitive advantage or creating enhanced products or services. In the third phase Contribution, an organization opting for IoT should study the practicality of the approach and the capabilities and resources available for the organization to implement IoT. In other words, does the organization own the knowledge and skills needed to succeed in implementing IoT.

Four types of IoT-enabled servitized business models were classified in [23]. Each business model was analyzed from three perspectives: the role of IoT, the firm's benefits, and the inhibiting factors. **Table 4** adapts from the study presents the four types of IoT business models and compares them based on the stated three perspectives. The four different business models have some shared features as the common role for IoT is adaptation, the common benefit is reducing operation cost, and the common inhibiting factor is the need for close relationship between different stakeholders.

IoT business models vary based on the type of deployment. Therefore, each industry has a different model that will fit with its value proposition. Seven IoT business models were reviewed by the researchers in [24]. Based on their analysis, six characteristics of the IoT business model were identified:

IoT Business model	Role of IoT	Firm's benefits	Inhibiting factors
Add-on business model	• Innovation • **Adaptation** • Smoothing	• Improve product-service offerings • Extend firms business • **Reduce operation costs**	• Privacy concerns • Data security • **Requires close relationship between different stakeholders in the network**
Usage-Based business model	• **Adaptation** • Smoothing	• Extend firms business • Generate steady income • **Reduce operation costs**	• Requires expertise in data management • **Requires close relationship between different stakeholders in the network**
Sharing business model	• **Adaptation** • Smoothing	• Improve service offerings • Increase resource utilization • **Reduce operation costs**	• Requires new ways of interactions with customers • **Requires close relationship between different stakeholders in the network**
Solution-oriented business model	• Innovation • **Adaptation**	• Extend firms business • Gain competitive advantage • **Reduce operating cost**	• Developing servitized offerings that aligns with customer's needs • **Requires close relationship between different stakeholders in the network**

Table 4.
Business model categorization based on role, benefits, and inhibiting factors.

- The ability to capture the transition between different business models.
- The possibility to connect IoT elements to the business model components.
- The ability to view the relation between business-centric and a network-centric approaches.
- The ability to map the Value Flows that involve revenue, costs, and assets
- The possibility to include the model patterns of digital business.
- The ability to balance between the actions and widening the rational thinking.

3.3 Analytical study of how IoT add-value to the industry

Given the potential impact and IoT devices' prevalence and ubiquity, one needs to understand how to leverage IoT technologies to realize the value-deriving benefits associated with them. For example, IoT can be used in the factory setting to make various processes more efficient. The IoT applications have noteworthy potential in value creation in terms of operation optimization and predictive maintenance. This can be achieved by monitoring, remotely tracking and adjusting the machineries, based on sensor data from different parts of the factory. It has been estimated that IoT has a potential to create value of $1.2 trillion to $3.7 trillion per year in 2025 by optimizing factory settings. This improvement in the working efficiency using IoT may also induce some security and privacy issues [25]. Moreover, technology does not automatically bring added convenience or value unless firms carefully consider the context into which it is introduced and how to derive any practical or monetary benefits. Mostly, add-value is related to performance enhancement. The latter can be improved through a variety of factors such as time saving, cost saving, and processing low-overhead to name but a few.

Table 5 shows some recent empirical research [26–31] on how to mitigate security challenges in an IoT industrial environment and different add-value. AI approaches are used more in access control, which is related mostly to the Network layer of IoT. Access control is a critical part of the system, which acts as a door for the factory to control authorized access to the recourses and the level of privileges. Due to the heterogenous and dynamic nature of the IoT networks, it will be significant to use AI approaches to enhance the access control.

The IoT add-value is constraint by several challenges and barriers. These can be categorized in two groups based on their domain as follows:

- Human limitations
- lack of social acceptance and knowledge
- lack of skilled workforce, technical knowledge
- Technology limitation
- the absence of technical accountability and regulation
- challenges related to data management and data mining
- privacy, security, and uncertainty

Ref.	IoT layer	Security mitigation approach	Performance (add-value)	AI used
[26]	Network	Graph-theory	Cost is not evaluated Different performances based on number of nodes	×
[27]	Data (access control)	Conditional proxy re-encryption primitive	Low overhead	×
[28]	Data (access control)	Context-aware analysis	Improve detection ration	×
[29]	Network	Deep Learning and Blockchain-Empowered Security Framework	Standard measures of latency, accuracy, and security	✓
[30]	gateways	Flexible rule-based control strategies	Cost and time saving	×
[31]	Network	A deep learning methodology for detecting cyberattacks	Improve detection accuracy of IDS	✓

Table 5.
Examples of AI usage in security mitigation approaches based on IoT layer.

- the immaturity of IoT innovations
- integration among networks and no standardization of regulations
- Business limitation
- difficulty in designing business models for the IoT due to a multitude of different types of connected products
- ecosystems are unstructured since it is too early to identify stakeholders and their roles

Uncertainty of how IoT will impact existing business models, organizational strategies, and return of investment, business models are considered significant barriers to implementation, where the add-value should be clearly identified.

4. Critical analysis of IoT security

4.1 Threat modeling

A threat model is an essential approach in defining security requirements. The goal of threat modeling is to understand how an attacker would be able to compromise a system, and then to ensure that proper mitigation techniques are in place to prevent such attacks. Threat modeling pushes the design team to consider the mitigations during the process of the system creation before deployment. In general, the threat modeling process consists of four steps.

- step 1: Model the application
- step 2: Recognize and Enumerate Threats
- step 3: Use countermeasures to Mitigate threats
- step 4: Verify and validate the mitigations

The most critical step is step 2 aimed at exposing the vulnerabilities and security challenges of the IoT systems. After properly classifying the threats, it will be possible to explore the mitigation techniques. For classifying threats in an information system, Microsoft introduced the STRIDE (Spoofing, Tampering, Repudiation, Information disclosure, Denial of Service and Elevation of privilege) threat model [32] Countermeasures are recommended and evaluated for each threat. The application of STRIDE for threat modeling in Industrial IoT (IIoT) has been studied before as discussed in [33, 34]. It also describes the adaptation of STRIDE for the Azure IoT reference architecture. After discovering threats, these should be rated according to their severity using some tools. The use of the DREAD (Damage, Reproducibility, Exploitability, Affected Users, Discoverability) model as one of commonly used tools to assign ratings to threats is mentioned in [35] .

Generally, each IoT system will have a multi-layered architecture consisting of various layers. These layers make use of diversified technologies, which introduce a plethora of challenges and security threats. As a result, the architecture of the IoT system plays a significant role in identifying the threats and attacks. However, there is no specific standard architecture because most of the IoT solutions are application-specific developed with explicit technologies, resulting thus in heterogeneous and fragmented architectures.

A secured IoT network architecture was proposed in [36] that would be using Software Defined Networks (SDN) for identifying the threats. It also summarizes how IoT network security can be achieved in a more effective and flexible way using SDN. Furthermore, studies, reviews, and analysis were conducted on some existing IoT architectures and a new architecture was proposed based on those architectures [37]. This new architecture includes a lot of the key elements of the other architectures, while fostering a high degree of inter-operability across diverse assets and platforms. Among the several IoT architectures reviewed in [38], it is found that the four-layer architecture (Application, Transport, Network, and Perception layers) is often being considered by researchers to address security challenges and solutions at each layer. Moreover, the most used IoT architectures are often three-tier/layer systems, including a perception/hardware layer, network and communication layer, and application interfaces and services layer. Additionally, the Open Web Application Security Project (OWASP) [39] identified attack vectors using the three layers of an IoT system: hardware, communication links, and interfaces/services layers. Thus, as shown in **Figure 5** at all layers of the IoT architecture, implementation of IoT security mitigation techniques should include security architecture [40].

According to the IoT security architecture, there are security issues and concerns at each of the three IoT layers. Because of their relative positions in the architecture, each of these layers has its own set of security needs. However, because they are all interconnected, if one is compromised, the others may suffer as well. The goal of IoT security is to protect customer's privacy, confidentiality, data integrity, infrastructure, and IoT device security, as well as the availability of the services. The following subsection discusses the IoT Security issues and threats at each of the layer.

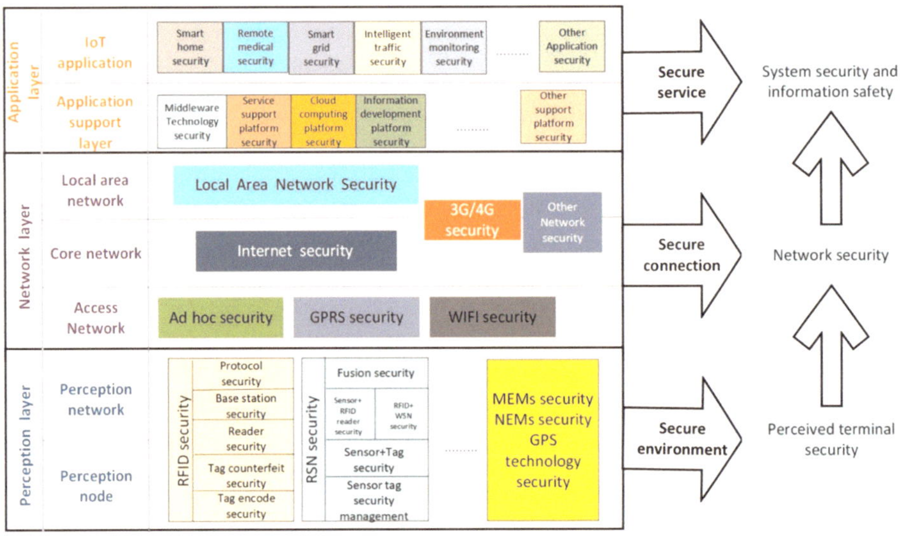

Figure 5.
IoT security architecture *[40]*.

4.2 Classification of IoT threats and attacks with solutions

Like in any other system, confidentiality, integrity, AAA, availability, and non-repudiation are some general security goals and requirements as already stated in previous sub-section. This section discusses about some of the most frequent threats and attacks at each IoT layer that might affect at least one of these criteria. Following **Table 6** provides an overview of the classification of the threats at each IoT layer and some proposed solutions corresponding to these threats [41–44].

4.3 State-of-art IoT security mitigations

The primary goal of implementing security mitigation is to ensure privacy, confidentiality, and the security of IoT users, infrastructures, data, and devices, as well as to ensure the availability of services provided by an IoT ecosystem. As a result, mitigation and countermeasures are often implemented in accordance with the traditional threat vectors.

In the above sub-section, some empirical based solutions have been listed in **Table 2** corresponding to the given threat or attack. Based on the studies performed in [11, 45–47], it is observed that some ubiquitous state-of-the-art technologies such as Blockchain, Fog Computing, Edge Computing, SDN, Artificial Intelligence can be used to enhance the security in an IoT environment. These technologies are vital and have enormous potential for addressing the IoT ecosystem's security concerns.

Blockchain (BC): A blockchain is a special kind of database. It differs from a standard database because of the unique approach in which it saves data. Data are, hence, saved in a series of blocks that are subsequently linked together to form a blockchain. IoT devices capture data from sensors in real time, and BC provides data security by establishing a distributed, de-centralized, and shared ledger [48]. Due to its critical operational properties, such as distributed functionality, de-centralized behavior, encrypted communication, embedded cryptography, and authorized access, it provides security solutions against a variety of threats across the different layers of the IoT such as disclosure of critical information, device compromise, malicious data injection, tag cloning, node cloning, unauthorized access, software modification, data manipulation, spoofing, session hijacking, false data injection, brute force attack.

Fog computing (FC): Fog computing allows processing, storage, and intelligent control to be close to the data devices themselves. Hardware failures, eavesdropping, device compromise, disclosure of critical information, leaks of critical information, node tampering, node capture attacks, node replication, battery drainages attack, illegal access, DoS and DDoS, MITM, etc. are just some of the threats and attacks that can be prevented by the vast processing, storage and management capabilities of the voluminous data that it processes, stores, and manages.

Edge Computing (EC): In edge computing, data are transmitted within the network or within the device. Data movement is reduced as compared to fog computing, which alleviates security concerns. Real-time services such as intrusion detection, identity recognition, access management enable edge computing to strengthen security against a variety of threats and attacks, including battery drain, hardware failure, eavesdropping, node capture, DoS and DDoS, SQL injection, jamming, malicious attack, virtualization, data integrity, cloud flooding attack, illegal access.

SDN: Software-defined networking is the preferred method of managing network security in a variety of application domains, including smart homes, businesses, and e-health care systems. The control plane and data plane refer to the two primary tasks

Layer	Threats/ attacks	Description	Solution
Perception Layer	Eavesdropping	An intrusion, also known as a sniffing or spying attack, occurs when someone attempts to steal information sent by the devices.	Deploying intrusion detection system.
	Replay Attack	An intruder listens to the transmission between sender and receiver and steals legitimate data from the sender.	Using one-time passwords and session keys and timestamps
	RF Jamming	RFID tags may potentially be exploited via a DoS attack, when RF transmission is disrupted by excessive noise signals.	Encryption & authentication.
	Node Capture	An attacker takes control over a key node, like a gateway node to use its resources.	Authentication and access control.
	Fake Node and Malicious	It is a kind of attack in which an attacker modifies the system by adding a node and injecting bogus data. This created node drains vital energy from genuine nodes and may gain control of them, thus destroying the network.	Authentication and access control.
Network Layer	Sybil Attack	In this attack, the attacker controls and changes the node in such a way that it shows multiple identities, hence compromising a huge portion of the system and resulting in misleading information about redundancy.	Trusted certificates that are based on a central certification authority.
	Sinkhole Attack	The attacked hole serves as a strong node, and so other nearby nodes and devices prefer it for communication or as a forwarding node for data routing, and thus acts as a sinkhole, attracting everything.	Intrusion detection system, strong authentication techniques.
	Denial of Service (DoS) Attack	This kind of attack causes the targeted system's resources to be exhausted, rendering the network inaccessible to its users because of an attacker's flood of useless traffic.	Configuring a firewall which denies ping requests or using AES encryption.
	Man-in-the-Middle Attack	Using a middleman attack, the attacker pretends to be the original sender, making the recipient believe that the message came from them.	Using high level encryption and digital signatures.
	RFID Spoofing	These attacks are designed to transfer malicious data into the system by gaining access to the IoT system. RFID spoofing, IP spoofing, and other spoofing attacks in IoT systems are examples.	RFID Authentication protocols.
	Unauthorized access	An unauthorized person may get access to the IoT device over the network.	Authentication and access control.
Application Layer	Malicious Code Attacks	This attack is done by executing the malicious scripts or code. It is a hacking method enabling the attacker to first insert the malicious code into the system and then data is stolen from the user by executing these malicious scripts.	Firewall is inspected at run time.

Layer	Threats/ attacks	Description	Solution
	Cross site scripting	Client-side scripts, such as javascript, may be injected into a trusted website by an attacker. An attacker may then totally alter the application's content to suit his requirements and illegally use original data.	Validating user input and the input by the web page.
	Phishing attack	The attacker spoofs the legitimate users' data to get their usernames, email addresses, and passwords. The attacker creates a false e-mail or website, and then the legitimate user logs in, stealing their data.	Using anti-phishing, prevention techniques.
	Botnet	By using a botnet, the hacker may take over a network of devices and control them from a single access point.	Using a secure router encryption protocol, such as WPA2.
	SQL injection	SQL script is used to log into the IoT Devices and applications.	Programming the log page using parameterized statements.

Table 6.
Common IoT threats, description, and solutions.

of switches/routers. The control plane determines where traffic should be routed, whereas the data plane routes traffic to a specific destination. The control plane and data plane are linked together in conventional networking, but are separated in an SDN architecture. The data plane runs on hardware, while the control plane runs on software and is logically centralized. SDN is capable of monitoring and detecting harmful activity on the network. It separates the compromised nodes from the rest of the network by identifying them. Flow statistics in SDN architectures was employed to detect anomalies through a variety of techniques, including DDoS attacks, port scanning, and worm spreading [49].

Artificial intelligence (AI): The use of artificial Intelligence is growing in cybersecurity because it can help protect systems from cyber threats in a more dynamic way. AI is most frequently employed in cybersecurity for intrusion detection, which involves studying traffic patterns and looking for activities indicative of threat. With the growth of IoT technology, AI has received considerable attention. As a result of this expansion, AI technologies such as machine learning, support vector machines, decision trees, linear regression, and neural networks have been integrated into IoT cybersecurity applications to detect threats and prospective attacks. AI is viable for IoT security, particularly for the four critical risks: intrusion detection, defense against DoS/DDoS attacks, device authentication, and virus detection [50]. The following section discusses the role of AI techniques and their comparative studies for IoT security.

5. Comparative study AI categories used to mitigate industrial IoT security

AI is a promising approach, which can be employed to mitigate the security challenges faced by IoT autonomous system. As per [51], the secure solution can be improved through AI approaches to predict future threats. The researchers point out generative adversarial networks (GAN) that are using generator and discriminator. The generator's scope is to add samples to the real data, whereas the discriminator's purpose is to remove the fake samples from the original data. The suggested AI-based solutions are from the data-driven type, which are support vector machine (SVM), neural networks (NN), artificial neural networks (ANN), recurrent neural network (RNN).

A framework has been proposed where AI based reaction agent is introduced [52]. The security enhancement is a combination between two intrusion detection systems: knowledge-based and anomaly-based. For network pattern analysis, Weka has been used as data mining tool and NSL KDD as dataset source and distributed JRip algorithm in which machine learning can be used for security enhancement. For anomaly-based IDS, the dataset is collected from real sensor data and the model uses library of python Scikit-learn.

The main finding of [53] is that AI can be used for IoT security mostly in intrusion detection system (IDS) in order to analyze the traffic and learn the characteristic of the attack. Naïve Bayes algorithm is mostly used to classify attack data where it is assumed this to originate from the independent events.

A two-tier framework is proposed by [54] for embedded systems such as an IoT system. The security mitigation is to improve the traditional host-based IDS. The machine learning approach used is of a pipeline method where a set of algorithms are involved which allow the flexibility of adjusting the ML processing and the link between different tiers.

From a comprehensive survey published by [55], it has been found that high-level encryption techniques are not advisable to be implanted in IoT systems due to resource limitation. Therefore, AI approach is a very strong candidate to enhance security in IoT system in addition to the other existing network security protocols. Consequently, to the nature of IoT-layered architecture, each layer has its specific security threats. It has been noticed that machine learning approaches are widely adopted in comparison to the knowledge-based expert systems.

Reference	Expert system	Machine-learning	Security mechanism to be enhanced
[51]	✓	✓	DoS, Sybil detection, intrusion detection, MITM, malicious node
[52]	✓	✓	intrusion detection system (IDS)
[53]	✓	✓	Mostly used in (IDS) and MITM
[54]		✓	Host-based IDS
[55]	✓	✓	Different layers of IoT system threats
[56]	✓	✓	IDS
[45]	✓	✓	Authentication mechanisms (Access control) and detection systems
[57]	✓	✓	False Data detection
[58]	✓	✓	IT trustworthiness (safety, security, privacy, reliability, and resilience)

Table 7.
AI branches used in IoT security solutions.

Another study published by [56] suggests that the machine learning based security approaches are used mostly to enhance the detection mechanism of IDS. The only approach that provides mitigation features is based on the techniques that utilizes deep learning such as Gaussian mixture, SNN, FNN, RNN or utilize supervised machine learning such as SVM. **Table 7** [45, 51–58] shows that machine learning is mostly used in the security mechanisms in IoT environment as there are a huge data to learn from.

As per the literature, AI-based methods are recommended to be used to enhance protection against IoT attack. However, most of them are not yet commercialized due to the difficulty of its implementation. The focus of proposing different IoT security mitigation is to introduce high-performance approaches with low cost in a real-time environment. Moreover, dataset preparation is a critical factor that affects the accuracy and efficiency of machine learning approaches.

6. Conclusions

As discussed in this chapter, industries deployed IoT technology to develop industrial applications to add values to their businesses and consumers in terms of performance and cost. Different business models are also reviewed to comprehend that the standardization of IoT business model is very difficult due to the different types of industries and their varied requirements. As such, it is critical for industries to ensure confidentiality, data integrity, availability to ensure data privacy, and security of the system. However, maintaining privacy and security emerged as a challenge in IIOT because of the sophistication of the IoT system. This chapter considered the most used three-layer IoT architecture to study and review the various possible threats and attacks and their conventional mitigation techniques. Conventional security mechanisms have a limitation in IIoT, particularly in predicting attacks.

The state-of-the-art technologies such as Blockchain, Fog computing, Edge computing, SDN, and AI have also been discussed to enhance the security levels in IIoT systems. But artificial intelligence (AI) has been emerging as a promising approach to secure the IIoT-based systems because of its ability to learn from the big data. It furthermore supports data analysis and enhances security mechanisms. AI techniques such as SVM, NN, ANN, RNN have been reviewed and recommended to design and improve countermeasures such as IDS. Data engineering is a critical phase to prepare the datasets required for machine learning. Therefore, it is highly recommended to consider this phase in order to achieve an effective AI deployment. Based on the analysis presented herein, it is the authors' view that this is an open challenge to enhance security mechanisms through AI-based mitigation techniques.

Acknowledgements

We would like to extend our appreciation to the Ministry of higher education and research and innovation for funding this research through the block funding program. This paper is aimed at contributing and further fostering the quality of research in the University of Technology and Applied Sciences in Oman. We extend our gratitude to the reviewers for their insights on the submitted manuscript that greatly improved the chapter.

Glossary of terms

AI	Artificial Intelligence
ML	Machine Learning
QoS	Quality of Services
ES	Experts System
B2B	Business to Business
STRIDE	Spoofing, Tampering, Repudiation, Information disclosure, Denial of Service and Elevation of privilege
DREAD	Damage, Reproducibility, Exploitability, Affected Users, Discoverability
SDN	Software-Defined Networks
OWASP	Open Web Application Security Project
Digital Twin:	A digital twin is a virtual representation of an object or system that spans its lifecycle, is updated from real-time data, and uses simulation, machine learning and reasoning to help decision making.

Author details

Shadha ALAmri*, Fatima ALAbri and Tripti Sharma
University of Technology and Applied Sciences, Muscat, Oman

*Address all correspondence to: shadha-alamri@hct.edu.om

References

[1] Bécue A, Praça I, Gama J. Artificial intelligence, cyber-threats and industry 4.0: Challenges and opportunities. Artificial Intelligence Review. 2021; **54**(5):3849-3886. DOI: 10.1007/ S10462-020-09942-2/FIGURES/3

[2] Shojafar M, Mukherjee M, Piuri V, Abawajy J. Guest editorial: Security and privacy of federated learning solutions for industrial IoT applications. IEEE Transactions on Industrial Informatics. 2022;**18**(5). Available from: https://ieee xplore-ieee-org.masader.idm.oclc.org/ document/9619939/. [Accessed: February 11, 2022]

[3] Swamy SN, Kota SR. An empirical study on system level aspects of Internet of Things (IoT). IEEE Access. 2020;**8**: 188082-188134. DOI: 10.1109/ ACCESS.2020.3029847

[4] Whaiduzzaman M, Mahi MJN, Barros A, Khalil MI, Fidge C, Buyya R. BFIM: Performance measurement of a blockchain based hierarchical tree layered fog-IoT microservice architecture. IEEE Access. 2021;**9**:106655-106674. DOI: 10.1109/ACCESS.2021.3100072

[5] Chao L, Peng X, Xu Z, Zhang L. Ecosystem of things: Hardware, software, and architecture. Proceedings of the IEEE. 2019;**107**(8):1563-1583. DOI: 10.1109/JPROC.2019.2925526

[6] Oteafy SMA, Hassanein HS. IoT in the fog: A roadmap for data-centric IoT development. IEEE Communications Magazine. 2018;**56**(3):157-163. DOI: 10.1109/MCOM.2018.1700299

[7] Arena F, Pau G. When edge computing meets IoT systems: Analysis of case studies. China Communications. 2020;**17**(10):50-63. DOI: 10.23919/ JCC.2020.10.004

[8] Mishra D, Zema NR, Natalizio E. A high-end IoT devices framework to foster beyond-connectivity capabilities in 5G/ B5G architecture. IEEE Communications Magazine. 2021;**59**(1):55-61. DOI: 10.1109/ MCOM.001.2000504

[9] Fagan M, Marron J, Brady KG Jr, Cuthill BB, Megas KN, Herold R, et al. IoT device cybersecurity guidance for the Federal Government. NIST. 2021; **800**:213. Available from: https://nvlpubs. nist.gov/nistpubs/SpecialPublications/ NIST.SP.800-213A.pdf. [Accessed: February 9, 2022]

[10] Almtrafi S, Alkhudadi B, Sami G, Alhakami W. Security threats and attacks in Internet of Things (IOTs). International Journal of Computer Science & Network Security. 2021;**21**(1):107-118. DOI: 10.22937/IJCSNS.2021.21.1.15

[11] Krishna RR, Priyadarshini A, Jha AV, Appasani B, Srinivasulu A, Bizon N. State-of-the-art review on IoT threats and attacks: Taxonomy, challenges and solutions. Sustainability. 2021;**13**(16): 9463. DOI: 10.3390/SU13169463

[12] Azrour M, Mabrouki J, Guezzaz A, Kanwal A. Internet of Things security: Challenges and key issues. Security and Communication Networks. 2021;**2021**. DOI: 10.1155/2021/5533843

[13] Dorsemaine B, Gaulier JP, Wary JP, Kheir N, Urien P. A new approach to investigate IoT threats based on a four layer model. In: 13th International Conference on New Technologies for Distributed Systems (NOTERE 2016). IEEE; 2016. DOI: 10.1109/ NOTERE.2016.7745830

[14] Russell S, Peter N. Artificial intelligence: a modern approach. Harvard: Prentice Hall; 2010

[15] Negnevitsky M. Artificial Intelligence: A Guide to Intelligent Systems. 3rd ed. Addison Wesley/Pearson; 2011

[16] Lantz B. Machine Learning with R: Expert Techniques for Predictive. Birmingham, UK: Packt Publishing Ltd; 2019

[17] Moustafa M, Choo KKR, Abu-Mahfouz AM, Guest editorial: AI-enabled threat intelligence and hunting microservices for distributed industrial IoT system. IEEE Transactions on Industrial Informatics. 2022;**18**(3). Available from: https://ieeexplore.ieee.org/document/9536391/ [Accessed: February 7, 2022]

[18] Gebremichael T et al. Security and privacy in the industrial Internet of Things: Current standards and future challenges. IEEE Access. 2020;**8**: 152351-152366. DOI: 10.1109/ACCESS.2020.3016937

[19] Chui M, Collins M, Patel M. McKinsey Report: The Internet of Things Catching up to an Accelerating Opportunity. Fluxus; 2021. Available from: https://fluxus-prefab.com/mckinsey-report-the-internet-of-things-catching-up-to-an-accelerating-opportunity/. [Accessed: February 11, 2022]

[20] Osterwalder A, Pigneur Y. Business Model Generation: A Handbook for Visionaries, Game Changers, and Challengers (The Strategyzer Series). John Wiley and Sons; 2010

[21] Dijkman RM, Sprenkels B, Peeters T, Janssen A. Business models for the Internet of Things. International Journal of Information Management. 2015;**35**(6): 672-678. DOI: 10.1016/J.IJINFOMGT.2015.07.008

[22] Cranmer EE, Papalexi M, tom Dieck MC, Bamford D. Internet of Things: Aspiration, implementation and contribution. Journal of Business Research. 2022;**139**:69-80. DOI: 10.1016/J.JBUSRES.2021.09.025

[23] Suppatvech C, Godsell J, Day S. The roles of Internet of Things technology in enabling servitized business models: A systematic literature review. Industrial Marketing Management. 2019;**82**:70-86. DOI: 10.1016/J.INDMARMAN.2019.02.016

[24] Mansour H, Presser M, Bjerrum T. Comparison of seven business model innovation tools for IoT ecosystems. IEEE World Forum Internet Things. 2018;**2018**:68-73. DOI: 10.1109/WF-IOT.2018.8355219

[25] Manyika J, Chui M, Bisson P, Woetzel J, Dobbs R. The Internet of Things: Mapping the Value beyond the Hype. McKinsey Global Institute; 2015

[26] George G, Thampi SM. A graph-based security framework for securing industrial IoT networks from vulnerability exploitations. IEEE Access. 2018;**6**:43586-43601. DOI: 10.1109/ACCESS.2018.2863244

[27] Fang L, Zhang H, Li M, Ge C, Liu L, Liu Z. A secure and fine-grained scheme for data security in industrial IoT platforms for smart city. IEEE Internet Things Journal. 2020;**7**(9). Available from: https://ieeexplore.ieee.org/document/9104725/. [Accessed: February 6, 2022]

[28] Tariq U, Aseeri AO, Alkatheiri MS, Zhuang Y. Context-aware autonomous security assertion for industrial IoT. IEEE Access. 2020;**8**:191785-191794. DOI: 10.1109/ACCESS.2020.3032436

[29] Rathore S, Park JH, Chang H. Deep learning and blockchain-empowered security framework for intelligent 5G-

enabled IoT. IEEE Access. 2021;**9**: 90075-90083. DOI: 10.1109/ACCESS.2021.3077069

[30] El Kaed C, Khan I, Van Den Berg A, Hossayni H, Saint-Marcel C. SRE: Semantic rules engine for the industrial Internet-of-things gateways, IEEE Transactions on Industrial Informatics. 2018;**14**(2). Available from: https://ieeexplore.ieee.org/document/8091285/. [Accessed: February 7, 2022]

[31] Iwendi C, Rehman SU, Javed AR, Khan S, Srivastava G. Sustainable security for the Internet of Things using artificial intelligence architectures. ACM Transactions on Internet Technology. 2021;**21**(3):73. DOI: 10.1145/3448614

[32] Microsoft, IoT Security Architecture Microsoft Docs. 2021. Available from: https://docs.microsoft.com/en-us/azure/iot-fundamentals/iot-security-architecture. [Accessed: February 11, 2022]

[33] Sven Schrecker H. Industrial Internet of Things volume G4: Security Framework. Industrial Internet Consortium. 2016

[34] Microsoft. Azure IoT Reference Architecture. Microsoft; 2021. Available from: https://azure.microsoft.com/en-us/resources/microsoft-azure-iot-reference-architecture/. [Accessed: February 12, 2022]

[35] Aufner P. The IoT security gap: A look down into the valley between threat models and their implementation. International Journal of Information Security. 2020;**19**(1):3-14. DOI: 10.1007/S10207-019-00445-Y/FIGURES/1

[36] Flauzac O, Gonzalez C, Nolot F. New security architecture for IoT network. Procedia Computer Science. 2015;**52**(1): 1028-1033. DOI: 10.1016/J.PROCS.2015.05.099

[37] ENISA, Baseline Security Recommendations for IoT — ENISA. 2017.

[38] Alshohoumi F, Sarrab M, AlHamadani A, Al-Abri D. Systematic review of existing IoT architectures security and privacy issues and concerns. International Journal of Advanced Computer Science and Applications. 2019;**10**(7):232-251. DOI: 10.14569/IJACSA.2019.0100733

[39] OWASP. OWASP IoT Security Verification Standard OWASP Foundation. OWASP; 2022. Available from: https://owasp.org/www-project-iot-security-verification-standard/. [Accessed: February 12, 2022]

[40] Zhang J, Jin H, Gong L, Cao J, Gu Z. Overview of IoT security architecture. In: 2019 IEEE 4th International Conference on Data Science in Cyberspace, DSC. 2019. pp. 338-345. DOI: 10.1109/DSC.2019.00058

[41] Sharma N, Prakash R, Rajesh E. Different dimensions of IOT security. International Journal of Recent Technology and Engineering. 2020;**8**(5): 2277-3878. DOI: 10.35940/ijrte.E5893.018520

[42] Ahanger TA, Aljumah A. Internet of Things: A comprehensive study of security issues and defense mechanisms. IEEE Access. 2019;**7**:11020-11028. DOI: 10.1109/ACCESS.2018.2876939

[43] Litoussi M, Kannouf N, El Makkaoui K, Ezzati A, Fartitchou M. IoT security: Challenges and countermeasures. Procedia Computer Science. 2020;**177**:503-508. DOI: 10.1016/J.PROCS.2020.10.069

[44] Ávila K, Sanmartin P, Jabba D, Gómez J. An analytical survey of attack scenario parameters on the techniques of attack mitigation in WSN. Wireless

Personal Communications. 2022;**122**(4): 3687-3718. DOI: 10.1007/S11277-021-09107-6/FIGURES/19

[45] Restuccia F, D'Oro S, Melodia T. Securing the Internet of Things in the age of machine learning and software-defined networking. IEEE Internet of Things Journal. 2018;**5**(6):4829-4842. DOI: 10.1109/JIOT.2018.2846040

[46] Hassija V, Chamola V, Saxena V, Jain D, Goyal P, Sikdar B. A survey on IoT security: Application areas, security threats, and solution architectures. IEEE Access. 2019;7:82721-82743. DOI: 10.1109/ACCESS.2019.2924045

[47] Noor MM, Hassan WH. Current research on Internet of Things (IoT) security: A survey. Computer Networks. 2019;**148**:283-294. DOI: 10.1016/J.COMNET.2018.11.025

[48] Miller D. Blockchain and the Internet of Things in the industrial sector. IT Professional. 2018;**20**(3):15-18. DOI: 10.1109/MITP.2018.032501742

[49] Giotis K, Argyropoulos C, Androulidakis G, Kalogeras D, Maglaris V. Combining OpenFlow and sFlow for an effective and scalable anomaly detection and mitigation mechanism on SDN environments. Computer Networks. 2014;**62**:122-136. DOI: 10.1016/J.BJP.2013.10.014

[50] Wu H, Han H, Wang X, Sun S. Research on artificial intelligence enhancing Internet of Things security: A survey. IEEE Access. 2020;**8**: 153826-153848. DOI: 10.1109/ACCESS.2020.3018170

[51] Puthal D, Mishra A, Sharma S. AI-driven security solutions for the internet of everything. IEEE Consumer Electronics Magazine. 2021;**10**(5):70-71. DOI: 10.1109/MCE.2021.3071676

[52] Bagaa M, Taleb T, Bernabe JB, et al. A machine learning security framework for IoT systems. IEEE Access. 2020;**8**: 114066-114077. Available from: https://ieeexplore.ieee.org/abstract/document/9097876/. [Accessed: February 04, 2022]

[53] Kuzlu M. Role of artificial intelligence in the Internet of Things (IoT) cybersecurity. Discover Internet of Things. 2021;**1**(1):1-14. DOI: 10.1007/S43926-020-00001-4

[54] Liu M, Xue Z, He X. Two-tier intrusion detection framework for embedded systems. IEEE Consumer Electronics Magazine. 2021;**10**(5):102-108. DOI: 10.1109/MCE.2020.3048314

[55] Zaman S et al. Security threats and artificial intelligence based countermeasures for Internet of Things networks: A comprehensive survey. IEEE Access. 2021;**9**:94668-94690. Available from: https://ieeexplore-ieee-org.masader.idm.oclc.org/document/9456954/. [Accessed: February 05, 2022]

[56] Jayalaxmi P, Saha R, Kumar G, Kumar N, Kim TH. A taxonomy of security issues in industrial Internet-of-Things: Scoping review for existing solutions, future implications, and research challenges. IEEE Access. 2021;**9**: 25344-25359. DOI: 10.1109/ACCESS.2021.3057766

[57] Aboelwafa MMN, Seddik KG, Eldefrawy MH, Gadallah Y, Gidlund M. A machine-learning-based technique for false data injection attacks detection in industrial IoT. IEEE Internet of Things Journal. 2020;**7**(9). Available from: https://ieeexplore-ieee-org.masader.idm.oclc.org/document/9084134/. [Accessed: February 11, 2022]

[58] Hassan MM, Gumaei A, Huda S, Almogren A. Increasing the trustworthiness in the industrial IoT

networks through a reliable cyberattack detection model. IEEE Transactions on Industrial Informatics. 2020;**16**(9). Available from: https://ieeexplore-ieee-org.masader.idm.oclc.org/document/8972480/. [Accessed: February 11, 2022]